TWO BRITISH CAMPAIGNS

IN

AFRICA.

MAJOR-GENERAL SIR GARNET WOLSELEY, K.C.B., ETC.

From a Photograph by Messrs. Window & Grove (by permission).

COOMASSIE AND MAGDALA:

THE STORY

OF

TWO BRITISH CAMPAIGNS IN AFRICA.

BY

HENRY M. STANLEY

SPECIAL CORRESPONDENT OF THE NEW YORK HERALD
GOLD MEDALLIST OF THE ROYAL GEOGRAPHICAL SOCIETY
CORRESPONDING MEMBER OF THE AMERICAN GEOGRAPHICAL SOCIETY
AUTHOR OF 'HOW I FOUND LIVINGSTONE.'

WITH NUMEROUS ILLUSTRATIONS FROM DRAWINGS
BY MELTON PRIOR (SPECIAL ARTIST IN ASHANTEE OF THE ILLUSTRATED
LONDON NEWS) AND OTHER ARTISTS, AND TWO MAPS.

Rediscovery Books

❖

Reproduced by kind permission of the
Royal Geographical Society

Published by
Rediscovery Books Ltd
Unit 10, Ridgewood Industrial Park,
Uckfield, East Sussex,
TN22 5QE England
Tel: +44 (0) 1825 749494
Fax: +44 (0) 1825 765701

To find out more about Rediscovery Books
and its range of titles visit
www.rediscoverybooks.com

Published in association with

**Royal
Geographical
Society**
with IBG

Advancing geography
and geographical learning

The **Royal Geographical Society with IBG** was founded in 1830 to
advance geographical science. Today it supports geographical
research, promotes geography in schools and through outdoor
learning, in society and to policy makers. Geography connects us to
the world's people, places and environments.
The **Rediscovery Books** series allow us to see how previous
geographers and travellers understood and recorded the world.

*In reprinting in facsimile from the original, any imperfections are inevitably reproduced
and the quality may fall short of modern type and cartographic standards.*

Printed and bound by CPI Antony Rowe, Eastbourne

PREFACE.

'COOMASSIE and MAGDALA : the Story of Two British Campaigns in Africa,' is the title I have chosen for this volume, written to record two grand successes gained by English soldiers in East and West Africa.

Before proceeding to Abyssinia as a Special Correspondent of the 'New York Herald,' I had been employed for American journals—though very young—in the same capacity, and witnessed several stirring scenes in our Civil War. I had seen Americans fight; I had seen Indians fight. I was glad to have the opportunity of seeing how Englishmen fought. In Abyssinia I first saw English soldiers prepared for war. What I think of them I have written frankly, and without malice. The story of Magdala was written five years ago. The reader will perceive it to be in a fresher style than the story of Coomassie. Much of what I formerly wrote was re-written several times, but the MS. was still reserved for an opportunity which I hoped would come. I now avail myself of it, and present it to English and American readers.

The story of Coomassie is dull compared to that of Magdala; but it is more heroic, and more worthy of our sympathy in behalf of those who fought, who suffered grievous sickness, and died.

Coomassie was a town insulated by a deadly swamp. A thick jungly forest—so dense that the sun seldom pierced the foliage; so sickly that the strongest fell victims to the malaria it cherished—surrounded it to a depth of 140 miles seaward, many hundred miles east, as many more west, and 100 miles north. Through this forest and swamp, unrelieved by any novelty or a single pretty landscape, the British army had to march 140 miles, leaving numbers behind sick of fever and dysentery. Five days' hard fighting ended the march, and Coomassie was at the mercy of the conquerors, to sack and burn to the ground. When this work was done, the commander of the force was compelled to march his soldiers back again to the sea, to save the remnant from perishing by flood and disease.

Magdala was a town planted on the top of a mountain about 10,000 feet above sea level, amid gigantic mountains piled one upon another, grouped together in immense gatherings—profound abysses lying between 2,000, 3,000, and even 4,000 feet deep—a region of indescribable wildness and grandeur. It was an almost impregnable stronghold, situate 400 miles from the point of disembarkation — a strange, weird country, full of peaks and mountains, and ruggedness lay between it and the sea. The scenes which flanked the march bristled with rocks and crags ; but they possessed the charm of novelty and picturesqueness, and the country was one of the most healthy countries on the face of the earth. The march was full of interesting incidents, more especially as we drew near the end. A battle was fought ; Magdala was taken by assault, then fired, and utterly destroyed. The Emperor committed suicide ; the captives were released ; and the conquerors returned to the sea, flushed with unequalled success, having suffered the smallest loss that could possibly follow an invasion of a hostile country.

A greater contrast could not be made than is presented by the story of the two expeditions which England undertook in Africa, in behalf of her honour, her dignity, humanity, and justice ; and more brilliant successes than attended these two campaigns in hostile countries are not recorded in history.

HENRY M. STANLEY,
Special Correspondent 'New York Herald.'

LONDON : *April* 1874.

CONTENTS.

—◆◇◆—

CHAPTER I.

CHAPTER XIII.

CHAPTER XIV.

LIST OF ILLUSTRATIONS.

MAPS.

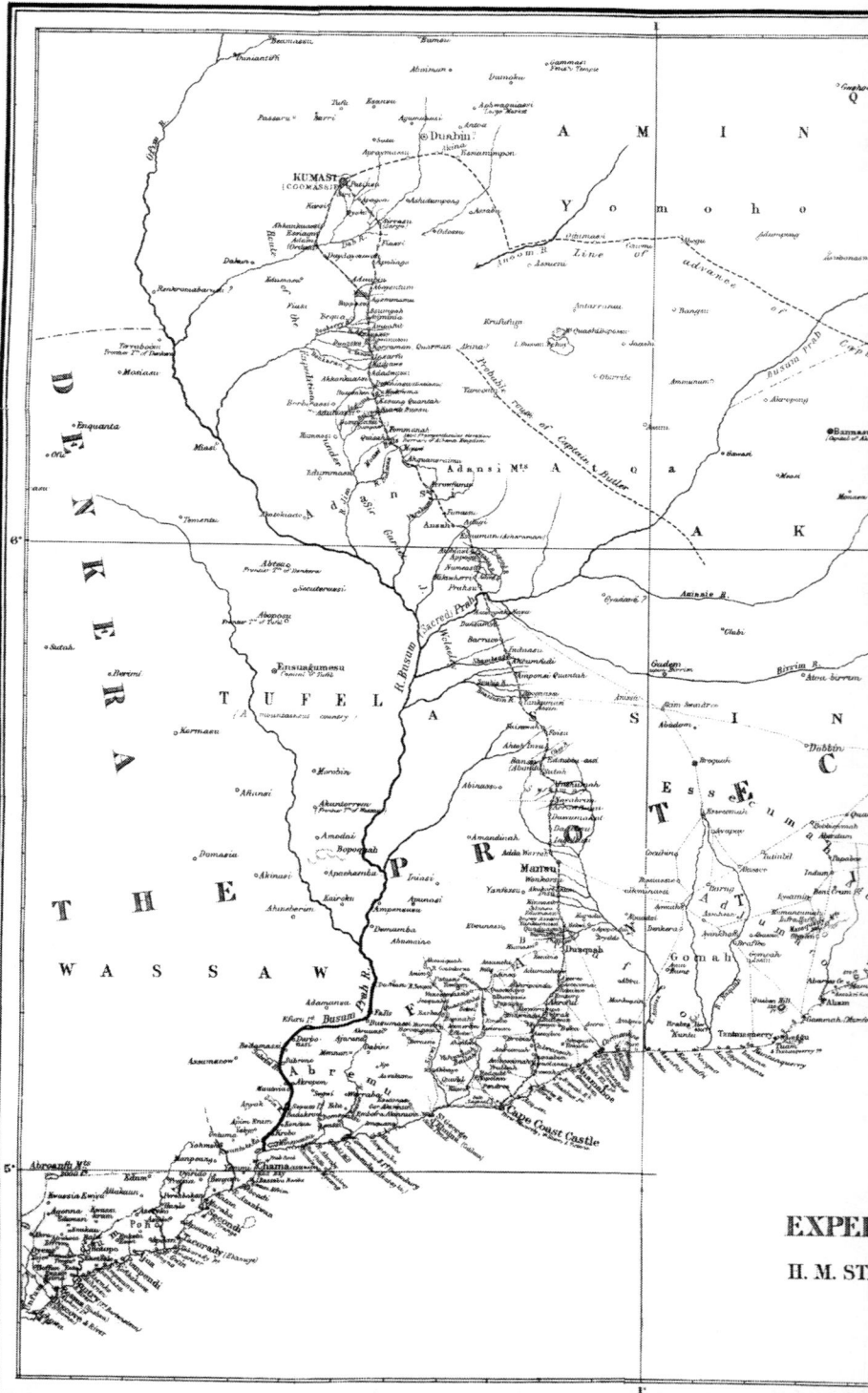

EXPE...

H. M. ST...

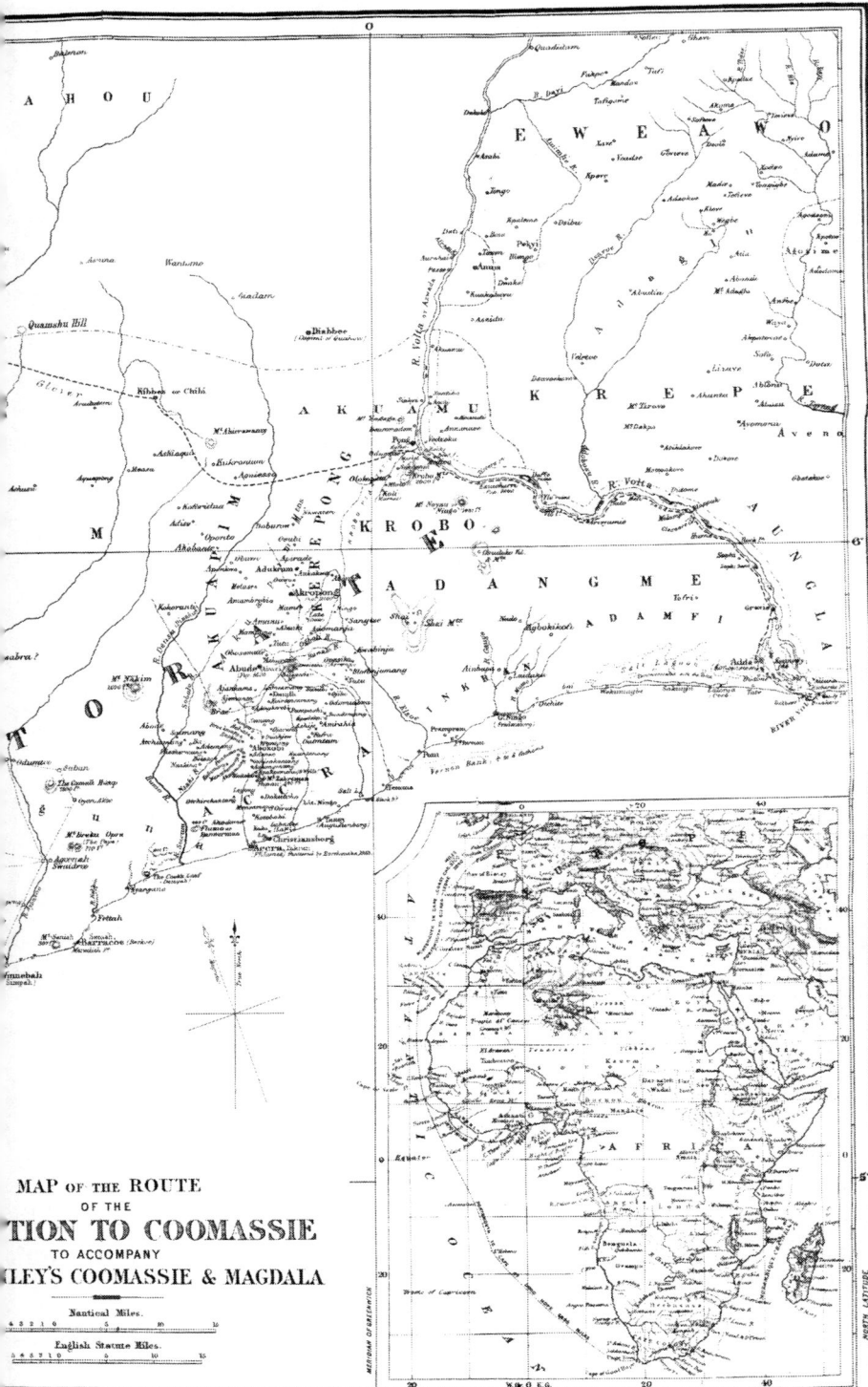

MAP OF THE ROUTE
OF THE
...TION TO COOMASSIE
TO ACCOMPANY
...LEY'S COOMASSIE & MAGDALA

Nautical Miles.

English Statute Miles.

WEST OR 0° EAST GR.

Beamassu

Duniantiffi

Bu

Aboin

Tufu   Ksansu

Passaru   Barri   Agamaua

Suta

Apraymassu

I

KUMASI
(COOMASSIE)   Patiasa

Sirisu

Karsi   Apogoa   Ashidu

Byoko   Sirrasu
(Sargu)

Ahkankuassi
Essiagu   Fiassi
Adasu
(Ordad)   Daydayawah

Dahun   Apoliage

Edumasa   Adwabia
Aboyentum

of   Ekoo   Agemmamu
Zappesas
Renkromabarudu ?   Fiasi   Esumgah
Bequa   Siminia

the   Soubere   R. Aferanassu   Amoahul

Dunsabu   Adwanassu

Terraboom   Denkaran R.   Korraman, Quar
Frontier Tn of Denkera   R. Yansu   Insarfu
Madjawe

Mosiasu   Ahkankuassi   Adadwassi

Hozainken   Dechiassu (Dacia
Mocowma

Borberassi   Essang Quanta
Enquanta   Adutiassi   Abara Buasu
(Dumpa)

Ofu   Dumpassi   Kommanah

Miasi   Humassi   Quisah   Mour

Ahquar

Edummassi

Edumassi

Akatokiado   Tementu

6°

Abteao
Frontier Tn of Denkera

Secuterassi

Aboposu
Frontier Tn of Tufel

Sutah

Berimi   Ensuagumesu
Capital of Tufel

T U F E L

D E N K E R

1°

Gammasi
Fetish Temple

Damoku

Gashoa
Q I

Aphwaguiassi
Large Market

Antoa
bin?
na
Essiamimpon

A M I N A A

g

Assabu

Y o m o h o

Odumassi
Anoom R. Line of

Abogu

Adumpeng

Man

Conomo

advance of

Assebanassu

Assueni

Antarranai

Bangsu

Krufufum

M. QuashiBopossu

Busum Prah

Captai

Akina?

L Busum Echuy

Jaashi

Yancoma

Obirribi

Amnunum

Akropong

Probable route of Captain Butler

Assini

Bammasu
(Capital of Akim)

cular elevation
ti Kingdom

Gawasi

mu

adansi Mts A t o a

Measi

fumtu

Monasa

unusu

Adagi

Essiaman (Asharaman)

A K

iast
Appoga
umeassu
wherri Ading
Prahsu

Prah

Aminnie R.

Assempah Naya

Gyadami ?

Dansamsu

Clabi

Barraco

Induasu

Shambensu

Ahcumfudi

Amponsi Quantah

Gadem
upon Birrim

Birrim R.

Soubin R.

Aponnassi

Boutinsin R.

Yankumasi
Assin

Anussu

Akim Swaidroe

Atoa-birrim

I N

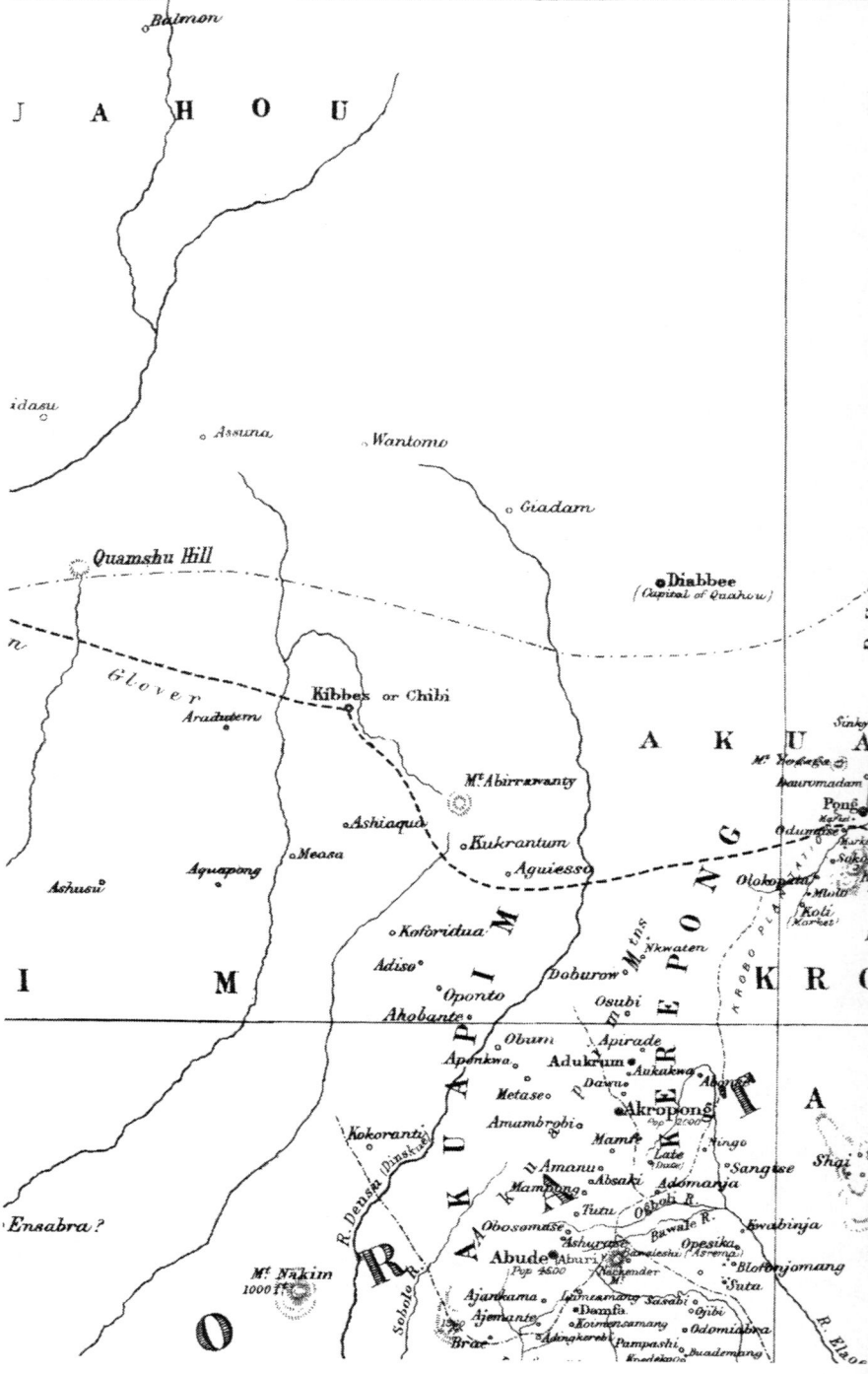

Balmon

J A H O U

O

idasu

Assuna

Wantomo

Giadam

Quamshu Hill

Diabbee
(Capital of Quahou)

Glover

R. Volta

Kibbex or Chibi

Aradutern

A K U A

Mt Yokafse
Dauromadam

Mt Abirrawanty

Pong
Market
Odumase
Sokben

Ashiaqua

Kukrantum

Aguiesso

Olokopua

Koli
Market

Measa

Ashusu

Aquapong

K R O

Koforidua

Nkwaten

AROBO PLATEAU

Adiso

Doburow

Osubi

Oponto

I   M

Akobante

Obum

Apirade

Apenkwa

Adukrum

Aukuko

Abura

Metase

Dawn

Amumbrobi

Akropong
Pop 2500

Kokoranti

Mamfe

Yingo

Amanu

Late
Date

Sangise

Shai

Mampong

Absaki

Adomanja

Obosomase

Tutu

Kwabinja

Mt Nakim
1000 ft

Abude
Pop 4500

Ashurase

Opesika

Blofonjomang

Ensabra?

Ajankama

Sasabi

Sutu

O

Ajemante

Koimensamany

Ojibi

Odomiabra

Brae

Pampashi

Buademang

R. Elso

(A mountainous country)

Kormasu

THE

Morobin

Afiansi

Akantorrem
(Frontier Tn of Wassau)

Amodai

Bopoquah

Domasia

Apachamba

P

Inias

Akinusi

Kairoku

Apum

Ahineberim

Ampensasu

Demamba

Abumain

WASSAW

Amim

Damasu R. Soup

Joo

Adamansa

Bas

Efuru Id   Busum

Falls

Busumassi

Mura
Boroques

Akranasi

Boras

Bedamassi

Darbo-
assi

Ajurand

Dabine

Sukrie R.

Mennan

Eje

Assamacow

Subreno

Assubiano

Abre

m

U

Akropon

Maubriad

Segwi

Werraba

Essaman

Anyak

Gwin R.

Ger Akennum

Apin Enim

Sepum Id

Edu

Embofra Akiunum

Yahyo

Badukrom

Domey

Amquana

Untuma

Kondua

Awassi

Amquania

Krobo

Yohmoh

Wantukeso

Minepuaba

Cormantin & FWandenburg

Brandy Hill

Manpeang

Prah Rock

R. Abrabi

(Akatey ski)

Yemmi   Chama

ESSIMARA

5°   Abroanfu Mts
2000 ft

Sand Bay

Onrido

Benyam

Bassebu Rocks

Edum

Preisia

Kwassia Ewiya

Attakaun

Kwassi Efrim

R. Ansnkwan

Agonna   Kwassi
Edumasi   Krum

Peresbakan

Aboadi

Hanlo

Katan

Akra

Snukau

Aseseka

Maraba

Secondi

Afrohota Bala

Enketa

Asaka

R. Orange

Oyene

Eson

Apoan   Apoasi

Tacorady (Ebanuye)

Poho

Shotopo

Gwin

Amansor

Akorady Jt

Boffin   Intu

Eketsko

Jua   Tompendi

Boutry (Ft Barkenstein)

Essan (Bushua)

Discove & River

Igwa

S Faisowah ◦Faisu

Ahtah Insu

Bansu ◦Eddubta-assi
(Abanta) Sutah

Abinasso Anshumah

ROO S ◦Nayabrim
◦Acrowhum
◦Dawumakot
◦Da

◦Amandinah Ingilasu

R Adda Warrah

Mansu

Wonkorsu

Yanfossu Akukorhamah
Insu

Kunnasu
Abinsu Kuradu
Edunassi
Imper Assam Bekwi
Ebounassi Yankumassi ◦Apopondu
Quaduagah Aryeldo
Murabboh
Thanassi
Kunasu Dunquah
B Escabio

Assanchi Adumachené
Falla Sinsa
Ashripinda Aperee
◦Dumassie Essecooma
Popagay Emsore
Abracarupa Prorak
Anamanah Bullayan
Effotoo Chipyaya Buka
Assaybo
Tibrobiah
Acproomah Senquah
Tokashi
Ameroon Essecroom
Abranser
Ambossimah Tugnabim
Yrubbah Yamolassu
Redoubt Prewaloe
Napoleon Fendro
Salt Aquafoe
Lagoons
Cape Coast Castle

SSINC

◦Abudam N

Dobbin
◦Broquah C

◦Essecomah
Esse Ecum ◦Quaman
◦Ayapan ◦Bobbiecomah
◦Aberdum
◦Cocubino Tutinbil ◦Papaboe
◦Akasoo Indum
Bissiassie Barng Beni Crum
◦Dominassi Eyeamin
Ad T Mumancumah
Ajemako Assahson Infra Baffia
◦Abuadei ◦Abassie ◦Obuen
Denkera Ju Braffee
Ayinkhan m Gompah
◦Assin
Gomah

Ayan Quaben Hill
Rume 580
ft

Brahra Tow Tantumquerry Aggu
Kuntu Tuam
Tantamquerry Pt
Mankassin

Amamo

EXPEI

H. M. STA

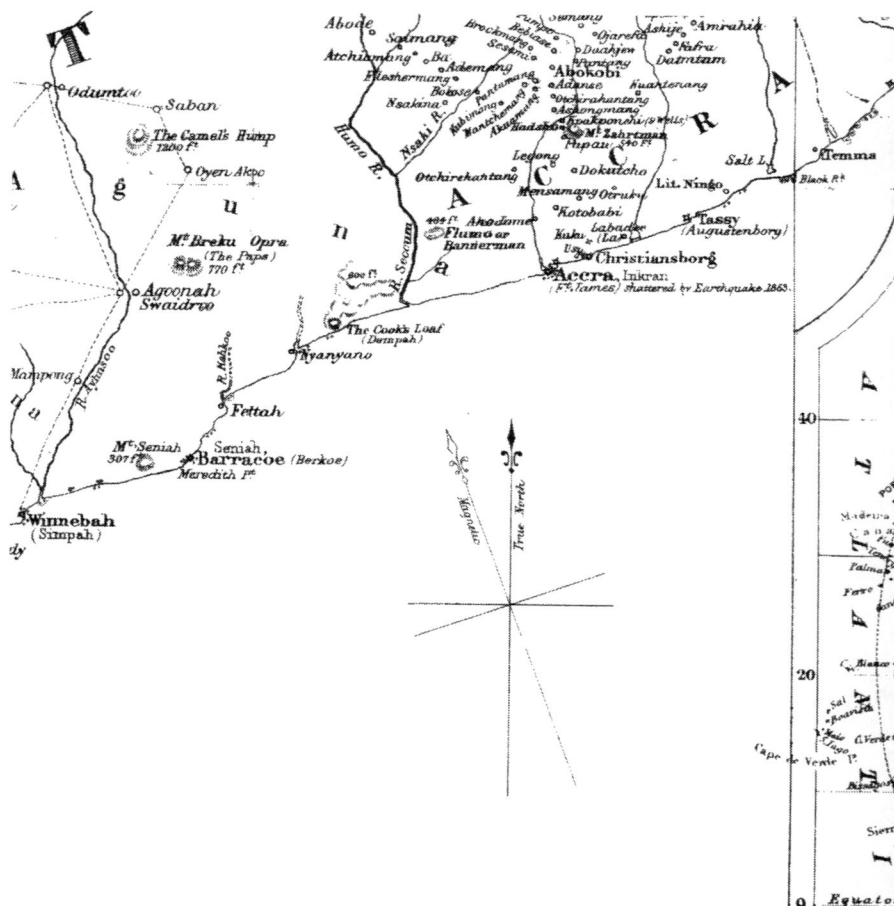

MAP OF THE ROUTE

OF THE

EDITION TO COOMASSIE

TO ACCOMPANY

STANLEY'S COOMASSIE & MAGDALA

Nautical Miles

5 4 3 2 1 0   5   10   15

English Statute Miles.

5 4 3 2 1 0   5   10   15

WEST GR. 0° EAST GR.

Marston, Low & Searle.

Map labels:

Odumta, Saban, The Camel's Hump 1200 ft, Oyen Akpo, Mt Breku Opra (The Papa) 770 ft, Agoonah Swaidroo, Mampong, R. Ainsoro, R. Nakko, Fettah, Mt Seniah 307 ft, Semiah, Barracoe (Berkoe), Meredith Pt, Winnebah (Simpah), Nyanyano, The Cook's Loaf (Detupah), Abode, Saimang, Atchiamping, Bi, Elashermang, Nsakina, Bobose, Nsahi R., Humo R., R. Seccum, Brockmang, Segom, Ademang, Pentamang, Montcho, Akodadeh, Duakjen, Pantang, Abokobi, Adanse, Kuartenang, Otchirahantang, Achomonang, Krakpomohu (S. Wells), Mt Zahrtma 340 ft, Dokuscho, Legong, Otchirehahtang, Mensamang, Otruku, Lit. Ningo, Salt L., Ahodome, Flumo or Bannerman, Kotobabi, Labadie (La?), Kuku, Acera (St James) Shattered by Earthquake 1863, Inkran, Christiansborg, Tassy (Augustenborg), Black Pt, Temma, Amrahia, Kafra, Datmtam, Ashije, Ojarefa, Samany, Madeira, Falma, Faro, Bisaco, Sal, Jago, C. Verde, Cape de Verde, Sierra, Tropic

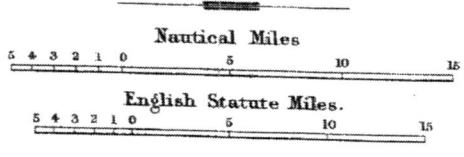

True North, MERIDIAN OF GREENWICH, Equator, Tropic

COOMASSIE AND MAGDALA.

CHAPTER I.

November 1, 1873.

WE STOOD on the deck of the 'Benin,' of the African Steamship Company, a mixed lot of passengers, gazing on the murky loom which marked where Liverpool stood, growing more and more indistinct as the good steamer shot bravely down the Mersey.

As Liverpool receded from view, and English land disappeared, we turned to one another for novelty, and mutual questioning elicited the fact that we were mostly all bound for Ashantee.

There were six doctors, five Control officers, one Staff officer, one volunteer, and one correspondent of us who were going to the war, but there were a few others destined for ports south of the equator. There were two Germans of the West African Expedition bound for the exploration of Central Africa; one of them was Dr. Falkenstein, a serious, earnest man, industrious, reflective, and fearfully copious in note-taking. The other, his companion, was a student. I was quite amused by the way in which the two evidently anticipated future delight in store for them when they should set foot on the shores of the classic Congo. If a fly buzzed about them a chase was made after him, and he was remorselessly prepared for the inspection of the savants of Germany. Two or three insignificant little shore birds were also added to the slaughtered and stuffed. I have been through all this myself, I have shared in this boyish

B

eagerness to grasp at once the delights which the fancy has painted in the perspective, and I can sympathise with men of robust health and untamed energy who, while they cannot at once stride into the enchanted land, clutch at anything, however trivial, which belongs to its outskirts, and regard it with wonder and undisguised admiration, whereas in a few weeks or so they will throw it away and wonder at their own simplicity.

Our doctors, six in number, were all young men, one or two of them with a high idea of their mission in the *fever*-field, while the others seemed better adapted for anything rather than a hospital. The Control officers—though what in the world they control I know not—were overgrown boys from Sandhurst College, airing their sub-lieutenant's uniform for the first time. They had plenty of stamina in them, and I have no doubt Africa will demand a large share of it, if not all. The Staff officer was Captain Butler, the author of the ' Great Lone Land ' and the ' Wild North Land,' who is going to join Sir Garnet Wolseley, and who perhaps will be able to pick up notes for a book he may well call the ' Great Black Land.'

Our destination in Africa is yet far ; many days will transpire before we shall see the Gold Coast, and in the meantime we shall amuse ourselves as best we may. We begin to tell each other our varied experiences, puffing vigorously in each other's faces the while at our cigars, or briar-woods ; but the tales cannot last for ever, so we borrow each other's books, and read listlessly. By this process of borrowing I procured a little red book called the ' Soldiers' Pocket Book,' edited to my surprise by Sir Garnet Wolseley.

Amid very much valuable matter, purely military, I came to Sir Garnet's opinions concerning newspaper correspondents.

If my memory serves me right, Sir Garnet calls correspondents a ' curse to modern armies '—and in other places he has bestowed opprobrious epithets upon the Press corps, among which is the term ' drones.'

Dropping the book on my knee, I picture to myself the kind of man the military author must be. I have never seen him, and I have only this excessive animus to the Press corps to guide me in my fancy portrait. A primmish man, of a Spanish cast of face, very stiff, formal, sour, crusty, vain, and afraid of criticism, conscious perhaps that he might commit faults and would fear being reviewed.

I ask a gentleman on board who are on Sir Garnet Wolseley's staff? I am answered—

'Captain Brackenbury for one.'

'Ah! yes. Brackenbury, of course, the "Times" man. A military correspondent, nothing would ever stop him from writing to the "Times." Strange that Sir Garnet, with his unconcealed hatred to newspaper people, should choose such an indefatigable caterer for the Press as Captain Brackenbury. Who else, pray?'

'Lieutenant Maurice.'

'What, Maurice the author of the "Wellington Prize Essay." He is another Press writer. I would bet anything he writes for one if not two newspapers. You need not name the others; I will take it for granted that they are mostly all newspaper writers in military clothes.'

Apart from his hatred of the 'Gentlemen of the Press,' as he facetiously terms them, Sir Garnet shows himself in his book as a man who would dare anything rather than brave defeat; he shows himself energetic and a master of all military detail, from the smallest minutiæ of an officer's outfit to the most difficult tactics for an army. Were the book reviewed bit by bit, there is many a place where Sir Garnet might be hit very strong on tender points; taken as a whole, it is a good instructor for officers, a treasury of knowledge for engineers, doctors, mechanics, drill-masters, generals, commissariats, for anybody you please connected with an army; in short, it is the work of a thorough soldier.

This exciting theme being over, we are now approaching Madeira, and the trade wind blows us southward and eastward at a cheering rate; and on the sixth morning, about the time or breakfast, the anchor was dropped in the harbour of Funchal, but we were not permitted to land. The Spanish Consul, with a malice I could not understand, had written in his own language, across the English bill of health, an account of the cholera case which has occurred at Liverpool on board 'a French brig;' and we were at once, as the phrase is, 'put in quarantine.'

Venting her spite upon the ill-mannered boors who refused

us permission to land upon the beautiful 'Isle of Woods,' the 'Benin' turned her stern indignantly towards the port of Funchal, and pointed her stem for Teneriffe, at which port we arrived in twenty-four hours. Nor at Teneriffe were we permitted to land. The yellow flag was at our fore peak, and we a thing to be dreaded. For the discourtesy with which we were treated shall I say anything good of these isles? Shall I when one cheek has been smitten desire the other to be smitten also? Shall I for a curse give a blessing? I have my private opinion of these islands and their people.

I will tell you a story about them. Occupying the ladies' cabin of our steamer, the 'Benin,' we had three passengers of the family of Jacob Doegling, of Chicago, the husband, wife, and daughter; the wife was in bad health, and the husband out of love for her left Chicago to give his wife and daughter the benefit of the salubrious air of Madeira. Arriving at Madeira they were refused permission to land, and as there was no place to receive passengers in quarantine in the port, they were taken to Teneriffe, in hopes that they could have accommodation there. When we came to Teneriffe the same refusal to land met them. They were compelled to journey on to Sierra Leone, on the west coast of Africa, 1,400 miles further still. Sierra Leone is the last place in the world to banish delicate women to, unless they are to be got rid of, when, if such is the case, Sierra Leone offers special advantages from its well-known insalubrious position. Fortunately, however, we met a steamer coming from the southward, and the sick woman, the daughter, and the loving husband were put on board, either to be landed at Madeira or to be taken back to England.

Three days after leaving Madeira the trade wind died away and the burning white Sahara desert hove into view.

Freiligrath's apostrophe to Africa comes to the mind as we sight the burning shores :—

> Oh zone so hot and glowing,
> Queen of the earth art thou;
> Sand is thy mantle flowing,
> The sun doth crown thy brow.
> Of gold, thou queenly woman,
> Are all thy clasps and rims,
> That fasten with fiery splendour
> The garment to thy burning limbs.

The sandy dunes and hills along the shore are bleached and forbidding. What a sad history has this shore. We all know the story of poor Captain Riley. His story a thousand times told would not complete the history of misfortunes which the inhospitable strand could relate were it able to speak. Besides the many many scores of foreign vessels wrecked on this shore, with their crews plundered, stripped, and hurried to the interior into a hellish slavery, how many a poor Canary fisherman has also found a home far inland in galling servitude of which we know nothing?

The sea along this coast swarms with fish for which the Canary fishermen have left the neighbourhood of their own snug little isles. Gladdened at heart at the prospect of soon filling their vessels, they approach too close to the shore and anchor; a storm rises, and the unlucky mariners are driven on the beach by the cruel gales, where the Moors lie in wait to bind and take them into life-long captivity.

While looking at the serrated line of sandy shore we cannot help fancying ourselves trudging across that torrid extent with blistering feet and backs, with the remorseless man-drivers behind us urging us to quicker pace with point of lance and snaky kurbash. But kind heaven forbid the mischance, and speed us on our way to more auspicious shores!

Imagination is assisted in the above doleful picture by the fervid atmosphere through which we move as we steer south towards Sierra Leone. We are impatient of the slightest vestments. The cabin is suffocating, for the thermometer is 88° Fahrenheit. On deck we have the cruel glare of the ocean, ruffleless all around. Sharks are frequently seen on either side of the vessel. I suppose their traditions delude them into the belief that here more than anywhere else man-meat is found. The 'Benin,' however, disappoints them, and as we glide ahead their black fins are seen cleaving the water at random, while they are looking for that which we cannot afford them.

On the sixth day from Teneriffe and the fifteenth from Liverpool the 'Benin' enters by night the mouth of the river of Sierra Leone, and the Lion Range obtrudes itself as the thick mist dissolves before the light of day. One glance at the position of the town reveals the cause of its insalubrity to the traveller.

It is placed in an amphitheatral terrace or shelf of ground, with an immense and lofty crescent of mountain at the back rising to a height of 1,000 and 1,500 feet above it. Between the town and sea a few spurs and outlying ridges prevent the full benefit of the pure sea breeze, and, as the breeze does not always come from seaward, when it comes from the interior the town becomes steamy, muggy, vaporous, a very hot-bed of sickness and fever. Nature has pointed to man the breezy upland above the town as the site whereon to build a house, yet the European has neglected the warning, and consequently has sickened and died; hence Sierra Leone has a bad reputation. English governors have endeavoured to mitigate evil sanitary regulations by industriously circulating clear fresh water through every street, but it has not been enough; the evil lies in the air, the air is the noxious exhalation from the leafy, umbrageous, close crescent around it. It is the want of that fresh air which bends the tall grass on the top of yonder mountain that has caused the fragile European to sicken, and fade, and die.

The first view of Sierra Leone is very deceptive. The stranger, on seeing the apparently strong, substantial, civilised-looking houses, rich in their luxuriant tropical surroundings, velvet-green palm fronds, rustling and waving over steeple and roof, papaw and bread-fruit trees, oranges and limes, and every tree that is delightful to the eye and taste, growing thickly about, would be apt to say, 'Oh, here is the fulfilment of my dream at last; this is the initial footstep of civilisation which by-and-bye is to penetrate to Central Africa.' He enters the town hopeful and sanguine, but the truth slowly penetrates into his brain that this town has been about a hundred years progressing towards its present pitch of prosperity. After a hundred years of occupation, the English are building a wharf! After a hundred years of occupation, the Episcopal church is but half constructed, and I should fear to say how much precious money has been spent on the rickety-looking edifice! After a hundred years of occupation, the zealous English missionaries have not been able to inculcate in the negro's mind that it is sinful to lie, to steal, and to be lazy! And this is the result of christianising Africa at Sierra Leone! If I were asked where I could find the most insolent, lying, thieving negroes, I should undoubtedly say at Sierra Leone. Through some strange caprice the English have permitted a colony of semi-civilised Africans to grow up in order to experiment, perhaps, how wild and how rank a colony of negroes can become when left to their own sinful and wicked devices, unchecked and uncurbed by the hand of law. The English will, perhaps, plead as an excuse that the climate is against the exercise of strong will; that no matter how valiant a man be in his intentions when he sets out to govern the blatant woolly-headed rabble of this colony, he will find his nerves and will prostrated before the unconquerable lassitude which the climate quickly engenders in him as soon as he sets foot on its shores.

The rumour of war with Ashantee had penetrated even here. One would have thought that Great Britain, after careful nurture of such a settlement, would have prepared a nursery of disciplined negroes as auxiliaries in time of need. Nothing of the kind; the only efficient force of Sierra Leone natives I have been able to see is composed of the few constables who perambulate the town, bâton in hand, to crack the heads of the unruly. What she has been able to effect in the way of a constabulary force ought to have suggested to her the raising of a faithful cohort of natives. It has not been done, however, and she has been obliged to despatch some of her West Indian legions to protect her possessions here; and not only those here, but her possessions all along the coast. Her West Indian negroes are fine tall fellows, marvels of discipline, showing what might be done with the raw material of Sierra Leone. Perhaps Great Britain, good nursing-mother as she is, has failed in this instance by proceeding too tenderly with her black children of Sierra Leone. It seems to me she has aimed at too high a standard of excellence in her mode of dealing with these people.

Mostly all of us on board the ' Benin,' destined for the war, wished to procure servants. But we found servants scarce, while applicants for clerkships far exceeded the demand. Indeed, I do not know that any of us were so effeminate, or so luxurious, as to demand a clerk to write our letters from the palace of King Coffee at Coomassie, while we really did need black servants, who could stand the climate, to carry our baggage to Coomassie.   The applicants of Sierra Leone said they could do anything, from reading the Bible to making long prayers for our success in the field; but when we asked them if they could carry a few pounds of canned meat for us at a shilling a day, they mostly all demurred at the ignoble task.

In almost every street in Sierra Leone I heard the voice of

A CITIZEN OF SIERRA LEONE.

praise and vocal prayer from the numerous aspirants to clerk-ships and civil service employ; but I am compelled to deny that I heard the sound of mallet and chisel, of mortar, pestle, and trowel, the ringing sound of hammer on anvil, or roar of forge, which to my practical mind would have had a far sweeter sound.   There is virgin land in the neighbourhood of Sierra Leone yet untilled, there are buildings in the town yet unfinished, there are roads for commerce yet to be made, the trade of the African interior yet waits to be admitted into the capacious harbour of Sierra Leone, for the enrichment of the fond nursing-mother of races who sits dreamily teaching her children how to cackle instead of how to work.

After a stay of twelve hours the 'Benin' turns from Sierra Leone, and at the mouth of the river passes the wreck of the steamer 'Nigritia,' three-fourths of which splendid steamer was lately sold for the sum of 400*l*. It is not too late yet for some enterprising man to charter a couple of ships to purchase the wreck of the 'Nigritia' at Sierra Leone, and that of the 'Monrovia,' which lies a quarter of a mile off Cape Palmas. The upper parts of both vessels are uninjured. Clever people setting out with plenty of appliances might make an immense sum out of the venture. The latter vessel was almost a new one, and cost 60,000*l*. Striking a reef off Palmas Point, she was at once run full speed into the beach where she now lies.

We steamed by Liberia's low wooded shores without the chance to observe how the sable republic flourishes by a personal view of things. Report speaks evilly of her, of her pride and her vanity, of the disinclination of her children to work and their pretensions to high-sounding titles and high places. Those on board the 'Benin' who have stopped at Liberia say that mostly every other man is styled an 'Honourable,' that the people are fonder of standing in groups in the street to discuss politics than of bringing the produce of the rich back country into the markets for sale, which if true is very disheartening.

Off Cape Palmas I had the pleasure of seeing one of the Liberia 'Honourables,' who introduced himself to me as the Honourable J—— M——, named after the Honourable J—— M——, Chief Justice of old V——. Said he, 'I was born in old V——, sir. A good old state, sir. I was named after old J—— M——, sir. You may have heard of him, sir. The Chief Justice, sir. I have been here seventeen years, sir, now, and we are improving little by little ; there is a promising future, sir. Oh yes, sir, I do not feel discouraged at all, sir ; rather have I, have we all, cause to regard the prospects of Liberia as being very hopeful, sir. Good afternoon, sir.

If you were to stop here a week, sir, I should feel honoured by your making my poor house your home, sir. Good day, sir. A pleasant voyage to you, sir;' and the pleasant-faced, simple-hearted old gentleman evanished into his canoe, in which he was rowed ashore by a parcel of naked Kru boys.

This Cape Palmas is said to be the most healthy place on the west coast of Africa, and looking at its position, exposed to the healthy winds of the Atlantic, it does not want much exercise of reason to be informed of its salubrity. The highest point on the Cape is 75 feet above the sea, and five substantial houses occupy the commanding sites, a graceful clump or two of palms adding beauty and life to the little rocky peninsula. The colony have called their town Harper as a tribute to Mr. R. G. Harper, of Baltimore, who has distinguished himself as a sincere philanthropist in the cause of the poor Africans, and have erected a capital lighthouse; but as the 'Monrovia' struck a reef or rock 500 fathoms off the extremity of the Cape, the passage on a dark night by this point is not without its dangers.

One of the most singular tribes, according to all accounts, on the West Coast of Africa is the Kroo, commonly called Krumen. They hire themselves out at a shilling a day, per capita, to every steamer and man-of-war destined for the southern ports. They are the most athletic people I ever saw; the masses of corded muscle one sees on the arms of some of them are absolutely astounding. The largest I saw seemed half as large as a 32-pound cannon-ball clapped on a human arm, while the amorphous muscles of the hips, and hams and calves give many of them a most ungainly gait in walking.

The Krus are remarkable for their good nature, and like all tribes in Eastern, Central, and Western Africa, lighten their labours with a rousing song, which though monotonous after a while, is not disagreeably so, the tones having somewhat the effect of a sedative on a white man. While they are certainly an industrious people, and valuable for the extension of commerce in the torrid regions, they are rightly considered the most cowardly. They are in many respects the very counterpart of the Wanyamuezi of Central Africa, with whom the Arab caravans are able to penetrate to the unknown regions north and south of the equator in search of the precious ivory. The Arabs never dream of arming the Wanyamuezi as escorts, as they are utterly unreliable. A Mgogo boy with a spear in his hand would be sufficient to make a legion of Wanyamuezi tremble.

The British General, in his search for dusky allies to drive back the insolent Ashantees from the Protectorate, has long before this time been informed that though the Krumen are invaluable aids on board British cruisers and mercantile steamers trading along the coast, they are useless as allies in war. The very name of battle is a terror to them, and they earnestly stipulate with all new-comers just now that they shall not be expected to go to the war.

The Krumen rejoice under names which prove who their sponsors have been. The sailor's genius is well known for the invention of marine nicknames. He has excelled himself in the multiform and mixtiform cognomens he has bestowed upon his smiling, large-mouthed, white-teethed, flat-nosed, dusky ship-mates of Kruland. I heard the muster roll of those on the ' Benin,' engaged at Grand Sestros near Cape Palmas, among which I distinguished such names as Pea Soup, Jolly Nose, Tar Bucket, Flying Jib, Salt Junk, Main Topsail, Jack Slush, Hard Tack, Tom Duff, Sheet Anchor, Cabin Boy, Star Gazer, Wool Pate, and other curious titles. The British sailors regard these as far more adapted to distinguish the Kru boys one from another than such names as Kalra Klemku, or Marponolongola.

Leaving the shores of Liberia behind, we come to the Ivory Coast, the French possessions, which extend as far as the Assinee river. Eastward of that river we sail along the Gold Coast, which has finally come entire through transfer and purchase into the hands of the English, extending from Assinee river to the Danoe, a coast line of about 290 miles, or from longitude 3° 20′ west of Greenwich to longitude 1° 10′ east of Greenwich. The town of Danoe on the Slave Coast serves as a boundary line to the British possessions and Dahomey. The Assinee river marks the boundary between the British possessions and the French possessions to the West of the Cape.

While the actual British possession comprises no more than a thin coast line, yet the Protectorate, according to the Ashantee Treaty of April 27, 1831, extends north along the parallel of longitude 3° west of Greenwich for a distance of nearly 100 miles, but in 1873 the proposed jurisdiction was to extend only to the southern bank of the Prah, or Boosum Prah. Within this extensive Protectorate are found the tribes of Denkera, Anka, Aowin, Amanahea, Ahanta, Wassaw, Fantee, Assin, Goomwah, Adjumacoom, Agoona, Aquapim, Accra, Akim, Adampe, Aquamoo, Crepee, Aveno, Kerapah. Besides these tribes there are many sub-tribes within the British Protectorate who were all formerly tributary to the famous despotic king Coffee Calcali, who holds court at Coomassie, a large and strong town situated north of the Boosum Prah.

We arrived at Cape Coast Castle on the morning of October 24. The first view of the place was one of a seven-hilled position, the hollows between the hills being occupied with houses white and coloured, which connected in neighbourly manner one hill with another. The mass of mortared walls and houses, in the foreground, close to which the big waves of the Atlantic rush with terrible fury over the rock that forms the Cape, is Cape Coast Castle.

The background is formed of three hills : Victoria Tower stands on the right ; Fort William—a round cheese-box affair about as large as the mammoth cheese we are going to exhibit at the Centennial Exhibition, 1876, at Philadelphia—is planted on top of a steep cone in the centre ; and Connor's Hill rises to the left, on the summit of which are some bell tents, officers' marquees, and a hospital. Connor's Hill is the sanatorium. The middle ground is filled up with hills covered with Government buildings, churches, and hollows filled up with long lines of mud houses, said to be occupied by Christian and Pagan Fantees. This is as much as you can take in at a glance from shipboard, of an obscure settlement just springing up into a notoriety which must make the place and its history a matter of common talk for months to come.

At anchor in the roads are a hospital ship, the ' Simoom,' commanded by Captain Piele, H.M.S. ' Barracouta,' Captain Fremantle, Senior Naval Officer at present, H.M.S. ' Decoy,' and a couple more ; the British and African steamer ' Bonny,' the African Steamer Company's ' Benin,' two American sailing vessels, and a brig owned by the great firm of Swanzy Brothers.

About 8 miles to the right of Cape Coast Castle, a glaring

white mass of buildings rises to view, which I am told is
Elmina and the Castle of St. George of Elmina.

This latter place is classic. It was discovered and settled by
the Portuguese navigator Diago d'Azambuja, as Aldea or the
village of two parts, but soon became more generally known as
Mina or the Mine; but when the castle, which took 86 years
to build, was completed, it was called St. George Da Mina.
The name Mina was derived I suspect from a family of that
name, who were in the middle of the fifteenth century exceedingly rich, and who advanced large sums of money to fit out
the exploring caravels. How it came to be called Elmina is
easy to explain; the prefix *el*, translated into English, means
the —El Mina—The Mine.

About the middle of the sixteenth century, Admiral de
Ruyter captured Elmina for the King of the Netherlands, and
left his bâton there as an insignia of authority for all succeeding
governors. That bâton of ivory and gold passed into the
hands of the British Governor Hennessey on May 6, 1872, the
time of the transfer of the Dutch territory into the hands of
the English, after its occupation by the Dutch during 235
years.

My object in drawing attention to the old Portuguese
castle, is because the cession of the place to the English is one
of the causes which led to the Ashantee invasion of 1873.

The history of the settlements is too complicated a matter
for a special correspondent to unravel while he is on the eve of
a stirring campaign, where his pen will be required any moment
to describe what transpires; but it would be unjust to expect
readers to feel interested in this campaign without revealing a
few salient points such as will furnish them with an idea as
to the purposes and objects of this war.

The Ashantees have been life-long enemies of England,
principally because her Commissioners and Governors have
never thought it worth while to cultivate their friendship
properly, such as their importance merits. It is true they
have always been polite to the king and his envoys, and have
sent him many rich presents.

Ashantee is as large a country as the whole of France or
Abyssinia, though not as populous even as the latter country; the
whole of it is united under one man whom every native is

bound to obey to the extent of his life and property. Abyssinia under Theodore in his last years was torn by contending factions and ambitious princes, and weakened ; but Ashantee can muster under the banner of its king a force of 100,000 warriors.

Time and time again has Ashantee entreated of England the right to come and go freely from the interior to the coast and *vice versâ.* This right has been refused. Yet England, great mistress as she is of the art of commerce, might have listened to the entreaties of Ashantee and granted what they asked, to her own lasting credit and enduring profit.

For nearly a couple of centuries Ashantee traded with the Dutch settlement of Elmina, eight miles from Cape Coast. When England purchased Elmina she sent conciliatory letters to the Ashantee potentate, telling him of her desire to remain on peaceful terms with him, and her hope that the trade might continue.

On May 6, 1872, she took possession of Elmina, with a blare of trumpet, cannon salvoes, and imposing ceremonies.

I now open the Parliamentary Blue Books, from which I extract the following news, which goes to show what was done by England from her occupation of the Dutch territory down to the time of the Ashantee invasion of 1873 :—

May 31, 1872.—Prince Ansah, a cousin of King Coffee, of Ashantee, being at Coomassie, writes to say that the king is pleased with the news which Governor Hennessey has sent, which is to the effect that the road will be opened for trade.

Some missionaries of the Basle mission, pursuing their religious profession quietly, had, about two years before, been captured by Adu Bofu, an Ashantee general, and Prince Ansah suggests that the payment of 1,000*l.* would release Messrs. Kuehne, Bonat, Ramseyer, Mrs. Ramseyer and child.

It should be noted that Governor Hennessey made a demand for their release, and offered to ransom them, though they were not British subjects. The British Foreign Office has already informed the German Government of their good intentions. Von Strausse, in the name of his august master, has thanked the British Government for its good intentions.

This was a diplomatic mistake, leading the Ashantees into the belief that they were people of importance.

The Ashantee General Adu Bofu, their captor, demands 1,800 ounces of gold, or 6,480*l.*

Later.—Governor Hennessey sends a present of pieces of gold-embroidered silk to the King of Ashantee.

In June a son of Adu Bofu is made prisoner, but is unconditionally released by Governor Hennessey, in the evident hope of stimulating a reciprocal release on the part of General Adu Bofu.

This month, also, Governor Hennessey makes a further diplomatic gift of a ring marked with the twelve signs of the zodiac.

At a later period Earl Kimberley, of the Colonial Office, advises Governor Hennessey to inform the King of Ashantee that if Europeans are detained, the annual stipend promised him will be suspended; and at the same time to warn Europeans that if they go beyond the limits of the Protectorate they do so at their own risk.

[A grave diplomatic mistake this on the part of Earl Kimberley, making use of such a threat when those Europeans were not British subjects, and when he was not called upon to do anything in their behalf. This fault, too, after Chinese and Abyssinian experiences!]

October 11, 1872.—The Ashantee king, lulling poor Governor Hennessey into a false security, writes to say that he is about to send to make peace, upon which Governor Hennessey states his belief to be that the release of the captives is of minor importance compared to peace with Ashantee.

Poor Governor Hennessey seems to have forgotten that he stated that peace could only be made by the release of the Europeans at Coomassie.

Peace with Ashantee means an influx of gold dust and native produce from Coomassie to Cape Coast and Elmina, which the conclusion of a sound peace would surely bring.

September 24, 1872.—The Ashantee king writes to say that he is willing to let the captives go for 1,000*l.*, which shall be paid half in gold dust and half in goods.

At this time a negro named H. Plange is the commissioner at Coomassie, negotiating for the release of the European captives, at a salary of 30*l.* per month. Mr. Plange writes to say that the chiefs, who overrule the young king, declare it their belief that if they only hold out they will obtain the whole sea coast in exchange for the captives.

[No wonder they do, after what Earl Kimberley and Governor Hennessey have written.]

Mr. Plange further says that the question at issue seems to be, shall the Ashantees come down to the coast to trade with British merchants without being made to pay to Fantees as middlemen? for by trading directly with the merchants they gain 50 per cent.

February 5, 1873.—There is a sudden rumour of an Ashantee invasion of the British protectorate.   Colonel Harley is at this time administrator at Government House, Cape Coast Castle.

February 10, 1873.—An Ashantee prisoner states that King Coffee is coming down to take Elmina, and Colonel Harley asserts that the King of Elmina has eaten fetish with the King of Ashantee against the British Government.   The opinion prevails at this time that there is a strong dislike to the British Government throughout the towns lately ceded by the Dutch to the English from Elmina to Assinee.

Governor Hennessey, at Sierra Leone, charges Colonel Harley with having quarrelled with the Elmina king; that he has also refused permission to Elmina people to practise certain indecent ceremonies. The only evidence of dislike to be found is in Governor Hennessey to Colonel Harley.

Middle of February.—Colonel Harley reports the first outbreak of the war.   The Ashantees have invaded Assin in the British Protectorate, and have burned nine villages.

A new cause of the war now discovered, through a letter which was mislaid somewhere at Sierra Leone, but which finally comes to light at the Colonial Office.   It reports the capture of an Ashantee chief, called Atjeimpon, who is uncle to the King of Ashantee, and his subsequent imprisonment at Cape Coast Castle.   He, however, has been started towards the Prah through the Assin country, which the Assins were loth to permit, but prevailed upon by Colonel Harley, they finally withdrew opposition.   It was on or about the 12th of December, 1872, when he started from Cape Coast.   It must have been about the beginning of January 1873, when he found himself across the border river, in Ashantee land, from which we may deduce the not unreasonable supposition that the time which intervened between his arrival at Coomassie before his nephew the king and the invasion of the British Protectorate was spent in the organization of the three great Ashantee divisions which rolled into the Protectorate at three different points—Denkera to the right, Prahsu in the centre, and Akim to the left, computed to be in the aggregate between 30,000 and 40,000 warriors.

Opposed to these splendid organizations there is a statistical number of 60,000 or more of the Fantee confederation, badly disciplined and organized, and incapable of effectual resistance, owing to their incohesion and want of a recognized chief.

Later in February.—King of Abrah volunteers the information to Colonel Harley that the invasion has been caused by the cession of the Elmina fort, and the Elminas having become British subjects, since King Coffee declares that from time immemorial his ancestors ate and drank at Elmina—in other words, that the fort should have been his, and that *he means to come and take it.*   It is also ascertained beyond doubt that the King of Elmina, having eaten fetish with him, was his sworn ally, and that he only waited the near approach of the Ashantees to declare himself and his friends from Elmina to Assinee for the Ashantees.

The King of Elmina is taken prisoner and shortly finds himself deported to Sierra Leone, having first positively refused to take the oath of allegiance.   At Sierra Leone he is to be retained until peace is declared between Ashantee and Great Britain.

The above are the notes I have taken from the valuable Blue Books ; and since the British Government and Press, up to the time of my departure, failed, so far as I could see, to perceive the causes which have led to this present Anglo-Ashantee war, I feel myself at liberty to suggest reasons for it.   From all I can see, war with Ashantee might easily have been avoided.   It has not been thrust upon the British Government.   The Ashantees have simply invaded the Protectorate, at which the British Government might have smiled undisturbed, for all that the Ashantees might have injured any of England's actual possessions along the seaboard.   Not a castle nor fort need have been surrendered had there been a million Ashantees encamped within rifle shot of the weakest.   British merchants might certainly have suffered diminution of trade.   This, however, need have been no cause for England taking an active part in the war or launching into a campaign to take Coomassie.   Slackness of trade and decrease of revenue were things to be deplored, but ought not to be considered just causes of war.

How, then, will you answer the question why England has taken up arms against Ashantee ?

If we are right in assuming that the Protectorate, which was never more than nominal, is of sufficient importance to England to maintain it as her own against Ashantee, of course we may then find a cogent reason for resisting the invasion, but to assume that such is the case is going beyond what the British Press has strenuously denied.   Regarding Great Britain only in the light of a great commercial Power, I should say that, whether there were logical reasons or not for accepting the gage of battle which King Coffee has thrown down, she has done perfectly right in entering upon the war, not that she was bound to do so, for that I distinctly deny—Earl Kimberley and Colonel Harley's letters being sufficient for my authority—but

c

because it promises to be a gain to her.   There is no reason at
all why she should not strive to make Ashantee tributary to
her.   King Coffee is too rich a neighbour to be left all alone
with his riches, with his tons of gold dust and accumulations of
wealth to himself.   Gold ought not to be hoarded, it ought to
be circulated freely.

Ashantee would be as rich an acquisition to the British
Crown as the Island of Cuba to the United States, for the people
are born traders, and it only requires a little careful management to be paid twenty times over for the cost of the expedition to Coomassie.   Those who fail to see this thing in its true
light, fail because of prejudice and bad taste.   A change of
masters would be a glorious thing for Ashantee.   Instead of
the despot who chops off a couple of thousand heads on the
burial of his predecessor, the people would have a rich and
generous nation to treat with, which is among the most skilful
and industrious in the world, and stands higher than Ashantee
even in its love for trade.   All Central Africa would soon be
benefited, and the inhabitants around Lake Tchad would in
time come to marvel at the palatial houses of the white merchants whose kafilahs ranged through the untrodden wilds of
the interior.   The people of the Protectorate think that this is
as much England's war as their own, though Colonel Harley has
often told them that the English have nothing to do with the
war, that it is the Fantee-Ashantee war ; but the British Government is now inclined to their way of thinking—hence the
expedition to Coomassie.

It is now nine months since the war began.   Another month
will probably roll by before the white soldiers will come and
take up their line of march.   England wasted time in talking
and negotiating ; but she might have learned a lesson of
promptitude from the savage Ashantees, who, quick as were
the Prussians to invade France, were quicker and more expedi-
tious than even they in their invasion of Fanteeland.

The English have been twice unsuccessful in their war with
Ashantee.   In 1823 Sir Charles McCarthy and 600 gallant
fellows perished before the furious onset of the Ashantees, and
that brave soldier's skull, gold rimmed and highly venerated, is
said to be still at Coomassie, used as a drinking-cup by King
Coffee.   In 1863–64 the English suffered severe loss.   Couran

marched to the Prah, 80 miles from here, and marched back again, being obliged to bury or destroy his cannon and hurriedly retreat to Cape Coast. I have not the information by me to give you the details of the disaster, but all the coast men speak of that expedition as 'ill-fated.'

It is now Sir Garnet Wolseley who is to try his fortune with the Ashantees. His antecedents lead us all to expect that it will be as successful as Napier's march to Magdala, though not so bloodless.

CHAPTER II.

November 8, 1873.

In the first chapter I endeavoured to inform readers how, and
why this war began ; I must now introduce them to Cape Coast
Castle, its authorities, and its people.

I could not go ashore with the other passengers, as I had
to superintend the lifting overboard of the little steamer the
' Dauntless,' which the proprietor of the ' Herald' had been
generous enough to let me have for the collection of news along
the Gold Coast.

About four in the afternoon, however, the 'Dauntless,' with a
large surf-boat loaded with coal in tow, steamed from alongside
the 'Benin,' towards the shore. The waves were uncommonly
high, and the surfy sea rolled its waves ominously in front.
Arriving within two hundred yards of the beach, the ' Dauntless' was anchored in eight fathoms of water, her engineer and
two Krumen were left in charge of her, with a caution not to
leave her until morning, by which time it would be supposed
I should find some quiet anchorage for her.

The coal boat was then rowed ashore. There were twelve Fantee
paddlers on board, strong brawny fellows, who had been buffeted
by the fierce surf of the Gold Coast for many and many a year.
We soon began to feel the influence of the big waves ; the
Fantees paddled gently, until one wave after another had
impelled the boat past the point of rocks on which the Castle
is built, when with a mighty shout and a mighty simultaneous
effort, the boatmen saw their chance, and sent her far on the
beach on the crest of a devouring wave.

Coal, however, was such a cargo that would take hours to

unload, and the booming surf still kept up its unceasing bluster and lashed itself into white foam all round the boat, which made it a most difficult job to unload; but about an hour after sunset we had got the coal on shore, and at this time, faint from exertion, dinned by the briny confusion and vocal noises of the Fantees, I was glad to hear the voice of the boy-servant of Captain Butler say—

'Please, sir, Captain Butler told me to tell you his room is ready for you, and dinner will be sent to you from Government House.'

A watchman had to be engaged to keep watch over the coal on the beach, lest the Fantees might carry the "black stone" away, and I then turned my steps towards Butler's quarters at the Colonial Surveyor's house.

Arriving at the house in anything but a presentable state, I found Captain Butler and another gentleman, who was introduced to me as Captain Charteris, A.D.C. to Sir Garnet Wolseley. Captain Charteris was the bearer of an invitation to dinner from the head-quarters mess. Captain Charteris had already impressed me most favourably, but the hospitable invitation considerably prejudiced me in his favour. After granting me time to dress, to change my surf-soaked clothes for the soft white flannels suitable for an enjoyable night in the tropics, we marched for Government House.

It is a barn-like building, about 100 feet by 80, mortared and whitewashed. It looks very imposing with its grand staircase buttressing the house, with a garden of bright flowers in front, in which may be seen the stuccoed forms of antelopes side by side with the broad gaping-mouthed mortars pointed threateningly at the innocent stars in the heavens. A couple of sentries promenading in front of the house add very much to the imposing appearance of the building. They gave effect to the house, in my mind, and smothered the half-formed desire to criticise Government House. We ascend the staircase, lights flash here and there, I get a peep at the pantry as I arrive at the top, my eye strikes off at a tangent to my right and catches sight of a well-laid dinner table; but to my left is the reception room of Government House, where the intending diners are assembled.

Captain Butler steps up with a gentleman to welcome his *compagnon de voyage*, and says—

' Mr. S——, Sir Garnet ! '

This stately little gentleman of proud military bearing, quick bright eye, broad high forehead, ardent temperament, a sparkling vivacious intelligence animating every feature—this then is Sir Garnet Wolseley—the pacificator of Red River, and the young hero chosen for the command of the British Expedition to Coomassie.

He is the very reverse of my conception of Sir Garnet Wolseley, who called the gentlemen of the press ' drones,' and a ' curse to modern armies.' If he had not been a soldier by his appearance, I should judge him to have made a first-class special correspondent, just the man to have seized an item and dared a general-in-chief to lay hands on him, just the man to be sent to any part of the world to collect news. His eager eyes betray the inquisitive soul and indomitable energy. Taking no offence whatever at his sharp-tempered criticism of the ' Necessity of the Age,' I admit at once that the British Government could have found no worthier man to entrust the castigation of the Ashantees to than Sir Garnet Wolseley.

At a luxuriously-spread table I found Captain Fremantle, of H.M.S. ' Barracouta,' with his left arm in a sling, from a wound received at the mouth of the Prah ; Major Baker, of the 18th Royal Irish, Chief of Staff ; Captain Brackenbury, military secretary of Sir Garnet, who is reputed to be a very able officer ; Captain Charteris, A.D.C. ; Lieut. Maurice, author of the ' Wellington Prize Essay,' who received the prize even above Sir Garnet himself ; Captain Butler, author of the ' Great Lone Land,' who is about to be sent on a special commission to the tribes of Akim ; and three other officers, whose names I forget.

Sir Garnet I found to have extensively travelled through the United States. He is said to be the youngest general in Her Majesty's service ; he is not forty years old yet ; and as he has not risen to his present rank through interest, it may be assumed that he has shown himself a capable and energetic officer, to have attained to his present position.

A glance at his ' Soldiers' Pocket Book ' will show the qualities of which he is possessed. Whatever its faults, every line is written like a soldier, and from a soldier's stand-point. There is no nonsense about him. The great aim he has in view seems to be to instruct young officers in their duties, and to inculcate the idea that everything should be sacrificed to success. He teaches them the art of diplomacy in their dealing with ignorant savages, as well as with European armies; he tells them how to avoid being interviewed by their messmates or press people, and how to govern their features when submitted to a too close questioning.

The General's dislike to press people does not arise from any black humour or any bilious acridity in him, for he is a most urbane gentleman; but it arises from some exaggerative conception he has formed of the Special Correspondent. I can imagine a meeting between this proud and haughty soldier and a 'special' from some of the unclassic districts of Western America, who has neither fear of danger nor reverence for good in him. I see the soldier struggling to be polite, and calm, and to be forbearing with the sneering and fleering man from the West, who is perpetually troubling him with impertinent and irrelevant questions about his age, the soundness of his teeth, the good character of his ancestry, until the choler of the soldier is kindled to white heat, and the impudent pen-trotter is seen fanning with his arms fifty feet in mid-air.

It must have been some sinister figure of this kind that Sir Garnet had encountered or imaged to himself as 'a gentleman of the press,' until he had conceived a morbid dislike for the whole tribe, and never omitted an opportunity to inveigh in terms unmeasured against them as 'a curse' to modern armies.

Fortunately, however, the representatives of the great London and New York dailies are of widely different material from the irresponsible reporters 'out West;' and Russian, Prussian, French, and English generals have found them not an hindrance but a valuable aid in their campaigns.

A gentleman on Sir Garnet Wolseley's staff during an argument with me relating to this very subject, and who thoroughly shares Sir Garnet's hatred for newspaper men, when it was suggested to him by me that if Sir Garnet in a European war merely trusted in a reporter's honour not to mention anything that would furnish information to the enemy, no *gentleman* of the press would disappoint him, blurted out—

'Trust in his honour, by heavens! I would trust to nothing less than his back. On the first publication of anything that I thought not proper, I would tie him to the triangle and trust to fifty lashes well laid on his bare back, not to do the like again.'

Whence one may infer that Sir Garnet is not singular among the military class in his hatred for the Press, and that it is rather a queer position a correspondent, however unobtrusive and considerate, occupies with this British Expedition to Ashantee.

I have been about Cape Coast Castle for some days now, and begin to feel able to write intelligently concerning this expedition to Coomassie.

I know not, however, who to blame for choosing Cape Coast Castle as a point of departure for Coomassie ; whether it is the British Government, the former governor of the Gold Coast, or Sir Garnet Wolseley.

Probably all three should have a share in the blame. Cape Coast was settled by the Portuguese, but was ceded to the Dutch in 1641, with whom it remained till 1655, when it was taken by the English and secured by treaty in 1667.

For 206 years this place has been in possession of the British, yet one knowing the character of this enterprising people, would barely believe it to be the fact, that the English have not improved the place since the first day they occupied it. They have kept the Dutch-Portuguese Castle which they received from the Dutch, in repair, they have built a Government House, post-office, and two or three other offices, and a church, and that is all during 206 years' possession !

The harbour they have left as nature made it, though a ship-load of concrete, laid down to a distance of 100 feet from the point of rocks on which the castle stands, would have secured an admirable landing-place for ships' boats and passengers.

What the Dutch have done at Elmina is in striking contrast to the apathy of the English at Cape Coast Castle. The Dutch made the Beyah River a little port, they walled it up on each side, threw a bridge over, until it looked like a miniature Thames ; they built a formidable castle on the hill of St. Iago, besides constructing the De Veers and the Beckenstein Redoubts. They cleared the downs and hills in the neighbourhood from forest, built charming residences in the suburbs, and cultivated farms whose buildings, embowered by orange and papaw trees, are pictures of rural felicity.

A view from the top of Connor's Hill will show you the wide prospect of bush which seems to threaten to bury the town itself in its wide-spreading arms.

The blame is not in the country ; even with very little system it has had always a surplus of revenue,[1] which one may see, by the Blue Books, has been squandered inadvisedly and fruitlessly. Instead of expending the surplus for the improvement of the port and the construction of roads into the interior, it has been spent in sending presents to the King of Ashantee and feasting his ambassadors, stipending the fractious tyrant instead of improving and disciplining levies of native troops against the emergency of an invasion which is now upon them.

The invasions of 1821–23, 1840, 1863-64, and 1869, have not taught the English that the time had come to demonstrate to the Fantees what object they had in holding any part or parcel of their coast. Probably that of 1873–4 will pass away without any lesson also being derived.

So little did the English Government know what Cape Coast Castle was, what had been done, what ought, and what might have been done, that Sir Garnet Wolseley was despatched before the troops to report on the situation, and find out what remained to be done.

Sir Garnet landed with one of the most efficient staffs that a general could be blessed with, and anticipating doubtless that difficulties would be encountered, from sheer despair of ever being able to do anything systematic in time, before the patience

---

[1] The statement issued of revenue and expenditure shows a considerable increase during the last twenty years in the public income of all the West African settlements, except the Gambia, in which, however, there has been some increase, though not so great as in the other settlements.

The revenue of Sierra Leone in 1853 was 26,659*l.* In 1871 it was 80,406*l.* The average revenue of that settlement for the ten years 1853–62 was 32,397*l.* The average revenue for the ten years 1863–72 was 64,066*l.*

The average revenue of the Gambia for the ten years 1853-62 was 15,645*l.* The average revenue for the ten years 1863–72 was 18,208*l.*

The average revenue of the Gold Coast for the ten years 1853–62 was 9,540*l.* That for the years between 1863 and 1872 (as far as the returns have come to hand) was about 20,000*l.* The income in the year 1853 was 13,249*l.*; that of 1872 was 42,745*l.* The estimated revenue for 1874 is 52,438*l.*

The settlement of Lagos was only annexed in the year 1862. The public revenue for the year 1863 was 15,837*l.*, that for 1872 was 41,346*l.*

The general increase of revenue which has occurred in the settlements is due to the growth of trade chiefly, not to additions to the burden of taxation.

of England was exhausted, determined to accept the difficulties as he found them without losing time in attempting to remedy them.

The first act that Sir Garnet did was to accept Cape Coast Castle as the place of departure for Ashantee, though, as I say, he would have done much better by making Elmina the port, dredging the Beyah River, and laying a railway from Elmina to Cape Coast along the smooth sandy beach, a work of about three days.

Sir Garnet's intentions are to send forward provisions and stores of war to the front, to stockade posts along the main road to the Prah River, then to send for the white troops, and on their arrival march them immediately to the interior without tents.

They say no animals are available for carriage here—all animals die as soon as imported. This may be true of the horse, the mule, and the ass ; but there is one animal not yet mentioned which would feel quite at home in this country, and that is the elephant.

Any one of England's troop-ships is capable of conveying 100 elephants here within 40 days *viâ* the Suez Canal from Ceylon or Bassein to Cape Coast Castle. The elephants, though timorous at the sound of battle, would not be so timorous as Fantees ; besides the Ashantees are not supposed to know that the ponderous brutes might be stampeded by the strife of battle. A certain lasting awe would be created in the Ashantees' mind at the fact that the white men could compel the service of so many lords of the forest.

Another benefit would accrue from the employment of elephants. They would tramp a road out for themselves and the use of the troops—a broad path, 3 feet wide—and thus the long delay of cutting out a road would be avoided. The road thus made might easily be widened to 10 feet, from the sea to Prahsu, the central depôt, and future outpost of the Protectorate. When I left England, it was reported that a railway was to be constructed from the sea to a point 40 miles inland. Such a project has been abandoned, being impracticable, and traction engines have been adopted instead.

The railway was impracticable from the nature of the land, which is very uneven ; but if it was ever seriously meditated—

as I have no doubt it was, otherwise the Government would never have shipped the materials—all I can say is, that the country was not surveyed for a railway; that if it had been, I have no doubt a route would have been found offering special advantages. The depôt might have been established near the

AN ELDER OF THE PEOPLE AND A WARRIOR.

Salt Pond, to the right of Cape Coast Castle, and the railway traced along the low valley that runs from it to the interior.

One of the traction engines has been landed on shore with a couple of trucks. The whole is made of steel, combining strength with lightness, the engine weighing about two tons and a half; but it is totally inadequate to the requirements.

After a feeble travel of about two miles, and twice tumbling into a ditch, it returned to the neighbourhood of the castle, where it is now employed in the more undignified labour of cutting planks for the Control Department. It was found that nearly as good, and as level a road, as would be required for a railway was required for the traction engine.

The General perceives that the great drawback is want of labourers. He cannot get the Fantees either to labour or to fight. Every Fantee man thinks himself qualified to fight, and has for long contrived to imbue the minds of the Cape Coast Governors with a high idea of his noble courage and martial worth ; and the Governors, consequently, have not been slack in supplying the Fantee with the means to exhibit his fighting qualities with the improved Snider and ammunition. But alas ! for their trustful confidence, the Ashantees have caused them to retreat precipitately before them, and in their perpetually recalcitrant flights have involved the death of many a noble British officer, who thought, by sacrificing himself, to make them more worthy of their leaders at a future time.

The more I ponder upon the long possession of this country by the English, the more I wonder that the qualities of the people whom they govern have not been perceived before, and that they have resolutely closed their eyes to the superiority of the Ashantees over the Fantees. With 5,000 Ashantees a British officer might be able to defeat and annihilate a force of 50,000 Fantees ; but with ' 70,000 Fantees ' any number of British officers are not able to drive 10,000 Ashantees from the neighbourhood of Cape Coast Castle. In the meantime, after advancing thirty-three miles into the interior, we are halted.

The Fantees lack confidence, owing to the immediate vicinity of the dreaded enemy, and are fearful of trusting themselves under the superintendence of a few officers to the labour of constructing a road, and we must either wait for the retreat of the Ashantees across the Prah, or for the arrival of 500 English navvies, who will shame the natives to work. Indeed, it is getting to be a most serious question this of transport and labour, and knowing the disadvantages under which a young general like Sir Garnet must labour, I begin to feel anxious for him. I doubt if he would have the courage to express the depth of despondency he must feel sometimes, when in spite of all his efforts, his wonderfully patient treatment of the people, and his long-suffering with their shortcomings, he is told by the Control Department, by the engineers and others, ' We can't get men, sir. They have all run away into the bush.'

Sir Garnet has the power to remedy this evil, but I fear it requires a more callous and obdurate soul than he is possessed of to employ the power his position and authority has given him. If, instead of sending out traction engines and railways, the British Government should send 500 sets of slave chains, and bind these runaways into gangs of fifties, each controlled by a non-commissioned officer with a long whip, the transport and labour question were for ever resolved, and success, a brilliant success, would be certain. When the expedition would be over, and the war ended, the British Government might compensate the people for the annoyance of being collared with iron bands, and apologise to them for the extreme measure they were compelled to resort to, to insure the independence of the country and lasting immunity from Ashantee atrocities. At first sight the proceeding may appear very cruel; but the other side of the picture should be seen before condemning the measure. Supposing that the British Government, seeing the utter impossibility of taking an expedition to Coomassie without the means of transport, withdrew from the war and surrendered the Protectorate to the mercy of the Ashantees, what would be the consequence? The Fantees, through their cowardly fears, unable to resist the fierce attacks of their enemies, would lay down their arms and be driven as slaves to Coomassie, to be butchered by thousands at the obsequies of some of the royal blood of Ashantee. On the one side is a temporary inconvenience; on the other side is an eternal servitude or death.

This expedition to Coomassie is to insure the future peace of the Protectorate; to prevent future invasions and their consequent relentless butchery of thousands; to restrict the Ashantees within their own territory; to visit them with some portion of the severity with which they have repeatedly visited the British Protectorate since 1811. All British expeditions to Ashantee hitherto have failed, and the conceit and arrogance of the enemy have grown in proportion. If, through fear of the condemnation of the Peace Society of Great Britain, the present expedition fails, because the Government dare not prove all measures to insure its success, all I can say is that the British Government will deserve the disgrace that will inevitably follow the final defeat, and to become the laughing-stock of the civilised world. Rather than suffer defeat for the sixth time, the British Government should seize upon every member of the sentimental Peace Society and deport them to the Gold Coast, to supply the deficiency of labour that is daily growing more alarming, to make the road to Prahsu through the jungle and convey the stores necessary for the subsistence of the British battalions now on their way out. This might probably be called an extreme measure, but I doubt if civilisation would visit the British Government with much harshness for it—nay, I think it would consider the act commendable and justifiable.

I have been wandering through Cape Coast Castle. Now behold the town of Cape Coast, like a smouldering volcano, blazing and burning and smoking in the hollows between the seven hills, domed by a sky of brass seven times heated by fire. In these smoking hollows, subjected to the heat of the sky of brass, are young English officers, fresh from the cool shades of the willows, the elms, and the oaks of England; from the verdant parks and breezy green hills; from the banks of salmon and trout streams. They are superintending what gangs of labourers they have been able to obtain, removing hills of commissariat stores from one spot to another.

Speak to any one of them about the heat of Cape Coast. You would be surprised at the depth of compassion in your soul when you would see the poor fellow's parched face, his blistered cheek, and the compassionate longing for a draught of cool water, which you see lurking in his eye. His tongue almost refuses to move, it feels so stiff and dead within the dry mouth, and finding it too great a task to talk, he gasps a deep deep sigh, which finds an echo in every recess of your soul.

Before long, rendered desperate by his sufferings, he seizes a cooler of water in the vain hope of assuaging the intolerable thirst, and drinks a long long draught of the tepid liquid, which soon transudes through every pore of his body in large drops of perspiration. He feels but a slight mitigation of the thirst which consumed him, and ever and anon during the day the friendly cooler is sought for.

When he finds the sun unbearable the half-scorched officer seeks the shade, where the breeze lulls him into the belief that the shade is far more comfortable than the glare and heat of the sun, but somehow or other there is a clammy chilly feeling at his back which he cannot explain. The air is warm, nay it is hot, sometimes so hot that it appears to scorch his very vitals, yet the spinal column feels uncomfortably cold, and a vague feeling of regret steals over his mind that he has allowed himself to drink so much water.

The next morning at breakfast he feels all right enough, accosts his messmates with a cheery good day, and sits down wondering at the ravenous appetite which seems to possess him, and while commenting upon it to his companions with a light laugh, declares it, as his opinion, that the Gold Coast has been very much slandered, that the sickness has been very much exaggerated, and ends generally by saying that he never felt better in his life, and that he wonders at people getting sick.

Suddenly, however, he pushes his plate away; an uncomfortable feeling has taken possession of him, even before the sound of his words has died away. He rises to his feet, stretches his arms, and yawns terribly, and with an effort represses the half-formed shudder that his body makes. He soon leaves the room, complaining that he feels chilled a little. If you would follow him in about half an hour, you would see him stretched on his bed with a pile of blankets over him.

With all the blankets, however, you will hear him complain that he is cold, and his chattering teeth are sufficient evidence that he does feel cold. He is suffering

> The bitter change
> Of fierce extremes—extremes by change more fierce—
> From beds of raging fire to freeze in ice.

He drifts from the equator to the arctic zone, and from the frigid to the torrid within a few minutes; he is one moment trembling and blue from cold, the next every pore of his body is a perspiring fountain, he shivers with quick-chattering teeth from raw waves of cold that seem to flow along his spine one minute, the next he tosses the clothes away as though he were being parboiled.

Finally the cold fits or aguish feeling is over, and a cracking headache follows with a steady profuse perspiration. When he has undergone about ten hours of this horrible experience, he feels as weak as a kitten, and with an abashed face, which has strangely enough assumed a jaundiced hue, he confesses that he has had the fever, as many a good man has had before him.

The next day he is down again, and on the third day also, when the Tertian departs from him, if he has been wise and taken the invaluable specific, quinine sulphate, in time—if not fever is not to be got rid of until the tomb closes over its unhappy victim.

This fever, as also the Indian jungle fever, is nothing more than a virulent ague. The agues of the Mississippi and Arkansas Valleys are mild compared with these agues of the African Equator. The former may be laughed at, may be taken and endured, to take and endure again, without seriously endangering the constitution, just as one would take a fit of sea sickness. The experience is bitter enough while it lasts, but the patient after a while finds in his secret heart a certain contempt for it. The African fever is a vicious sickness, deluding one into the idea that it is not very dangerous by its mild preliminary symptoms, but it sometimes changes suddenly, drops its intermittent form and becomes remittent, or continued, and from three to five days only are required to kill the stoutest men. It is the remittent type that has been so fatal to travellers in Africa, that slaughtered the members of the Zambesi Mission, that terminated poor Mrs. Livingstone's life at Shupanga, that decimated the unfortunate marines at Elmina. Even now I see that its brother, the yellow fever, has appeared at Shreveport and those towns along the lagoons of the Mississippi, reaping a fearful harvest of lives. The malignant remittent fever of the Equator is a distinctive name without much difference from 'yellow jack.' The symptoms are almost the same, except that the African remittent is not contagious.

The intermittent or ague of this continent insidiously saps the best constitution. Its effects are either seen in an early decay of the whole system or in a weakening of the mental faculties. A man hitherto noted for his excessive energy and industrious habits soon begins to feel a heavy languor or torpitude restraining every impulse and faculty.

One of the most remarkable features in a British Expedition to me is the quantity of varied but excellent advice which each officer gives his fellow on first arrival, whether it is in Abyssinia or Ashantee Land.

In Abyssinia the advice was, 'Fortify the system, sir, drink brandy, soda, and bottled beer,' often pronounced by the educated classes in England bottled ' beah.'

In Ashantee the advice is more varied, but as excellent. ' Live generous, sir. It is the best thing in the world, sir, depend upon it. Patronize beer, claret, and good wines, wear a cholera belt about the loins, wear flannel next the skin, bathe twice a day, avoid the sun as you would ——, exercise moderately, but not in the early morning or at night. Beware of late dinners and suppers. Avoid excess in eating, and all unripe fruits. Should you feel a fainting in the stomach don't rush to the bottle to sustain it, but take a wee wee glass of sherry and bitters. Take a glass of wine every morning as a stimulant.'

With all of which excellent advice, however, men suffer from the fever, confess to languor, lassitude, feeble appetite, and failing of strength, and in a few weeks are compelled to be invalided and sent home to regain their bloom and brightness of manhood, which had almost been wrecked during their short stay on the Gold Coast.

It is a common mistake with many new arrivals here to suppose because a week or so elapses before they have been attacked by fever, that the climate on the Gold Coast has been exaggerated; but before long it turns out that these robust people, who pride themselves upon their strong constitutions, invariably suffer more in the end than those who have had to undergo on short notice a seasoning fever.

Yet it is so difficult to conceive that Coast Castle is so very unhealthy, especially to those who have their head-quarters at the castle, exposed to the genial sea breeze, or to the dwellers on Connor's Hill, in the cool chambers of Government House, or at the Colonial Surveyor's office. They

D

are wonderfully exempt from fever, I admit, and so was I while I enjoyed the refreshing sea breezes ; but owing to the crowded state of the castle, and the continual arrival of new people, I was obliged to seek quarters elsewhere ; and as few eligible quarters offered, I sought a house in one of the valleys between the hills in the native quarter. Through this hot and oppressive valley a mere thread of a rivulet meanders sluggishly from the base of Connor's Hill to the beach near the Cape. On its banks a thin greenish scum rests, slightly covering the earth. In front of my house, close by the stream, is a grove of palm-trees. Under the cool damp shade the humid atmosphere has begotten the malarious fungi which cover the soil, as with a green veil. A little to the left of the palm grove, not a hundred yards from my house, there is a deep excavation, about thirty feet in diameter, and lower than the little rivulet flowing by its mouth. The excavation has drained itself full of water ; the surface of the excavation, now a pond, is covered over with weeds and plants. Out of this pond, the sluggish little stream and the slime under the palm grove, is exhaled every evening and throughout the night the miasma which causes so much misery in Cape Coast.

What follows? What followed with me after coming to live in the neighbourhood answers the question. On the second day I was ill with fever ; on the third day my servant had an attack ; on the fourth my native cook came to me, with swimming eyes and throbbing forehead, to ask for medicine.

The fever was defeated again, and left us all with only its bitter memories ; with abated energies and strength ; with frames susceptible to a new attack of the disease, when it has once more gathered force sufficient to be aggressive. When I summon resolution to leave the fetid valley for the heights, crowned by the Government offices, my lungs inhale new life from the pure sea breezes which are constantly being wafted ashore from the wide Western Ocean, and a spasmodic energy is felt for a time, until the stagnant vapours of the valley, where I live, shall sap and destroy it again, and lay me once more prostrate.

In addition to this valley, with its filth, which poisons the whole of Cape Coast, the late rains have demolished over 300 mud-built houses, which crushed to death, as they fell on the occupants, about eight human beings ; and what the mud huts buried besides is not known. The ruined heaps are very suggestive, as you look at them with eyes constantly seeking for the causes of the disease which has given Cape Coast its unenviable reputation, though you would hardly hazard the statement that death lurks under the brown ruins in the shape of putrid garbage and corrupting humanity ; yet I know it would make one feel more comfortable if the ruins were cleared away, and the spaces given up to light and air. No one can deny that the sanitary condition of the place would be improved.

On a hill proudly rising above the bush-clad terraces and hollows in its neighbourhood behind Cape Coast stands Prospect House, whence your eye may command quite an extensive view. The whole town of Cape Coast seems to slumber at mid-day in the warm haze which floats above it. Fort William stands silent and sentry-like before Prospect House, as the stranger looks towards the ocean ; Victoria tower, away to the right, seems solitary and alone ; and Connor's Hill to the left during the hot hours of day, crowded as it is with white tents, looks deserted. The view from Prospect House is worth seeing, and one would imagine that when the breeze blew strong, the veranda at Prospect House would be very enjoyable.

Yet, strange to say, Prospect House has proved to be very unhealthy. Every inmate of it since the landing of Sir Garnet Wolseley has been laid prostrate through fever. The cause of it lies in the bush-choked hollows around it, and the fetid valley which runs a straight course from foot of the House to the beach. The wind bears on its wings the exhalations and drops the genesis of disease with which its wings are clogged right at the portals and on the veranda of the house, and human lungs respire and inhale the poisoned air. Major Baker Russell, who occupies the house with other officers, seems to be a favoured child. He is tall, stalwart, and strong ; and when we landed was a very picture of robust manhood. He was happy at the prospect of living at Prospect House. Men envied him the breezy height which seemed to ensure health, and said, ' What a lucky fellow ! ' Of all the officers which the steamer ' Ambriz ' brought to

Cape Coast Castle with Sir Garnet, Major Russell suffered from
fever first.

Already the hospital ship 'Simoon' has been the residence
of many a fever-stricken patient.   Every man however con-
fessing to ill-health is not suffering from fever.   The disease
in some cases manifests itself in severe attacks of dysentery,
and three officers of the staff are at present laid up with this
sickness on the 'Simoon.'

What is the element which produces so much sickness
amongst Europeans in Africa?   Opinions are divided.   Some
doctors say it is the water, some doctors say it is the poison
exhaled from stagnant water, while others again declare it to
be a deficiency of ozone in the atmosphere to disinfect the
malaria in the air.   It has been proved that if a man spent
twelve hours in a hot-bed of malaria, and lived the next twelve
hours where the ozone is pretty well developed, the quantity
of ozone inhaled into the lungs would be sufficient to neutralise
the malaria taken into the system.

When a person cannot move so quickly into an ozonic atmosphere, twenty grains of quinine should be taken, which is
said to contain the same quantity.

I have not the slightest doubt myself but that the savants
are correct, and the quantity of ozone in an atmosphere should
be seriously considered by people when about to choose residences.   Fortunately this is easy to discover by an ozonometer,
When this instrument indicates over 8° in the atmosphere, the
resident might accept the comfortable assurance that if the
malaria was not in excess the intending resident might enjoy
tolerably good health.

But what can the traveller or the soldier effect in the way of
preserving himself from a deficiency of ozone?   What can we
do in the expedition to Coomassie?   What can the traveller in
the wilds of Africa do?   What comfort could be found by the
traveller reading on his ozonometer, maximum of ozone during
the day 2°, maximum during night 1°.   He must do as others
have done before him—grin and bear it, and take quinine.

Do you who live in healthy countries, who are able to preserve your health with a fair amount of exercise and a little
attention to diet, realise what it is to be constantly suffering
from fever, to bear about with you the seeds of that which may

at any moment lay you prostrate in delirium, to be perpetually anxious about the due taking of quinine and cathartics?

Our condition is best described by the melancholy reply given to the traveller when crossing the Pontine Marshes:—

'Nous ne vivons pas, nous mourons,'
(We do not live, we die.)

In a town on this coast where statistics are kept, records prove that within eight years the deaths numbered 1,649, while the births were only 812, showing that the deaths exceeded the births by 837. These people, with the poor marines who lately went into action 105 strong, and returned aboard ship only 25 strong, might well say with us who have come to the Gold Coast, 'We do not live, we die.'

If the administratorship of the Gold Coast were given to some one who, combining great experience in office with enough energy in him to carry out what his experience suggested was necessary for improving this unhappy state of things at Cape Coast, the evil name now attached to the place might soon be eradicated. Such a man is Captain Lees, now Colonial Secretary at Cape Coast. He has had a great experience at Lagos, and is the witness of what energetic administratorship can do. He has seen the success of good work, and no doubt would be induced to attempt to convert the principal settlement on the Gold Coast to a habitable place. The climate agrees with him better than with most men; he has studied the great necessity of the settlement; and none better could be employed than such a man.

The Anglo-Ashantee war threatens to be, for a time at least, until the arrival of British regulars from England, as full of insignificant details and puerile prognostications as the Carlist-Republican war now raging in the Iberian Peninsula. Even then I fear that we shall have nothing to write about, unless it be about the immediate return of the troops to their native shores.

From all I can hear or see, the whole expedition depends upon the caprice of King Coffee Calcali. If this sober gentleman sees fit, after repeated blows given to his army now in the Protectorate, to eat humble pie, to acknowledge that he was wrong to disturb the peace of Western Africa, it is reported and believed that Sir Garnet has received orders to close with any proffers of peace or settlement that may be received from the stricken king. As a soldier proud of his reputation, and ardent for fame, I presume that such a tame conclusion to what promised at first sight to be such a fair opportunity for glory, would not be very palatable to Sir Garnet, but then Sir Garnet is bound to obey orders.

Almost every officer on the Gold Coast expresses a hope that England will not withdraw from the war now that she has armed herself for an earnest conflict with her old enemy of Ashantee, without retaliating for the long series of enormities against her, or her Protectorate, which Ashantee has committed since the beginning of this century, by burning Coomassie, the capital, and reducing it to a smoking ruin like Magdala in 1868.

Considering the extent of injury so many thousands of Fantees have suffered at King Coffee's hands, in the utter desolation and havoc wrought in the Protectorate by his armies, it is somewhat assuring to those who have come out to engage in the war to think that King Coffee will never be able to pay one twentieth of the demands that would be made should he hesitate to prosecute the war and show a desire to avoid the stroke of retributive justice by sending propositions for peace.

Meanwhile we wait patiently the course of events, though it is a dreary task imposed upon correspondents to write letters upon such uninteresting topics as the preliminaries of an expedition that may never take place. The task is made much drearier by the tantilizing mystery with which Sir Garnet has seen fit to involve his intentions. I fear the days of the garrulous and good-natured official who told one everything that was in his heart are passing away, and that we are about to have the insufferable officer who will tell one nothing, but will delight to wear an air of mystery that he may hold the world in suspense.

Lest perhaps you should deem me exaggerating, I will give you an instance of the new method of military tactics which Sir Garnet has initiated by means of that extraordinary volume of his, the ' Soldiers' Pocket Book.' Three or four days ago a rumour, soon followed by official intelligence, came to Cape Coast Castle of a battle that was being fought between the

Ashantees and the Allies of Abracrampa, which had already lasted several hours. At 3 A.M. Captain Brackenbury, the military secretary, was sent to the Senior Naval Officer from Sir Garnet with an order to hold himself and a detachment of marines and sailors in readiness to march to the front. The head-quarters mess sat down to breakfast about 9 A.M., and from some casual remarks which were made at table, Lieut. Maurice, the private secretary of Sir Garnet, gleaned sufficient to make him suppose that the General intended making a move upon the besieged villagers some time during the day. Half an hour elapsed, and Sir Garnet was being drawn in his carriage towards Abracrampa, but his own private secretary, living in the same house, was not aware that Sir Garnet had departed until an hour later.

Some military readers may see much to commend in this excessive reticence and silence on the part of a general, as auguring well of his conduct when engaged on more important expeditions. I fail to see anything commendable in it. Sir Garnet is now engaged in an expedition against savages on the West Coast of Africa, far removed from the influence or the enterprise of the public press, and any intelligence he would care to impart could not injure his mission whatever, but rather benefit him. It takes twenty-one or twenty-two days for a letter to reach England, it will take as many days for the newspaper publishing it to return to Cape Coast Castle, I therefore fail to see any plausible reason for Sir Garnet's reluctance to speak of his intentions. I suspect, however, that Sir Garnet does not desire to deprive the Colonial Office and the Home Government of the honour of being the first to receive any news there may be to transmit.

Sir Garnet's reluctance to speak of his mission and its purpose has found many imitators on his staff. It is carried to such an extent that they may all be compared to a lot of wise-looking owls, and it is evident that they share in their chief's distrust of special correspondents and 'gentlemen of the Press,' as Sir Garnet courteously calls them. Whether these silent, distrustful gentlemen of the staff are possessed of any items of news that the world would care specially to know, I know not, but if each one carried in his breast the fate of a continent they could not look more important. For my part,

I only hope that Sir Garnet holds the fate of a continent in his hands, that he will employ his well-known prudence and discretion in arbitrating wisely and well, and that in the course of a few years at least he will see fit to let the admiring world know what cares and responsibilities weighed down his great soul.

Having traced the history of the Ashantee War down to the date of the arrival of Sir Garnet on the Gold Coast, I should proceed to relate what I know of the preliminary little battles in which he has participated.

When Sir Garnet arrived here, he heard of an immense force of 40,000 Ashantees being in the neighbourhood of Abracrampa and Dunquah, as a menace to the weak garrisons occupying the posts. While staying but long enough at Cape Coast to gather information concerning their tactics, their plans, and their mode of fighting, he heard that a large force, evidently bent upon fulfilling King Coffee's threats of capturing Elmina, had suddenly diverged from the main road and marched past Abbaye and Napoleon towards that place. Sir Garnet arrived upon the scene next morning, and aided by an auxiliary force of marines and sailors, plunged upon the foe, which proves him possessed of energy and daring, and deserving the honour conferred upon him of leading a British army to Coomassie.

The enemy lay hiding in wait in the middle of what they call here a ' bush,' but should be more appropriately called a jungle. It is so dense in some places that one wonders at first sight how naked people can have the temerity to risk their bodies in what must necessarily punish their unprotected cuticles most painfully. This jungle probably covers many thousands of acres, literally choking the earth with its density and luxuriance. It admits every kind of shrub, plant, and flower into a close companionship, where they intermingle each other's luxuriant stalks, where they twist and twine each other's long slender arms about one another, and defy the utmost power of the sun to penetrate the leafy tangle they have reared ten and fifteen feet above the dank earth. This is the bush into which the Ashantee warriors creep in on all fours, and lie in wait in the gloomy recesses for the enemy. It was in such a locality as this that Sir Garnet found the Ashantees, and where he suffered such loss in his staff and officers. Until the sonorous sounds of Danish musketry awoke the echoes, few of the English suspected the foe so near; until they themselves betrayed their presence, the English might have searched in vain for the hidden enemy. Secure, however, as they were in their unapproachable coverts, the volleys of bullets which their loud-mouthed challenge evoked, searched many a sinister-looking bush, and in a couple of hours had effectually silenced their fire, but not before Colonel McNeil, the Adjutant-General, had the muscles of his fore-arm torn off by a discharge of iron slugs, and Captain Fremantle, R.M., and other officers had been wounded.

This was the first affair in which Sir Garnet was engaged, where he showed himself by his activity and energy and disposition of forces to be what he really is, a dashing, brave, and skilful officer, who, if he lives, has a splendid future before him. Undoubtedly he is possessed of high soldierly qualities. One half the encomiums which have been passed upon his activity would suffice for even the most ambitious. Indeed I have heard no general more highly spoken of by his own officers after so short an acquaintance, than Sir Garnet has been.

After this little brilliant affair the Ashantees seem to have contented themselves with procuring supplies for their starving numbers, cutting roads through the bushes towards Abracrampa and Dunquah, and avoiding the scouts of the allied tribes. Rumours floated down occasionally, that they were suffering greatly from famine, that in order to satisfy craving hunger, they lived on early yams and unripe plantains, which disagreed so much with them, that they were fast perishing from dysentery and famine fevers. This was borne out, it seems, by the reports of the scouts, who, it was said, frequently met the poor worn-out frame of an Ashantee who had died of starvation.

Meanwhile, in the quiet interval that elapsed after the repulse which the Ashantees suffered at Elmina, Sir Garnet was not idle either in the bush or at Cape Coast. The white man's tactics sorely perplexed the Ashantee chiefs. His scouts —mainly consisting of Arabs, Kossoes, Winnebahs, with a sprinkling of Houssas—were everywhere now along the line of the main road, between the garrison posts of Mansu, Dunquah, and Abracrampa, and soon along the path between Elmina and Napoleon. His labourers, under the command of a gallant officer named Gordon, were extending their work beyond Mansu, cutting a road Prah-ward. Capt. Gordon was not to be deterred by ever so many reports of a wild nature, and kept his unarmed and defenceless squad manfully at their toils.

Stores were also being landed at Cape Coast by many hundreds of tons. Huts for commissariat stores were put up, and hospitals, prepared in England, were being erected for the use of the troops, and being conveyed by male and female porters, in long lines to the front; and as day after day passed by, without molestation from the enemy, the materials of war were being safely transported and housed at Dunquah and Mansu, two days' journey from Cape Coast.

While Sir Garnet and his staff were thus busy at Cape Coast, the former never lost an opportunity to glean information of the intentions of the enemy. The scouts and the commandant of the advanced posts kept Sir Garnet well-informed. The force of the enemy was variously estimated at from 10,000 to 40,000, much too large a force to operate against in the bush, with his detachment of marines and blue-jackets; but at the beginning of this month he heard that the Ashantee general had been recalled by the king, and that the army was now separated to operate in three columns. One of these columns was reported to be in the neighbourhood of Mampon, a small town not far removed from the main road to the Prah. The British general, observant of his opportunities, hastily collected what force of sailors and marines he could, consisting of about 200, and marching from Cape Coast Castle, encamped at Assayboo, ten miles distant. The next morning he proceeded to Abracrampa, three miles from Assayboo, and ordered a reconnaissance to be made by Lieutenant Pollard with the irregular native troops, who succeeded in making a few Ashantee prisoners, and creating some confusion in the enemy's van.

While Sir Garnet was resting this day at Abracrampa, a reconnaissance was made by Lieut.-Colonel Festing from Dunquah, a fortified post situate about twenty miles from Cape Coast Castle. This was a very successful affair, inasmuch as the enemy was surprised in his camp while at breakfast. Several prisoners were taken, and a quantity of powder and camp furniture was destroyed before the enemy recovered himself.

Out of nine officers engaged with Lieut.-Colonel Festing, five were wounded. The other losses were four wounded out of the 2nd West India Regiment, forty-two wounded native allies, and one killed.

The next day Sir Garnet moved from Abracrampa towards Dunquah, hoping to encounter the enemy; but taught by the signal defeat they had suffered the day before, the Ashantees did not dare to make their appearance.

The Commandant at Dunquah, Lieutenant-Colonel Festing of the Royal Marine Artillery, had been requested by Sir Garnet to meet him on this day at a village half-way between Dunquah and Abracrampa, but the native troops' commander dared not leave the friendly shelter of the palisades, showing plainly that no reliance could be placed in the Fantee auxiliaries.

Hitherto the Ashantees had boasted that the white men were afraid to meet them in the bush, but the manner in which they were met, and forced from their hiding-places by Festing's people, has dissipated the illusion.

Finding that nothing more could be done against the retreating enemy, Sir Garnet, after leaving a force of fifty bluejackets for the defence of Abracrampa, returned to Cape Coast Castle.

Soon after the British general's return to the coast, the Ashantees once more appeared on the scene. The Ashantee chief Amanquatiah had sworn to the king that he would destroy the village of Abracrampa, and with this view no doubt he directed his energies to cutting a road through the jungle towards this fort. His people were said to be busy as ants upon the work.

On the 5th of this month the enemy appeared out of the jungle in force in front of Abracrampa, and at once opened fire. But the station was well able to sustain such siege as the Ashantee general meditated, or any attack that might have been made upon it. Besides a strong palisade that completely surrounded the village, a deep trench had been excavated and the jungle cleared for a space of 250 yards around.

The attack began at 4·30 P.M. and continued without intermission until 5·30 P.M., accompanied by vigorous tom-toming at all points, and much inharmonious chanting from lusty throats. But the native troops ensconced behind thick palisades, with superior weapons in their hands, were not to be daunted by ever so much tom-toming or inharmonious vocal music, and replied with their Sniders with such good effect that the enemy were obliged to withdraw for a time.

This pause was taken advantage of by Major Baker Russell, 13th Hussars, who commanded the station, to despatch word to Assayboo that the Ashantees were in the neighbourhood, evidently determined on a persistent effort to reduce Abracrampa. The commandant of Assayboo despatched the news to Sir Garnet, who received it about midnight.

Whether the British General expected to receive more hopeful news is not known, but he delayed his departure from Cape Coast until 9 A.M. next morning.

Meantime the Ashantees had soon broken the truce, and recommenced the attack after dark, keeping up a determined firing until 11 P.M. At 4 A.M. they renewed their attack, and continued it until the morning of the 7th. Sir Garnet had arrived at Assayboo about 4 P.M. of the 6th, having marched his troops during the hottest part of the day, thereby causing much suffering to the blue-jackets under Captain Fremantle. Numbers of these poor fellows fell out on the wayside, completely prostrated by the heat, the glare, and thirst induced by the fierce sun, which so enfeebled his force that he was unable to proceed to the rescue of Abracrampa. Before Sir Garnet's arrival at the besieged station the enemy's firing had ceased—Russell had driven the enemy away.

It would be difficult even for those who saw the action, which I did not, to render any account of the loss inflicted on the enemy; the circumstances under which the siege occurred, though favourable to the besieged, were specially disadvantageous to the besiegers. The besieged were armed with first class breech-loaders and Enfields, and were protected by their defences; while the Ashantees, hidden by the jungle, were betrayed by the puffs of white smoke which, in the breezeless air, rested above their heads after they fired; and frequently these localities became the special regard of many a cool rifleman. It may be presumed that above a hundred Ashantees lost their lives before Abracrampa, while only one white sailor and about a dozen natives were wounded of the defenders. But even the great loss inflicted on the foe hardly compensates for that inflicted on a reconnoitring party that set out from Dunquah on November 4.

Colonel Festing, adjudged by Sir Garnet to be active and unceasing in persecuting the enemy, had left the garrison station of Dunquah for the Ashantee camp situate a mile beyond the barriers, with a mixed force consisting of Houssas, Kossoes, Arabs, and Annamaboes. Festing attempted to attack the camp from the rear, but he was not as successful as he hoped to be. The Ashantee scouts had already alarmed their people, and each warrior had immediately hastened to the bush, and taken his post in the vicinity of the abandoned camp. When Festing's force came up they found it had been deserted, and their presence known; and barely had they arrived at this conclusion when the bullets and slugs of the ambuscaded foe came hurtling amongst them, informing the astonished Fantees of their presence. Lieutenant Eardley Wilmot of the Royal Artillery received a severe wound in the left arm as, with a revolver in his left hand and a sword-bayonet in his right, he was attempting an advance against the enemy, and, by placing himself at the head of his men, hoped to encourage the Annamaboes under his command. The young officer—he was only twenty-four—fearing to discourage his men by going to the rear to have his wound attended to, never moved from his post in the van, but simply told his servant to run to the fort to get lint, and continued shouting to his men to advance, plunging foremost himself sword in hand. When the servant returned with the lint young Eardley Wilmot was dead, with a rifle bullet-hole through his heart. The spot where he fell became dyed soon after with the blood of his fellow officers. Lieutenant-Colonel Festing, while lifting the body of the dead officer, was hit, and Dr. Gore, coming up to offer his services, also received a wound in the groin. Lieutenant Paget was scarred with a slug, and a non-commissioned officer of the West India regiment had a pebble imbedded in his neck at the angle of the jaw, which inflicted a ghastly wound. The cowardly allies deserted the fatal spot, and a retreat was sounded; but the body of Wilmot was taken away by Festing so that it might not be decapitated by the savages. The victory was undoubtedly on the side of the Ashantees at Dunquah, but those who attacked Abracrampa paid dearly for it next day.

The body of Lieutenant Eardley Wilmot was conveyed to Cape Coast, where it was buried in the presence of the General and his staff and a large concourse of sorrowing officers.

The camp of the enemy in the neighbourhood of Abracrampa was attacked after Sir Garnet's arrival with reinforcements. So well was the attack planned, and so well was it made, that the Ashantee general, Amanquatiah, barely escaped capture; one of his wives and two of his nephews fell into the hands of the British, besides his own chair in which he was accustomed to be carried by his slaves. If reports are to be believed, Amanquatiah is in deep disgrace with his own people, who complain that they have been deceived, for, when they set out in February on this invasion, they understood that they were to fight against the Fantees, but they now find themselves opposed to white men whom, they say, it was never their intention to fight. Captain Grant and his company of 2nd West Indians distinguished themselves nobly and gallantly in this action.

The Cape Coast Volunteers have fallen into disgrace with the British authorities. They were despatched, soon after the defeat of the Ashantees at Abracrampa, in pursuit of the flying enemy, along with the Abras, Kossoes, and Houssas. The latter, over eager to harass what they considered to be a demoralised force, hung so close upon the skirts of the retreating Ashantees that they compelled the enemy to turn on them, so that the Abras, Kossoes, and Houssas were compelled to return faster than they had advanced. As they returned they were met by the Cape Coast Castle Volunteers, who, in their fright, mistook them for enemies, and they were fired upon, by which two of their number were killed and six wounded. Sir Garnet, on hearing of their cowardice, disarmed the Cape Coast Volunteers, and transformed them into labourers and commissariat porters. Indeed, it were a wise thing if Sir Garnet transformed the major part of the allies of the Protectorate into pack-bearers and commissariat store carriers, the most useful employment he could give them, and one quite in accordance with the opinion of all Europeans who have seen them display their hare-valour in battle.

Sir Garnet Wolseley has paid the penalty which Africa inflicts on the European who treads her shores. His audacity in braving the hot sun at mid-day has been punished by an attack of fever.

Others are sufferers as well. Two officers only, however, have been buried since my arrival—Eardley Wilmot, already men-

FANTEE VOLUNTEERS.

tioned, and Harry Mount, of the Control Department, who died of dysentery. Colonel M'Neill, Chief of Staff, has been invalided home; Captains Brackenbury and Charteris have been on the sick list; Major Baker, Acting-Adjutant-General, vice M'Neill, has suffered from dysentery; Deputy-Surgeon-General

Home has sore throat; Surgeon-Major Jackson has had the intermittent fever ; and almost every officer who came with Sir Garnet to the Gold Coast has been compelled to succumb to the general unhealthiness for a time. Out of ninety-two labourers who arrived from Sierra Leone lately, forty-two have been sent back unfit for duty. When these facts are summed up and digested, it will perhaps be admitted that the Ashantee king has terrible allies to fight his battles for him in shape of fever and dysentery which attack the highest as well as the lowest, and penetrate even to the comfortable chambers of Government House.

CHAPTER III.

THE FANTEES—CAPE COAST CASTLE—FETICH—THE ASHANTEES AND THEIR CUSTOMS—HISTORY OF KING COFFEE'S DYNASTY—THE KING DESCRIBED.

November 20, 1874._3

NEXT to my surprise at the unaccountable apathy of the English on the West Coast of Africa, is my surprise that the English should have adopted for allies the Fantees, the miserable and worthless tribe occupying the country lying between Cape Coast Castle and the boundary river flowing between Ashantee and the Fantee Land. It may be said—with justice —by the English, that they could not help themselves; that they found the Fantees in possession of the country and of Cape Coast itself, and that they had no right to displace the Fantees for the Ashantees or any other people; that such a course on their part would have been the grossest injustice.

It is true the English could not have done this arbitrary act without incurring great guilt. But, after the cession of Elmina to them by the Dutch, good policy would have indicated what course they should adopt towards furthering the wishes of the opulent and commerce-loving Ashantees.

Elmina is the only place suitable for the residence of an Administrator-in-Chief or Governor on the whole shore of the Gold Coast from Assinee river to the Volta. It is centrally located: nature has furnished it with a river which might be made a capital harbour for small craft; there is a good anchorage outside for larger vessels. The Dutch have furnished it with two castles, infinitely superior to the miserable castles or forts at Cape Coast. Elmina was a large, thriving town, in a healthy situation, exposed to the pure sea breezes; the country around was well cultivated; numerous plantations studded the eminences behind the town; plenty reigned everywhere; and though the Elmina natives were formerly a branch of the Fantee nation, they were far more industrious, and were able to supply

E

the necessaries of life to the garrison at one-fourth the cost those around Cape Coast Castle could afford.

In every way Elmina was a desirable locality for the principal station of the Gold Coast; and had the English at once transferred their offices of administratorship hither they would not have been under the deplorable necessity of reducing to dustheaps the native quarter eighteen months after the exchange.

Having acquired this rich territory from the Dutch, and aware of the anxiety of the Ashantees to possess a small portion of coast line which should be theirs and none other, the English out of their abundance might have set apart a strip of ten miles immediately adjoining Elmina, and proffered it to the Ashantees, and they might have even gone so far as to assist the Ashantees in the construction of a road from Prahsu to the new Ashantee settlement, thus proving to them that they were as ready to enter into a friendly alliance with them as the Dutch had been. The fear that the Ashantees would soon have made themselves masters of the country need not have deterred the English, for the coast line would always remain in their power, and always within reach of the guns of their African naval squadrons. Elmina and Cape Coast also would ever be protected by the guns of the castles and the forts, so that such a policy as the one indicated above might have been prosecuted to a very desirable end—viz., that of populating an uninhabited part of the coast with a portion of an opulent tribe desirous of opening direct commerce with the English, and securing the friendly alliance of a nation which, in a commercial sense, was worthy of being allied with.

The traveller, after observing the tremendous energy of the English in other lands, and the keen intelligence which always directs their policies, finds himself compelled to condemn the extraordinary apathy which characterises the British authorities on the West Coast, and the unwisdom of their mode of government. From Sierra Leone down to Accra, ruin, decay, and neglect are visible in every settlement. Thirty years ago Sierra Leone was a thriving and prosperous commercial mart. A forest of masts reared their tall heads in the unrivalled harbour, the soul of business seemed to animate her streets, the natives appeared to possess some idea that they were made to work, and that their limbs were formed for labour.

Now, however, an occasional steamer may be seen in the harbour, homeward or outward bound, and the natives are utterly spoiled for work, and seem to think that celestial state has already arrived when humanity ceases from toil and betakes itself to perpetual psalm-singing and hallelujah chanting. Apollonia, Dix Cove, Chamah, and Elmina are heaps of ruins. The Dutch left them prosperous and happy ; the English took it upon themselves to govern them, and within eighteen months ruined them, laying the settlements in mortar and dust-heaps.

The Fantees have not found out that their alliance with the English has been a curse instead of a blessing, because they have not yet been taught the art of deduction. But should they ever be schooled to logic, I have no doubt but they will say :—' If we had not been friends with the English the Ashantees would not have burned our villages, or consumed our produce, or laid waste our plantations.' To which the English may perhaps answer :—' We went to war with the Ashantees for your sakes ; if we had not been fools, we should not have rejected the Ashantees and accepted such miserable cowards as you have turned out to be for our allies.'

I have already informed you that the Fantees occupy the country between Cape Coast Castle and the Prah, though in reality the Fantees proper inhabit the town of Cape Coast and its immediate neighbourhood. Those of the Fantees who dwell in the villages of the interior are distinguished after the names of their villages. The whole of the Fantee territory may be about 2,000 square miles. It is not a kingdom. Fantee Land is more of a republic of villages, whose chiefs are banded together for military protection against the common enemy— the Ashantees. Cape Coast has its king, or chief. Annamaboe has another, though it is but seven miles distant. Dunquah, sixteen miles distant, has another chief, and Abracrampa has its chief. When any chief has a grievance against any other, the British Governor of the Castle is called upon to arbitrate, and his word is supposed to settle the ' palaver.'

Confining my remarks, however, to the Fantees of Cape Coast, and taking it for granted that the Fantees of the interior are no better, I must say that the town presents the strongest evidence that could possibly be produced of the apathy of the

English on the West Coast of Africa.   Two hundred and five
years has this town been in possession of the British, and yet
out of its population of 20,000 souls I doubt whether there are
1,000 of them who understand English.   The people are as
barbarous, untutored, and superstitious as though they had
never seen an Englishman's face.   They are as wild in appearance, as naked in body, as filthy in their habits as any tribe of

FANTEE WOMEN.

savages I have ever seen.   Their fetich idols, and medicine
and Mumbo Jumbo rites, are visible in the streets of
Cape Coast.   The bizarre fantastics in which women caper
about nude, and ancient hags disagreeably remind you of the
bonyness of the human body in old age, and little conscience-
less children prank and hop, involved inextricably between
the pedal extremities of their grandmothers, form a picture

utterly unsuited, I should say, at least, for any portion of the British colonies.

If I have to echo the opinions of all white men who have come out lately to join this expedition, and declare that these people are the most indolent, toil-hating tribe it has ever been my lot to see, I am bound also to say that the administrators, the administrators-in-chief, the governors and sub-governors, sent out to rule them must have been selected by the British Government for those very qualities which we are all so ready to condemn in the Fantees of Cape Coast Castle.

Consider the thing ; 205 years in possession ! Why, Prussia was made a kingdom only 172 years ago ! The battle of the Boyne was fought only 183 years ago ! Just think, then, how many things have transpired during the 205 years since first John Bull laid his paw on this spot on the Gold Coast, and yet the Fantees, the population of this government station, have to be described as the most shiftless and indolent of the negro race !

I must charge the British also with their usual fault of raising the cost of labour and necessaries of life to about twenty times what it ought to be. Englishmen are as much sufferers as I am myself. They suffered as much from the same folly as I did in Abyssinia. When we might have obtained fowls at the rate of twelve for $1 in Abyssinia, Sir William Merewether commanded three fowls only should be given for $1. There has been as yet, to my knowledge, no such order from Sir Garnet respecting the number of fowls to be given for $1, but I do expect some such order will be issued soon, as proclamations are being issued almost daily. In the meantime, however, the prices of everything have risen, until I should much prefer to pay the high prices of living at the Fifth Avenue Hotel rather than at a miserable Fantee hut.

Two chickens cost $1 here, and mutton is not purchasable under 25 cents per pound. Much of anything else cannot be had for love or money, except rice, which sells at sixteen pounds for $1. The household expenses are enormously high. First, your hammock-bearers cost you at the rate of $45 per month ; your porters, who are to carry your effects and provisions to the ' bush,' cost you as much more ; your house rent costs you $20, and your cook $18 per month : so that if one finds his bills at the end of a month to be under $200 one must consider himself specially fortunate.

I did not intend that my remarks about the Fantees should be abusive of them when I began this chapter, so that I think I had better bottle up my indignation for the present.

I have already remarked that the Fantees are very super-

FANTEE SERVANT BOY.

stitious. Their fetich consists of a wooden image of an arm-less and legless man, which is placed upright by the side of some public street, and at its back is generally found a medi-cine heap, covered over with either a turtle's back, or a huge stone. If the fetich is not formed of wood, a figure of clay, or a padded, doll-shaped thing answers just as well. This image, made of wood, or clay, or old rags, is endowed by the fetich priest with extraordinary powers. A white man, educated at a university in all the secrets of medicine and surgery, is not gifted with one-half the powers commonly ascribed to and commonly believed to be possessed by this miserable idol. The fetich man carries the image to the bedside of a fever-stricken patient, mumbles formulas into the ear of the sick one, waves the fetich over his head, and departs satisfied that a cure has begun already. But should the British surgeon come up in time to arrest the progress of the disease and effect a cure, the fetich man curses the surgeon for making his fetich to lie, believing that the failure of the image was caused by the underhanded and false machinations of the surgeon.

If I were to describe all the ceremonies which I have noted under my verandah I should fill many columns ; but there is one which I have seen, which causes intense disgust to the coloured missionaries of the town. It is the exhibition of a virgin arrived at years of puberty to the rude gaze of the rabble of Cape Coast. Before the British authorities interfered, the scene was really disgusting and downright savage ; but at present, though it is a singular one, the immodesty of it is veiled somewhat. The young girl, in former times, was accustomed to be stripped naked, and obliged to march thus through the streets, loaded with gold and silver ornaments. The wealth of the family was sometimes seen on her person, amounting in value to several hundred dollars' worth, while her friends and relatives, her playmates and acquaintances, followed her with all the fanfaronade which delights the ears of savages. Several *kinkassis*, or drums, were beaten in the most energetic manner ; fifes were blown, and the vocal noises, inharmonious at the best of times, were absolutely deafening. She was thus paraded round the town, until every household had enjoyed the ineffable felicity of seeing the virginal maid. At dusk, she was conducted to the bridal bed of her husband, who had probably paid dowry to her parents to the amount of $40 or $50. The earlier hours of hymeneal joy were honoured with an increase of din, at which all the friends and relatives participated, and a bestial orgie concluded the singular ceremony. In these modest times, however, the maid is shown about properly clothed and bejewelled, and since the absence of the Fantee males in the ' bush ' the noise and din of the festivity is not so unbearable and soul-distracting as formerly.

Lest this chapter may be intolerably long—dull as I know it to be from lack of war news—I propose to give a few remarks concerning the Ashantees.

To begin with the King, then. The present chief or king of Ashantee is called Coffi or Coffee Calcali, the son of Coffee Tutee, an Ashantee noble of great power, by Princess Effnah Cobee, of the blood royal. He was born about the year 1837, and, is therefore, at present in his thirty-sixth year. He ascended the throne of the kings, August 26, 1867, as the eighth monarch since the dynasty was established and the Ashantees became a nation, which occurred about the year 1700.

To attempt to trace the past history of any savage African nation which has no written record for its basis, would appear too much like romancing; but such facts as the sages, and courtiers, and elders could store in their minds and transmit from father to son we have been able to obtain. No doubt a further acquaintance with Ashantee chiefs would enable us to add still more to our *résumé*; but what we have at present suffices for the ordinary reader, whose desire would be mainly to have an idea of what the Ashantees are, and whence they came.

The special correspondent in a peculiar land like this feels a melancholy pleasure in being able to glean somewhat from the past of this now powerful nation, and transfer some portion of its history to the pages of a book ere the traditions which now inspire the elders at sunset, under the patriarchal trees of their villages, to recite the glories of the great founder of the nation, and the deeds of the heroes and conquerors who succeeded him, should pale and fade, and become lost in oblivion.

If any reader, after coming to this land, should become interested in any tribe or nation, and attempt to write its history, he would soon find how difficult a task it would be to distinguish fact from fiction, and be able to understand the pleasure I feel in writing the following remarks upon the Ashantee nation.—

In the days of Cada Mosto, Fernando Gomez, D'Azambuja,

Vasco da Gama, and other heroic navigators, who flourished about the latter part of the fifteenth century, the Ashantee nation was not known. If it were a concrete people, organised under a despotic king, as renowned in war as the Ashantee nation is to-day, there is no doubt but that some of the enterprising Portuguese explorers and writers, especially De Barros and Cada Mosto, would have heard something about its existence and noted it in their journals with the same exactness and accuracy with which they penned their notes of other kings and nations whom they came in contact with or of whom they heard report ; for, while looking eagerly over their quaintly-written journals of discoveries along the coast of Guinea, we are struck at the ease with which the steps of the Portuguese navigators may be traced to-day. It is true that many names of tribes and rivers have become lost, but the major number are still to be recognised.

Cada Mosto writes in 1469 of the success of Fernando Gomez in his trading for gold dust at the port of Mina, which we at once recognise to be Elmina of the present day, where, during peaceful times, the same trade in the precious metal is still carried on. Again, he says that the first exchange in gold dust occurred at a village called Samma, a few miles further up from Mina, which we know to be the place called Chamah, where Commodore Commerell and his boat fleet received such a check lately. It may not appear singular that Mina, now called Elmina, should have retained its name, since it has been in the possession of a European Power ever since ; but it is singular that a small village like Chamah should have retained its native name in a country like Guinea, where everything is so mutable and where changes transpire so quickly, and nations and tribes, once prominent in local traditions, so soon sink into oblivion.

My opinion is, that the Ashantees at the time of which I write were divided into small village clans or sub-tribes, acknowledging no ruler, save their local chiefs, or elders ; that wars for supremacy and right of way to the coast were frequent among them ; that defeat of one village served to add power and subjects to the conquering sub-tribe, which, probably, in the course of time was augmented by other victories over weaker neighbours. There is a tradition among the Fantees that, in former times very far back, the Fantees and Ashantees composed one tribe ; but that while engaged on a war expedition into a distant region they suffered severely from hunger ; that one portion sustained itself by feeding on a shrub or plant called Fan, while another portion found sustenance in a shrub called Shan, from which the eaters of Fan became known afterwards as Fantees, while the others in like manner became known as Shantees, to which imperfect hearing has added the prefix A, so that the tribe is now called Ashantees by the English. This is, of course, fable, but only a more thorough knowledge of the roots of the language would enable us to decide whether the fable is correct.

We know, however, to a certainty, by the records of the Portuguese navigators, that gold dust in considerable quantities was brought down from the interior to Elmina, by people who were not distinguished under any distinctive title during the latter half of the fifteenth century ; and we know, through official records by the Dutch governors, who succeeded the Portuguese at Elmina, that the same trade was carried on between Elmina and the interior during the whole of the sixteenth century.

But at the beginning of the seventeenth century there struggles slowly into the minds of the European authorities on the Guinea or Gold Coast a knowledge of the existence of a nation growing in opulence and power where the gold dust comes from, called the Ashantees, and of a chief or king who governs them whose name strikes terror into the hearts of the sub-tribes on the coast, who were accustomed to convey palm oil and cloths into the interior to exchange for gold dust. This king's name was Sy Tutu. During his lifetime the sub-tribes, the clans, villages, and towns throughout a great extent of country were consolidated into one kingdom, with Sy Tutu, the conquering chief, as a sovereign. In the attempt to extend his power on this side of the Prah river, in the territory of the Akims, at the great battle of Coromantee, fought on a Saturday, the conqueror, Sy Tutu, was slain.

As in more civilised countries, a glorious era of conquest and power was not forgotten by the nation he had founded and made great by a series of victories, and the memory of Sy Tutu is kept green to this day by the Ashantees. The most binding oath known to-day among this people is the one which is made by ' Coromantee Saturday.' The present king, Coffee Calcali, has taken this solemn oath that he will either drive the British from Elmina, or he will make a prolonged and bloody war.

It was natural that, after a glorious reign such as Sy Tutu's, his successor should be his son, and accordingly the throne devolved upon Apuku, his eldest son, who during his reign maintained the supremacy of the Ashantees, conquered the Akim territory and added it to Ashantee. It is said that Apuku, like the Plantagenet John, desirous of curtailing the power of the sub-chiefs, exasperated the chieftains and caused them to band together and drive the monarch from Coomassie, the capital; but, more successful than his British prototype, he was enabled to rally to his standard an army of warriors who had probably shared in his father's conquests, with whose aid he retook Coomassie and subdued the fractious and turbulent rebels.

Apuku died, and was succeeded by his son, Akwassi. This king, after extending his power towards Denkera, in the west, turned his attention to a powerful country on the east called Dahomey. The fact that Dahomey is still independent, and is understood to be as powerful as Ashantee, proves that Akwassi obtained little or no success in his wars against the Dahomans.

Akwassi, dying about the year 1752, was succeeded by his son, Sy Kudju, a warlike prince, whose memory is venerated for the numerous successes which attended his arms throughout a long and prosperous reign.

Sy Kwamina, the fifth monarch since the dynasty was established, succeeded Kudju; but he was shortly dethroned by the chiefs, and was replaced by a younger brother, Sy Tutu Kwamina, about the year 1800.

It was during the reign of this prince that the British became personally acquainted with the prowess of the Ashantees.

The first war was that of 1811. Then came the disastrous campaign of 1823 and 1824, during the Governorship of Sir Charles M'Carthy. Major Ricketts, a member of his staff—one of the very few who survived the fatal battle of Esmacow—has left us a very interesting account, simply written, but full of pith and matter. When the Ashantees were marching up to attack Sir Charles they advanced singing and beating a wild alarm of drums and tom-toms, as is their custom. Sir Charles, not to be outdone by noise, ordered his band to strike up 'God save the King.' However, the music had no effect on the enemy, and soon both forces were engaged in desperate strife. After a battle of some hours the British allies were without ammunition, and a mistake committed by a commissariat officer, who despatched biscuits instead of ball cartridges, proved fatal for the allies and their gallant commander. Sir Charles and all his staff, with the exception of Major Ricketts and a Lieutenant Jones, were beheaded, and the heads were carried in triumph to Coomassie, where, it is reported, they are even to this day held in high honour, being decorated with gold bands and jewels.

Sy Tutu Kwamina died soon after the battle of Esmacow, and was succeeded by his brother, Sy Okoto.

After the victory of Esmacow the Ashantees advanced upon Cape Coast Castle, when there was but a garrison of about fifty men in the fort. Even the streets of the town were invaded by the enemy, and the slaughter which ensued was terrific. It is calculated that about 20,000 Fantees lost their lives in that campaign, while the miserable state into which the protectorate was plunged is indescribable. It may be said that until 1826 the Ashantees were masters of all the protectorate up to within a mile of the seacoast forts. The tables were reversed in 1826. The British collected a force at Accra, and met the Ashantees at Dudowah. The battle that ensued then was most sanguinary, but nearly at the commencement of it the Ashantees were stricken with a panic and fled, and a great many of their captains, rather than meet the king, committed suicide on the field. The Fantees and their allies on this day amply avenged the horrors of 1823 and 1824, and an even balance of losses was thus struck.

The successor of Sy Okoto was a prince called Kwaweda, who departed this life in 1867. Kwaweda invaded the protectorate in 1840, and waged a bloody and relentless war on the Fantees; but the English were not summoned to take up arms for the latter in this campaign; they simply advised them, and supplied them with arms and ammunition. On the death of Kwaweda, the present King Coffee Calcali ascended the throne, in his thirtieth year.

The present king of Ashantee is said to be an intelligent man, with the appearance of a mulatto, slight but sinewy in figure, hospitable and gracious to strangers. He resides at Coomassie, the capital, which is described by those who have seen it as a populous and well-regulated town. The palace of the king is a structure of stone, large and capacious, with great squares, where reviews and assemblages are held. The whole is walled around, having an outer and inner enclosure. The rooms are lofty and commodious, and those occupied by his Majesty are furnished with European and native articles. The presents he has been continually receiving diplomatically have assisted in supplying him with unusual luxuries. Pictures adorn the walls; sumptuous sofas are ranged round his walls; thick carpets cover his floors; his tables are loaded with a thousand costly knick-knacks.

Next in importance to the Palace among the public buildings is the Bantammah, which is the Tower or the Louvre of Coomassie, wherein are stored the more precious gifts it has been the good pleasure of the kings of Ashantee to receive from the British, the Dutch, and the French Governments. It is the treasure house, filled with the costly ware the kings have gathered from time to time : the Bank, wherein lie the specie and the gold dust, ingots and bars of solid gold, and many a chestful of rings and chains of gold, gorgeous silks and satins, and much else of wealth that goes to make up the treasures of a rich and powerful despot like the Ashantee king. In the Bantammah, side by side with the crown of the kings, is laid the gold decorated cup fashioned out of the skull of the unfortunate Governor of Cape Coast Castle, Sir Charles M'Carthy. It is only on days of state and high festival that this ghastly cup is used. I fancy, if the English take the Bantammah, that this state cup will be considered one of the most valuable trophies that could be secured.

The king possesses a numerous harem, like the sable monarch of Dahomey. He is privileged to marry as many as he pleases; for even in Ashantee the king can do no wrong or trespass on any law. I should fear to say how many wives he has married, for we shall probably know the exact number by-and-by; but he can take his pick out of the noblest, the fairest and best in the land, after which it is certain death for any other man in Ashantee to look on her face, for she is the king's. The harem is jealously guarded in a quarter of the palace overlooking the palace gardens by a body of 150 eunuchs. It must not be supposed, however, that the rights of ordinary women are curtailed thus; in the households of all but the king the women are at liberty to stare and be stared at, to talk with any man or be talked with.

The king, with the usual privilege of monarchy, may stay at home during war, while he may command every soul capable of bearing arms to proceed to the war. He has his Prime Ministers, his Commander-in-Chief, Keeper of the Treasury, Chief of the Eunuchs, Generals of Divisions, and Nobles of the Council.

The Prime Ministers, two in number, the most important of his household, manage all public business, and proffer advice to the king upon all matters of vital importance to the state.

The Commander-in-Chief is generally some ancient warrior, whose deeds in his prime form never-ending subjects of social chat among the old and young. When he assumes this high post he does not venture into battle any more; his work seems to be limited to portioning the levies to each noble, to keeping the army up to a certain standard in numbers, to organising, and appointing the officers and troops for active service. Next to the Commander-in-Chief comes another venerable old soldier, whose post seems to be that of strategician-in-chief, who plans campaigns and wars, and whispers military advice into the ears of the generals about to proceed on active service.

The Keeper of the Bantammah or Treasury of Coomassie is a most important man. He is generally a noble, and a distinguished soldier. The present keeper is the chief called Amanquatia, already mentioned. He is reported to be in our front seeking honours in skirmishing with the British allies.

The generals of divisions are generally great and powerful chiefs, who have supplied the king's army with hundreds of their retainers and slaves, and are appointed for their influence.

As may be seen from the above, the Ashantee nation is composed of four classes—the monarch; the nobles, or feudal chiefs; the free Ashantee yeomen, land proprietors, small village chiefs, all free-born Ashantees; the slaves taken in war or born in bondage form the fourth class.

The great nobles own vast territories and are lords of thousands of vassals. They may be said to be petty kings, subsidiary to the central government at Coomassie. Foremost among these, like another Warwick, is the Prince of Jwabin, in whose territory is the populous city of Selaga, with a population of 400,000 (?) souls, situate on the upper Volta. This prince can furnish the king with a force of 20,000 vassals for his army. He seldom ventures out himself into active service unless the King of Ashantee proceeds in person to the war.

The Princes of Becquah and Mam-pon are each able to furnish the national army with contingents varying from 5,000 to 10,000 vassals. The Chief of Fomanah, between Coomassie and the Prah, enters on service with a force of 1,000 of his own retainers. But there are several large cities beyond Coomassie, both affluent and populous, of which we may know more when we shall have entered Coomassie as conquerors.

The Ashantee army in the field bears flags for distinctive signals or ensigns of vassalage, but they have no importance in the eyes of soldiers, save as distinguishing pennants or banners. The loss of any of them involves no disgrace or misfortune. The chiefs' umbrellas represent in the Ashantee soldiers' eyes what a regimental flag does in a European army. The gorgeous sunshade which an Ashantee chief holds over his head, ornate with vari-coloured pieces of silk, rich with its appendages of silver and gold, embodies in an Ashantee's eyes the honour of his chief. The loss of one of these in battle involves disgrace and defeat.

The chiefs are borne from place to place by hammock or chair bearers. The chair which the British captured at Abracrampa was a specimen of Ashantee manufacture in that line. It appeared to be exceedingly cumbrous, though it was not of great weight. In shape it was exactly like the body of a light phaeton lifted from its wheels. Eight stout fellows bearing it on their heads, protected by thick pads resting on each bearer's cranium, sufficed to take Essamanquatiah, the Ashantee general, from camp to camp. The woodwork of the chair was studded with an infinite number of brass tacks, and specimens of what skill in tracery and ornate woodwork of which the Ashantee artisans are capable were visible on the sides and front of the chair.

The king appears in the field only *in extremis*—when his presence is necessary to inspire the flagging enthusiasm or devotion of the army to his cause, when discouragement has followed repeated reverses in battle. Then every soul capable of bearing arms, every chief owing allegiance, every officer of his household, arms himself for battle, and a force of 100,000 men is thus speedily raised.

His state umbrella is borne over or before him by a chief wherever he moves, the sight of which rouses the interest and zeal of every member of his army. This umbrella costs a small fortune—about 3,000 or 4,000 dollars. It is made of alternate pieces of crimson and black velvet, while hundreds of gold pendicles form a fringe around it, and a large boss of pure gold surmounts it above.

The dress of the king on such occasions is exceedingly costly. A tunic of crimson velvet covers his body, his loose Moorish pantaloons are made of the same stuff, a broad band of gold encircles his waist, a cap or turban of silk, richly embroidered, covers his head ; his weapons are decorated profusely with the precious metal.

The king's revenue is derived, apart from his own private estates, which are vast, from the gold mines. The dust becomes the property of the miners or proprietors, but every nugget, great or small, throughout his dominions, belongs to the king. Whatever person is guilty of reserving any of the king's portion is liable to the death penalty, which is invariably carried out.

The arms of the Ashantees consist of long five-foot Danish guns (flint locks), though many are also found with blunderbusses. The cartridge boxes serve as girdles, the leather or wooden cups into which the powder is poured being sewed on a belt, the two ends of which are either tied with a leather thong or buckled in front. The bullets or iron slugs are in a small leathern pouch, slung over the shoulder. This pouch, sometimes found on the dead Ashantees, is generally found to contain as miscellaneous a set of articles as may be seen in any Jack tar's box or bag. Bark thread, bark waste (probably for wadding), iron pins, queer-looking stones, an assortment of dark-coloured beans, a stale piece of yam or manioc, a piece of chew-stick, a handful or so of small snail shells and other extraordinary articles, represent what an Ashantee's pouch contains on the battle-field. Any reader who may have seen a Greek brigand's, an Albanian's, a Kurd's, or a Bedouin's cartridge box may guess at once what kind of a cartridge box the Ashantees use, and they undoubtedly derived their notions of the utility of such a thing from the trading Tuaregs of Timbuctoo. One need not wonder, then, at the very few severe wounds inflicted upon the English, despite the very many splendid opportunities the ambushed Ashantees have had to inflict instantaneous death. The powder is thrown loosely into the barrel, and, being often without wadding material, the bullets or slugs of iron, or handfuls of snail shells, or a piece of unsmelted iron ore, are dropped on the loose powder, and the loose charge is thus fired when only a few feet from their foes. Hence, at least, we need not wonder that so few wounds received by the English have proved fatal.

There is not the least doubt but the Ashantees have improved immensely since they first became a nation. They can now point to eight kings who have worn the crown of Ashantee, who have carried terror and dismay to the hundred nations around, and have spread the reputation of the Ashantees far and wide. The certain consciousness of the fact that their Ashantee forefathers were warriors, and their kings were conquerors, and that they themselves have not degenerated, has imparted dignity of bearing and an air of superiority to the meanest Ashantee soldier.

The Ashantee warrior, miserable as he may appear compared to the gaudily-dressed West India British soldier, is his equal, nay, I may say superior to him, in the battle. The West Indian, confronted with the almost nude warrior of Ashantee, may, with a serene smile of self-satisfaction, tap his breech-loading Snider rifle and dare his foe to the battle, sure that his weapon, his own superior knowledge of military laws and tactics and the discipline in which he has been trained will carry the day. On the other hand, the Ashantee—full of the traditions of his race, conscious only of the power of his king and that he is the king's warrior even to the death ; that, until he saw the British soldier, the Ashantee went to the battle only to conquer;

that in four campaigns, at least, against the British, the British were worsted; that in the Bantammah, the king's treasure house at Coomassie, there lies a bleached skull of a British Governor as a trophy of the Ashantee's might—can tap his cheap Brummagem gun with its old-fashioned flint lock, his bad powder and his cast-iron slugs, and, relying on his superior craft in the bush, may with a smile of proud serenity accept the combat which the bedizened and tasselled negro of the West Indies proffers him.

Give the Ashantees the same weapon, the breech-loading Sniders, with the proper ammunition, with which the allies of the English have been furnished, and even the white troops would find that the Ashantees are a foe worthy of their best efforts.

It is a thousand pities that the British Governors have not succeeded in winning the confidence of the Ashantees. Yet, who knows but this war may prove the greatest blessing that could befall these people? If the English are wise they will deprive their present enemy of their king, attach to themselves these brave and formidable warriors, and through them open the whole of Central Africa to trade and commerce and the beneficent influences of civilisation. The Romans would have been delighted at such an opportunity of extending their power for the benefit of themselves and the world at large. Two thousand Ashantees, under the leadership of an intelligent British officer, would soon extend the power of the English from Cape Coast Castle across the Thogoshi mountains to Timbuctoo, and from the Mandingo land to Benin.

CHAPTER IV.

A VOYAGE PROPOSED—ADVENTURES AFLOAT—THE SURF OF THE GOLD COAST —SCENES ALONG THE GOLD COAST — ACCRA — TRADE AND TRADERS— CAPTAIN GLOVER.

November 22, 1873.

It was dreary waiting at Cape Coast Castle for news, and expecting the arrival of the white troops and the soldiers of Her Majesty the Queen, who are said to be coming to destroy Coomassie, the capital of the ferocious king of Ashantee. Inactivity and ennui were rapidly sapping every ounce of energy and vitality.

Many, like myself, found it hard work to live week after week in a Fantee hut, stationed in a malarious valley, doing nothing but smelling the abominations around us, or counting the minutes which must elapse before being prostrated by another attack of fever, listening to the tweedle-dee and tweedle-dum of Fantee music, staring in a kind of lack-lustre-eyed way at the antics of sireless little negroes who seem to have been born no one knows how. If one went in a desperate state straight to head-quarters seeking news, it was ever the same answer, 'Nothing new.' If one met an officer, and judging by his face that he would be likely to impart something, one would say 'Good morning, good morning. How do you do. Glad to see you look so well. Anything new this morning?' the answer would inevitably be, 'Good morning, good morning. Quite well, thanks. No, nothing new. Ta-ta, old fellah!'

The question 'Anything new?' as soon as uttered seemed to annihilate all courtesy, cordiality, and good feeling. Mind you, not that there was really anything to impart; not that the officer was in possession of state secrets and regarded the man who had the audacity to ask him for any with distrust;

but because General Wolseley has so thoroughly imbued his
subs with his own peculiar notions of special news-gatherers,
and stamped his own individuality on almost every member
of his staff.   Socially, then, there is not a disagreeable man
here—at least I have not met any.   Meet any of them without the face of inquisitiveness, and the nose of a news-
gatherer, and the eye of an itinerant newspaper man, and you
will find them as genial souls as any who ever called themselves
English gentlemen.   The general himself is chiefest and best
of them; as hospitable and kindly-mannered as he can be.   But
on the matter of furnishing news to newspapers he entertains
most peculiar ideas, which are by this time patent to all the
world.

In the meantime it was dull dreary waiting for the arrival
of the troops: one's eyes became quite jaundiced looking
at the same objects; one's ears became tired with the meaning-
less hum and babble of people on the verge of inanity.   Even
the stirring items of ' Ashantees in retreat,' ' two Houssas killed,'
' four of the allies wounded,' and news such as another officer
down with the fever, one of the staff has a dysenteric affection,
a marine has had a sunstroke, failed to attract.   If letters
had had to be written upon such things as these, I fear that
the Anglo-Ashantee war, with its tedious, small, uninteresting
details, would soon be placed on a par with the ever-enduring
Carlist war, as the most uninteresting campaign ever planned.

Well, I said it was dull dreary waiting for the troops: it
was becoming positively sickening, for already about fifty per
cent. of the officers who had been attached to the prospective
expedition were down with fever.

I proposed one evening that we, the Correspondents, should
proceed down the coast to Accra in the ' Dauntless,' and find
out how things were progressing to leeward with Glover's expedition.   The time was after dinner, stomachs were more
than usually full of a good dinner, eyes were a little brighter,
and faces were a little more animated with good cheer.   There
were four of us ; one was the agent of a great London house;
one was special correspondent for a London daily, a man of
some importance ; a third was the partner of the agent afore-
said ; the fourth was your obedient servant.

The London special replied kindly : ' My dear fellow, this

is really very kind of you ; I should have thought you would have considered me a rival.'

Answered I :—'I just happen to have a little steamer here, and you, like myself want to go down the Coast to gather news. Permit me, then, to offer you a passage. If I considered you as a rival, I would not.'

Then spoke the agent : ' Oh, we can manage it easily enough. It will only take us ten hours to go to Accra; thence six hours more will take us to Adda, at the mouth of the Volta. It will be impossible to cross the bar with the steam launch, but we can hail a surf-boat, and go ashore. If Glover is there, our journey is at an end ; if he is not there, we shall have to manage to enter the river and go after him : one day will be enough for that.'

The partner chimed in with all the agent said, and the London special hailed the idea with pleasure.

We all felt uncommonly brave, unusually energetic, youthfully sanguine. What we were going to do, and what we were not going to do, were themes which were fertile and talkable. It was left for me to provision and coal the ' Dauntless ; ' they were simply to consider themselves as guests. It was finally resolved that we should set out on our adventurous cruise at evening of the next day.

The morning of the day dawned, noon came, and the afternoon was rapidly drawing towards evening, when the gentlemen who were to be my passengers were seen advancing towards my house.

When they had come in, and been greeted, I said : ' I suppose you are all ready to start ? '

' No-o-o,' drawled the agent, with not one-twentieth of the assurance or the audacious fire which lit his face the night before. ' You see, I have been reckoning up my time pretty closely, and I don't see how I can spare eight days, for it will take eight days and no less this trip of yours, and my ship is coming ; she may be here to-morrow or any day, and it would not do for me to be absent. If you were only going to Accra now, and coming straight back, why then I think I should go.'

' What do you say, Mr. Correspondent ? '

' Well, you see,' he replied, ' if Mr. Agent was going to Addah, where Glover is, I would not mind going because he knows the whole coast. But going to Accra doesn't tempt me at all, besides I fear I should be fearfully sick, your launch is such a small thing. No, I don't think I can go.'

'Very well, Mr. Agent, we shall go down to Accra, and if I can get the information I need there, I shall come back at once. Will that suit your book?' The new arrangement suited Mr. Agent, though it evidently was a labour to him to get his mind up to the travelling pitch, as he begged me to defer starting until next morning.

It must be understood the agent of the great London house is a kind of king on the Gold Coast. He has sub-agents everywhere, all of whom keep very hospitable houses. He is the man whom we expect will cash our bills by-and-bye, when ready money has gone, and he is the man who could furnish me with more coal if I required any; it was therefore a matter of policy to submit to the caprices and humours of such an important man.

At 7 A.M. we proceeded on board the 'Dauntless,' and were about to weigh anchor, when I asked the agent if he had remembered the oil-cans.

No, he had not; he would send a man immediately after the oil, which he did. It was no light matter to have forgotten the oil for the engines, neither was it a light matter to send a boat back through that awful surf, which the English have left to be a curse and an annoyance to all who seek this part of the Gold Coast. We had both in coming through the frantic surf been drenched to the skin by a vicious wave, which had simply wetted us for spite, though it might have drowned us had we submitted the boat to its power.

Patience, mixed with a little undercurrent of fuming and fretting, obtained its reward finally; the oil-cans came on board, we sounded our whistle for a parting signal, we weighed anchor and steamed to leeward, that is towards the east, along the African shore, at the rate of nine knots per hour.

I was falling—not falling, but gradually sinking—into a delicious stupor, a soft quiet ecstasy, lulled into uncon-sciousness, or listlessness rather, by the gentle cradling of the deep, by the soft, pure, healthy breeze which stole whisperingly under our awning, by the luxuriant scenery of the tropics, which was banked up on our left, in the palm-clad hummocks and low ranges.

We had reached Annamaboe, about 10 miles from Cape Coast, within a very short time it seemed to me, when I was awakened from the repose of mind and body, and the influence of the wind, the sky, and the scenery, by a gurgling sound proceeding from my friend the agent, through whose good offices I expected to have my note-book full of interesting notes respecting the war.

'What is the matter, my friend?' I asked. 'Sick, eh! it will do you good.'

'Oh! much good; bile, you see, collects so quickly in these tropical countries, that really when one inhales fresh sea air the whole system seems to revolt at the tax laid upon it by the vicious bile when ashore, and gladly frees itself from that which choked and impeded the machinery of—ugh-ugh-ugh—the body. I feel much better already. But what in the name of fortune is the matter with the engine?'

'Eh! what, of course we have considerably slackened, haven't we? what can be the matter, I wonder.'

The engineer was at once questioned imperiously upon the subject. He answered that he himself had been wondering what the matter could be, and he had been anxiously asking himself the question, besides examining carefully the engine and machinery. The cause was far beyond his ken, but he feared the boiler was not adapted for salt water. He had started with 80 pounds of steam, but it had gradually got lower and lower until he had no more than 15 pounds left, and though there was a splendid fire he could not increase the steam-power.

The yacht was going at the rate of about 2 knots an hour. My friend the agent was perfectly miserable from the disagreeability imposed on him by Neptune. The sky lost its cerulean aspect and assumed a gloomy colour, which, I was assured by the agent, who knew the coast well, was the forerunner of the tornado.

'My God,' said he, 'I would not be overtaken in this little thing for the world. Nothing would save us. We are only 500 yards from the beach, and if she can't do better than this,

she would be driven ashore in no time.   I vote we go back
before it is too late.'

I turned her head round, and inwardly vowed that once I
landed him I should continue my voyage without him.

In a couple of hours we were abreast of the fleet we left
some three hours before full of happiest anticipations.   Everything seemed changed in that short time.   Our own thoughts
were sullen, baffled hopes soured the mind, the sky was of
an inky blackness, the sea was dark, the land appeared overhung with a murky pall, there was no pleasantness in anything,
looked we inwardly or outwardly.

Suddenly the funnel gave a roar, and before we could recover from our astonishment the engineer shouted through the
rising breeze, ' It's all right now, sir ; the damper was up, that's
all, sir !   She will go now, sir.   She will beat anything about
this neighbourhood.'

' Shall we go back, Mr. Agent ?   Would it not be better
for us to continue the voyage ?   The engineer has found the
secret out,' I asked.

' All right,' he answered ; ' if you think there are no more
secrets that may wreck us, I am willing to prosecute the
voyage.'   Considering that the gentleman was miserably sea-
sick, and advancing tornadoes are not to be sneered at on this
coast on board a tiny steamer like ours, the answer was plucky,
and worthy of the best John Bull that ever was afloat.

The little yacht was now throbbing through every beam in
her, the steam sizzed and swished vigorously, the funnel roared
in concert, while the fan kicked up a foaming wake behind in
quite an exhilarating style.

We were going a steady eight-knot pace, and soon sighted
Annamaboe, an ochrish-coloured fort situated on a spur sloping
from a palm-clad range to the sea.

As we continued our journey, glimpses of unrivalled shady
groves of palms, shrubs, plantains, and tall arrowy silk cotton
trees, met the ravished eyes.   It was such a change to see real
nature once more after all the dreary waiting for active life at
Cape Coast Castle.   The tornado clouds vanished over the
horizon, the sky regained its tropic tints, the shores emerged
out of the gloom, the sea vied with the sky in its blueness, the
surf, ever frantic, ever roaring, ever plunging, reared its myriad

crests of snowy foam and dashed itself on the sinuous beach
with a noise almost deafening.

The 'Dauntless,' as trim and taut a little yacht as ever
bent to surging wave, raised her head proudly and gracefully,
and, despite the uproar on her lee, cleaved the Gold Coast's sea
as if she cared nought for rock, or reef, or wreck.

By hugging the shore closely we were enabled to detect
beauties that are never seen by passengers travelling on the
steamers. Tiny nut-brown villages modestly hiding under a
depth of green plantain fronds, and stately silk cotton trees,
which upheld their glorious crowns of vivid green foliage more
than fifty feet above the tallest palm tree; depths of shrubbery
wherein every plant struggled for life and breathing-space with
its neighbour, through which the eyes attempted to penetrate
in vain beyond a few feet; tracts of tall wavy grasses, tiger,
spear, and cane, fit lurking places for any wild beast of prey,
varied by bosky dells, lengthy winding ravines literally choked
with vegetation, and hills on the slope of which perhaps rested
a village of a timid suspicious sub-tribe. And if the eye, ever
in search of the picturesque and novel, ever roving for strange
scenes, chanced to fall on the long line of surf, and the ear
listened to the sonorous deep-sounding thunder, the mind
recurred to the Psalmist's expressive words: 'They that go
down to the sea in ships, that do business in great waters, these
see the works of the Lord, and his wonders in the deep.'

The surf on the African coast is ever a wonder and a
danger. There is no other coast in any part of the world a
thousand miles in length without possessing a cove or harbour
where a ship could anchor secure from being rocked by the
surf waves. Try along the whole of the Grain, Ivory, the
Gold, and Slave Coasts, and there is not one port. But fortunately for ships trading to these places, there is seldom a
hurricane or a gale blowing, so that they are able to anchor
about a mile from shore. There is never any dead calm.
The sea is ruffled in the morning, by the breeze from
oceanward: during the night it is moved by the land breeze,
so that ships anchoring in the roadsteads are ever to be seen
rolling uneasily; they are never at rest. Unceasingly the long
line of waves is to be traced, rolling onwards towards the shore,
gathering strength as they advance nearer, until, receiving

the ebbing waters flowing from the beach from preceding seas, there is a simultaneous coiling and rolling, and at once the long line of water is precipitated with a furious roar on the land. Where the water meets a rock a tall tower of spray and foam is suddenly reared, the wave line is broken and is in mad confusion. Where the beach is smooth sand, you may trace a straight unbroken line of foam nearly a mile long.

One may easily understand then the trouble and annoyance trade undergoes on such an inhospitable coast. An ordinary ship's boat is useless; it would be a drowning matter for a crew of sailors unaccustomed to the surf to attempt to land anywhere along the coast between Sherbro and Lagos, a distance of nearly 1,200 miles. The mouths of the insignificant rivers which feed the sea along this distance offer just as dangerous impediments to an ordinary ship's boat as the beach does. A long line of mighty breakers run across the mouth of each river and form a bar, which is almost certain death to cross, except in boats especially constructed for the peculiar work, and native canoes. These surf-boats have no straight stems or sterns, the keel is in the shape of a bow, which allows the advancing wave to be well under the boat before its crest lifts it aloft to precipitate it in the deep trough left in the track. A straight-stemmed boat cuts the wave, and its divided crest as it falls unites and swamps it. Many and many a poor sailor, ignorant of these things, has lost the number of his mess on this coast, and the history of trade on it is fraught with many a doleful tale.

Beyond Annamaboe, Accra is seen nestling comfortably in the lap of palm-clad hills, a small unpretentious village, and soon the five hill cones of Cormantine rise to view. A mass of white buildings glisten in contrast to green palm fronds; an old fort, and a ruined tall tower, also catch the eye. Cormantine was evidently a place of some importance once; its position is admirable, the range behind it augurs well for the salubrity of the climate; and the five tall singular hills just aback of the range would have been well adapted for sanitoria if the authorities of Cormantine were acquainted with sanative principles.

Cormantine has a history. I wish I could give you the plain simple story that one of the stones in the ruined fort could tell, had it a tongue to speak. I daresay it would be one of the most pathetic stories that could be told connected with the Gold Coast. It has the merit of having been the first settlement which the English made on this coast. Admiral de Ruyter wrested it from the English in 1663, and called it Fort Amsterdam, after his beloved natal place; it was a great slave mart once, human cattle have filled the square of the fort, after tramping long weary leagues from far Gaudo and Sokoto on the eastern bank of the Great Niger. Descendants of the Moorish trader and his Fellatah slave now inhabit the town of wicked memories, and in the olive bright-eyed mulatto of Cormantine you may recognise the seed of the Dutchman.

Beyond the old town a few miles, the range dips down into a broad alluvial valley formed by the Amissa and the Nacqua rivers, which, however, in the dry season never, and seldom even in the rainy season, empty themselves into the sea. This is another feature peculiar to the West Coast of Africa. There are numerous small rivers within the bank of comminuted shell and sand which the Atlantic waves have forced up, but very few of them have current enough to withstand the force of the restless surf.

Rivers such as the Assinee, the Prah, and Volta, have volume sufficient to clear to a superficial depth the sand which the action of the sea heaves up, but the smaller rivers, with their feebler currents, find themselves imprisoned by a bank of sand averaging from a few yards to a few miles in breadth; consequently finding their natural debouchures barred up, they form lagoons, like those of Grand Bassam, Volta, and Lagos, many scores of miles in length. Lagoons of limited extent are here called salt-ponds, such as the salt-ponds near Cape Coast Castle, Annamaboe, Cormantine, and Accra, and a score of other places. The alluvial deposits conveyed by the small rivers settle down in the lagoons, the bottom of which in the course of ages will become of the same altitude as the beach; the rivers will then overrun their barriers, eat out a channel to be again imprisoned by a new bank of sand to again form lagoons, and be again created land; and thus the process of land-making, I have no doubt, has been going on for ages here.

The soil from Cormantine to the distant low-lying cape of Tantamquarry, for which we now point our little steamer, must be very fertile. Large groves of cocoa-nut trees stand apart, apparently but a few hundred yards from the utmost reach of the surf, embosoming villages with their ample and verdant leafage.

Rounding Tantamquarry we saw another ruined fortlet—another evidence, if more were wanting, that the Gold Coast was considered to be an important acquisition, and that a foreign Power was determined to guard what they had acquired at all cost, at one time. And right in sight, a few miles away, we saw still another, that of Assam, once held by the Dutch, but destroyed by the Ashantees in 1811.

A truncated conical hill of unusual height above its neighbour, possibly 1,000 feet above the sea, marks the neighbourhood of Winnebah, an important settlement once, and still the seat of some trade in palm oil, oil-nut kernel and other customary articles of the African trade. The whitewashed houses of the merchants were soon visible, situated on a terrace overlooking the bend of the shore, called its harbour : a river empties into the sea close by, said to come from a great distance inland.

Barracoe Cape was reached a few hours later, and on rounding it a stretch of country was visible, which extended to Accra, the second most important settlement of the English on the Gold Coast.

At two in the morning, after a run of over 70 miles, we arrived opposite Accra, and hailing a mail steamer, the 'King Bonny,' Captain Hamilton, just arrived from leeward, were soon on board, recipients of large-minded hospitality from the kind old Scotch skipper.

At seven o'clock I was wakened by a hideous din of human voices jabbering alongside the ship in their surf-boats an unintelligible jargon of words such as no Christian like myself could stand without nerves getting unstrung. I therefore looked upon what was called Accra with a sullen face and in no amiable frame of mind.

The scene ashore was that of a straight beach backed by a mud terrace, which stretched to the right and left and rear of Accra for many miles, singularly open and clear as seen from shipboard. Accra itself straggled for nearly a mile on the edge of a terrace overlooking the beach, many pretentious houses, whitewashed, attracting attention from their prominence above the clay-brown huts amongst them. Almost to the extreme left was the Commandant's house, aloof and exclusive; its wide verandah denoted luxurious coolness, its wide space around it informed you that at one time or another some occupant of it had been assiduous to procure unpolluted air. Away to the extreme right was another large house with wide verandahs, and abundant grounds about it. This was the Basle Mission House, occupied by a singular community of religious Swiss and Germans who have banded together for the very sensible purpose of teaching the natives and making money by them by honest trade in palm oil and gold dust. In the very centre of the town was the port and lighthouse of Accra. Between these houses the body of the town of native and European buildings jammed itself. Some three miles to the east of the Basle Mission is the village of Christianburg, a picturesque mass of whitewashed buildings, consisting of a ruined castle, a ruined Martello tower, and another large establishment of the enterprising Basle Mission.

You would hardly believe it perhaps, but what I have described to you as Accra was in the possession of two foreign Powers at one and the same time, Great Britain and the Netherlands. The two portions of what appears to be one town were known respectively as Jamestown and Dutch Accra. Each town had its own native king. King Kudjo was under the protection of the British, King Pakki owned allegiance to the Dutch, and at Christianburg there reigned King Dawmah. Indeed, the same excessive confusion reigned throughout the Gold Coast until the British purchased the Dutch out-and-out a couple of years ago.

After breakfast I proceeded ashore in a surf-boat manned by native Accras. Please bear in mind what I have already related of the difficulties of landing on the Gold Coast; at Accra, the difficulties are terrible. The boatmen urge their heavy short-built British boat with an admirable nonchalance; though looking ahead at the awful confusion on the beach, I could not tell, to save my life, how or where they proposed to land. I saw only long lines, apparently unbroken, of dark, damp rocks, and the confusing, bellowing, thundering, foaming waves, in which, unless my men are steady, I shall be presently engulphed and swept high on those dark rocks a battered corpse. Pleasant, is it not? Destiny is inevitable, however; I must go on. The boatmen's paddles are lifted high and dashed emphatically into the waves, their clear, sweet, healthy-toned voices rise in sonorous chorus, and the outer line of rocks is reached. It is only now that I am made aware that there is a passage running obliquely through the rocky series, and now the steersman astern deftly threads the way, wave after wave, roaring and menacing, lift us higher, still higher — Ah! God of mercy!—We—No—By George that was a narrow escape! The boat's keel bites the beach of Accra, and we are safe.

I will forbear giving another version of my sentiments or opinions respecting what I think of the British Government for leaving this place in the state of nature it has been left in, as every other place on the West Coast of Africa. Referring to the boatmen, I will say that they valued their labour of bringing me ashore at the very same rate I valued their deftness and skill; but I would not like to risk a frequent passage among those Accra rocks, much as I admire those Accra boatmen, nor, indeed, have I seen a European who did care to risk it often.

When we were freed from the boat-gang, my chaperone conducted me to the old fort of the slave-traders. Almost the very first thing I saw that was interesting was a large mural tablet near the gate, dedicated by a man to the memory of his friend whom he had, to his eternal sorrow, shot to death by accident. Both were English officers.

Accra's fort, like all forts on the West Coast of Africa, is a mere stuccoed, or lime and mortar, imitation of one. Its walls, comparatively speaking, are as thin as cardboard, but I presume it answered its purpose in the antiquated times of 24-pounders. It serves now as a prison for refractory Accras and coloured culprits, who are fed with *kenke* [1] balls.

On mounting the stone stairs leading to the battlements, I am enabled to appreciate the change which has come over the fort since the town of Accra became wholly a British possession. The old cannon have been tumbled into dishonourable heaps

---

[1] Kenke—a stiff-doughy mixture of corn-meal and water.

over the walls, and are now the prey and sport of the briny surf. There were probably 500 old pieces of artillery in the forts of St. George da Mina, Cape Coast Castle, Annamaboe, Mumford, Apam, Winnebah, Accra, Christianburg, at the beginning of Sir Garnet Wolseley's administratorship of the Gold Coast, and it was seriously intended to collect and ship them to England to be recast; but when the authorities came to calculate the cost, they found that the cost of shipment would be greater than the value of the material, and the guns were accordingly tumbled out of the forts as rubbish.

Having completed an examination of this Dutch slave fort, we proceeded through the intense heat into the town. The huts of the natives have been established anywhere, without regard to order or to any symmetrical arrangement. The consequence is that the streets are uniformly narrow, crooked, and oppressive from the filthy habits of the natives. The principal merchant of the town is a Mr. Croker, the agent of the great mercantile house of F. and A. Swanzy Brothers, sons of the famous Swanzy who founded the legitimate West African trade. Mr. Croker's house raises its tall head above a mass of dingy grey thatches, so that the outlook from the windows of the second storey takes in, perhaps, a hundred of these thatched roofs, in all stages of decay and native improvidence. From the third storey, however, the eyes are delighted with views of sea, shipping in the roadstead, the area of the ill-planned town, the houses of the European residents, and a vast stretch of plain country, covered with cactus, and gums, and thorns, and grass, and a winding lagoon of grey-green water, on which Mr. Croker, who is an intelligent and energetic Briton, finds pleasure sometimes in paddling his Rob Roy canoe, to the shame of half-a-dozen sallow-faced and moribund European clerks.

The trade in which these Europeans under Mr. Croker are engaged is that of purchasing palm oil, gold dust, and gum-copal; while the Basle Mission buys not only palm oil, gold dust, and gum-copal, but black monkey-skins, cotton, india-rubber, gum, and almost everything that can be turned into money remuneratively in Europe.

The currency of the Gold Coast is gold dust, and in some parts cowrie shells are still used, though they are being rapidly

superseded by British silver coin. An ounce of gold dust is
sold for 3*l.* 12*s.* The natives frequently exchange among
themselves the weight of even a small bead in the precious
dust, which they call a pessua, a trifle as insignificant to the
Accras as a dime or a groat would be to us.

The Basle Mission House, on closer inspection, is a large
roomy building of two lofty storeys, surrounded by broad
verandahs. The yards about the building resound with the
hammering of the coopers, making up puncheons ; with the
tattle of the gossipy traders in palm oil, who have just arrived
from the interior somewhere, with scores of black earthen pots
full of the yellow, buttery palm oil. Puncheons are being
rolled about, palm oil pots are being shifted, black people are
in the height of business in the heat which threatens to strike
the first European who dares it unprotected dead. I see that
a little store also is kept in the lower storey, and I am thinking that these Swiss-Germans are making much money by
these African pupils and converts of theirs. The superintendent informs me that the principal educational establishments are at Christianburg-on-the-Sea, a few miles lower
down, and at Akropong in the interior. After a look at his
own breezy quarters upstairs, I proceed to Christianburg in a
carriage drawn by fourteen half naked Krumen on the run.

It is a very large establishment at Christianburg which
the Basle Mission has put up for trade with, and conversion,
and education of the natives—very large indeed. In its yards
are numbers of men at work upon puncheons and palm oil.
Trade, I fear, is uppermost here also ; though the superintendent tells me he has 80 children in-doors of both sexes
under the tuition of male and female teachers. I did not see
them being instructed.

Two buildings made in Germany and shipped to and
erected here a little behind the village of Christianburg, contain the pupils now under process of Christianising. Their
very aspect promises that within their walls the divinity of
Christ is being taught the students ; and taken in connection
with the fact that at Akropong the college of the mission, with a
model farm, &c. is situated, the Basle Mission, let us hope, have
a more brilliant future than the past has been, and that the

members of it will go on doing a vast amount of good, spiritually and temporally, to all concerned.

When I come to speak of the political and war news collected at Accra, I am reminded, by a glance at the number of my page above, how very brief I must be.

Captain Glover, formerly of the Royal Navy, but for some years past connected with the Civil Service, has imparted a degree of confidence among white and black in his powers of administration by fighting, which is quite extraordinary.

He must be a Wellington in strategy and the very model of an administrator. A white trader on this coast does not feel comfortable when he meets a newspaper man until he has conversed about Glover's genius, his Clive-like abilities and energy. On the other hand, I have noticed a faint attempt at depreciating him on the part of the military officers, who acknowledge Sir Garnet as chief. Either the man must be a great and skilful leader or be one of ordinary calibre, unfit to be a leader, but capable of conducting a small insignificant business tolerably well—which is it? The British have entrusted to him a most important duty, second only to that of Sir Garnet, if not as arduous. I believe it to be, myself, a very important and a very arduous task, and I hope sincerely he will bear out the high praise I have heard spoken of him.

Captain Glover has been entrusted with the duty of conquering the contumacious tribes of the Volta, and of collecting a force many thousand strong to take the Ashantees by the right flank, by marching north-west from the Volta river. He has succeeded very well so far. For weeks he waited at Accra for the 5,000 men promised him by the king of West Akim, until in high displeasure he sailed with some 1,800 men to Addah, at the mouth of the Volta, where he has made his camp of rendezvous. The 5,000 reinforcements of his force which he had waited for so patiently drew near Accra at last, only to be tempted by the offers of Sir Garnet J. Wolseley. It is not quite decided yet whom the king of Akim will join. If he joins Sir Garnet, then Captain Glover is considerably weakened, and the full measure of success which he expected may not be given him, though his energy is such, it is said, that he would not stand passive, and see himself robbed of all opportunities of winning glory, but would march upon Coomassie with the 2,000 native troops he had been able to secure.

Before starting on his expedition he has to distinguish himself by marching upon the Aquamoos and Awoonahs on the right bank of the Volta and crush them, lest when he is absent on the expedition to Coomassie these warlike tribes should overrun the countries of the tribes who are his allies.

The Aquamoos were formerly allies of England, but they invited the Ashantees to attack the British Protectorate in 1867. The Duffas and the Ulos are also to be summarily suppressed, in the same manner and for the same reason, by Glover. He has a steamer, the ' Lady of the Lake,' and several steam-launches, to assist him in his enterprise along the Volta. He has also several able young English officers to help him. Nothing as yet, however, has been done, though he will attempt something shortly. Both Glover and Sir Garnet, it seems, are waiting ; a dreary thing enough in this unhealthy climate, as I found it to be at Cape Coast Castle.

The first cruise of the ' Dauntless ' ended, I have just returned to Cape Coast, to wait another opportunity of getting news.

CHAPTER V.

THE TOWN OF ELMINA — CAPTAIN HELDEN, 2ND W. I. R. — SIR GARNET
WOLSELEY'S EXPEDITION.

November 29, 1873.

I AM writing from the old Dutch town of Elmina, or the 'Mine,'
whence the Portuguese, who farmed the revenues of the Gold
Coast in the sixteenth century, obtained ingots and gold dust
enough to stir the kings of Portugal to mightier and grander
achievements, which finally culminated in the discoveries of
the Eastern and Western Indies. I love the old town for its
associations and its history, for the impetus it gave to civilisation, for the fair promise that it held out to Don Juan and his
brave captains of an empire to be established, of the redemption
of Africa by means of the Catholic missionaries, a dream, however, that has not yet been realised.

The Castle of St. George, a tall mass of white buildings,
stands on a spit of sand and rock to my right; the Castle of San
Iago, situate on a hill which commands the town of Elmina and
Castle of St. George, stands on my left; the harbour of Elmina,
formed by the river Beyah, is between the castles. The town
of Elmina groups itself at the base of San Iago, and then
straggles in two irregular lines along the shore of the bay of
Elmina. It probably contains a population of 5,000 souls in
time of peace, but just now the warriors have been mustered in
as auxiliaries of the English against the Ashantees, while the
drones have been captured by press-gangs and enrolled as
carriers to convey material of war from Cape Coast to the front.
The people do not like this state of things a bit; it is in such
striking contrast to what they have been accustomed to under
the benign rule of the Dutch. They feel aggrieved that they
are compelled to do service under a flag which they never acknowledged, and that they are forced to fight against the

Ashantees, a people whom they were accustomed to look upon as allies and friends in the old days.

The Elminas see a vast difference between their treatment by the Dutch and the English. The Dutch indulged them in long-winded palavers in the castles, treated their chiefs with deference and respect, were merciful to their foibles, their old customs, and their ignorances ; the English are brusque and peremptory in their dealings with them, absolve them from attending at palavers, content themselves with giving them orders and to seeing that such orders are carried out by armed police, pooh-pooh their antiquated and ignorant objections, keep watch and ward over them vigilantly, drive their warriors to fight against old friends, and press men to transport service without so much as asking the consent of their chiefs. All these peremptory doings of the English are causes of grievance.

Nearly all the Elminas can speak Dutch fluently, because the Dutch were accustomed to keep two schoolmasters at the fort for the instruction of their subjects. There are several clever mechanics and sailors there, also, whom the English find exceedingly useful at the present busy time.

The superiority of this town over Cape Coast, as the headquarters of the British on the Gold Coast, is at once seen by anyone coming here for the purpose of comparing the advantages of the two towns. The people, in the first place, are very much more advanced than those of Cape Coast; the Dutch have mingled more with them, and inculcated in them industrious habits, and their intelligence has been very much more expanded. The castles are stronger and better adapted for defence; the country is healthy and might be made more so, until, in fact, fevers would be as rare as in any part of Europe. Then, as a last argument in favour of Elmina over Cape Coast, the former has a harbour which is accessible to boats at a high tide, and with a better outlay and a small dredging machine, schooners under 100 tons might enter with ease and safety; a very important thing for the Gold Coast, which has not a single harbour where a ship's boat can land with safety. I should say that an outlay of 20,000l. would make Elmina a first-class harbour. The sand and mud at the bar and in the river are the only impediments ; these, however, are easily removable. Large schooners once floated in the harbour, there is no reason why they should not float again, even in deeper water.

The civil commandant, Captain Helden, of the 2nd West India Regiment, is doing a good work, however, and in the right direction. It is a step which if followed up by the Colonial authorities would lead towards establishing Elmina as head-quarters, and promoting the commerce of the Gold Coast. The ruins that have disfigured the neighbourhood of the Castle of St. George since the destruction of the native town last June, by the launches and boats of the English fleet, are being removed by convict labour, and a wide embankment is being constructed along the right bank of the Beyah, which will confine the waters to narrower limits, and enable the swift ebb-current to deepen the harbour. When the ruins have been cleared, a noble promenade or plaza, worthy of a European watering-place, will be left exposed to the cool breeze of the sea, which will ensure to the European quarters, or the town of Elmina as it is now, an unpolluted atmosphere and a healthy future. The prospect is encouraging; and if only the authorities abandon that nest of malaria and sickness, Cape Coast, remove to Elmina, and continue the good work of improvements, which would be a pleasure to most people, on considering the vast advantages to be derived from them in a sanitary, social, and moral sense, Elmina will have been the means of removing from the Gold Coast much of the evil character it now bears.

The news from the front is tame and uninteresting. A skirmish between the advanced forces and the retreating laggards of the enemy happens now and then, which has a local interest for the weary souls at Cape Coast. The last account we have received is about Colonel Evelyn Wood. It seems he was making a reconnaissance beyond Sutah near Fasua, with a force of 100 Houssas and a few hundred of native allies, when a large body of the enemy suddenly attacked him on all sides. Though taken somewhat by surprise, he instantly formed a square, putting the transport porters in the centre with their burdens, and in this position fought, it is said, for three hours, after which he retreated in order to Sutah, the point he started from. He lost but few men comparatively, while the enemy's audacity was punished severely.

Hardly a day passes, however, but a few are wounded on either side. The Ashantees keep well under cover of bush, and the British allies stick to the protection of the palisades, until they emerge out in strong parties to reconnoitre. This state of affairs will end only on the arrival of the white troops.

Sir Garnet informed me the other day that as great a force will ultimately move upon Coomassie as Napier marched upon Magdala. This force will be made up of three battalions of the line (white); one battalion of marines and sailors (white); one battery of Artillery; one company of Royal Engineers (white); two battalions of West Indians; one battalion of Houssas; two battalions of native allies: total, about 4,000 men. You may imagine then, with what interest we all wait the arrival of the white troops. Their very presence will revive every sick man at Cape Coast, and inspire every weary correspondent who finds the task of feeling interested in these petty skirmishes that now take place a work of real difficulty. Meanwhile we sicken more from inactivity than from the malaria. It is to the inactivity to which we are compelled that I ascribe the sickness and overpowering lassitude which has already prostrated so many. There is really nothing to write about, nor hardly anything to talk about. We make the most of every little accident that comes to our notice, and in the absence of anything of real importance we are inclined to think skirmishes great battles.

Sir Garnet must have felt that he was bound to do something to revive his fading energy when he last week proceeded to the front to examine personally into the state of the advanced stations. He travels slowly from place to place, as he has found out that it is highly dangerous in this climate to expose himself too freely. Some have expressed an opinion that it is very unlikely that he will see the end of this expedition, simply because he was too advanced in life to begin an expedition into Africa and that the climatic influence will tell severely on his frame. With this opinion, however, I disagree. Sir Garnet is not over forty years old, and is the youngest General in the British service. To choose a younger man, therefore, the British Government would have had to entrust the command of this expedition to a young Lieutenant-Colonel. While in England, Sir Garnet looked remarkably well and hearty, and his indomitable energy promised everything that could be required in a man to fill his position. If Sir Garnet becomes invalided through fever or other sickness, it will be because he exposed himself too freely to the sun and climate before active operations began, and to tell the truth because the doctors of his staff have coddled him too much.

We have lately had auxiliaries from Bonny. Prince Charles Pepple, and Prince John Jumbo, have arrived with a force of 100 men, from King Pepple, and Oko Jumbo, the rival of the famous Ja Ja Jumbo, of Bonny, to fight against the Ashantees.

The officers of the Control Department are as busy as ever in their duties. The stores are being rapidly conveyed into the interior, and there is every prospect that by the time the troops arrive Prahsu will be in the hands of the English, with abundance of stores safely housed therein. The advance guard is now within twenty miles of the Prah; and the Ashantees, it is reported, barring a few of the rear-guard, who exchange skirmishing shots, have crossed the Prah.

CHAPTER VI.

TRIP DOWN TO ADDAH—COMMISSIONER J. H. GLOVER AND HIS EXPEDITION
—CAPTAIN GOLDSWORTHY—BIOGRAPHICAL SKETCH OF GLOVER.

December 16, 1873.

EVENTS at Cape Coast Castle had become tame and uninteresting ; they lacked the charm of novelty. Sir Garnet Wolseley
and his intentions unconsummated, indeed as yet not begun,
had been written about threadbare. There remained actually
nothing upon which anything more might be said connected
with Sir Garnet Wolseley's expedition. Before rendering myself up to utter despair, I bethought me of the other expedition to leeward, about which I had attempted to write at a distance, but of which few people of Cape Coast, the authorities
included, knew anything rationally and intelligently. This
was Commissioner J. H. Glover's expedition to Ashantee-land,
viâ the Volta.

The ' Dauntless ' was lying peacefully at anchor in the lagoon
of Beyah, close under the frowning white castle of San Iago
Teredos, and saline particles gradually corroding her iron works
below, and the hot sun blistering her bulwarks and scorching
her decks after her troublous trip to Accra. The ' Dauntless '
offered me the means to proceed in search of Commissioner
Glover, and I at once availed myself of her. But it appeared
to me to be a very selfish proceeding, this availing myself alone
of the only means to hunt up Glover, when there were so many
of my itinerating colleagues hungering after such information
as would be derived from the trip to Addah, the reputed location of Commissioner Glover's camp ; and I offered to take
down any of the correspondents who chose to venture within
the limited confines of the tiny cockpit of the ' Dauntless.'

One only was found brave enough to accept the offer. ' It is

very tempting, I confess,' said my colleague of the ' ———,' 'but
I would not venture in such a tiny boat on the Atlantic for a
thousand pounds!' 'Oh no, by George,' said another, 'you
don't catch me on board. If I lost my life on the trip, my
people would say, "Well, what a fool the fellow was to be sure,"
a very scant reward, as you will admit. I like the sound of
the surf and waves, when I am safe on shore; but I do not like
the idea of being tossed about on big waves with the thought
prevailing in my mind that the launch would very probably go
down, and I should immediately be gobbled up by the sharks.
You are very kind and obliging, sir, but I must really decline.'

The brave man who accepted my offer was my colleague
Mr. George A. Henty, of the 'Standard,' whom I first met in
Abyssinia. As soon as he had made his determination known,
he and I were set upon by all, with a doleful chorus of voices
predicting direst calamities, such as 'Sure to go down, battered
by the surf, a certain shipwreck in store. The bar of the Volta
is awful, plenty of crocodiles about. Think of the bar at Lagos;
the launch will never cross those breakers; vessels three times
larger have been upset on that bar. You won't find any one
at Addah; Glover is certain to be gone to the front. If he is
marching, how will you catch him? The white troops will
come in your absence, and you will have lost the opportunity
of seeing them landed.' These and a hundred other uncomfortable suggestions were thrown out for our benefit. We
treasured them in our memories, but declined to act upon them,
knowing as we did, that if we fell in the cause of our respective journals we would fall gloriously.

The 'Dauntless' awoke from her slumberous repose in the
lagoon of Beyah about midnight, and steamed down the coast
at a rattling rate, heedless of the burden of forebodings which
had been heaped on the heads of her passengers. The beaming
stars and silver moon shone over our heads brilliantly, and lit up
our course, and in the morning the sun arose, and with the sun
the wind came and blew the sea into a million million uneasy
wavelets, which made the tiny yacht hop up and down merrily.
On our left was the Gold Coast of Africa, darkly green and
beautiful; now rising into gentle hills and anon subsiding into
valleys, or stretching away a level plain, scantily tufted with
cactus and thorn clumps. Clusters of conical huts appeared at

almost every headland, and in the recesses of every beach-lined bay. Umbrageous depths of shrubbery filled the hollows and depressions of the coast-line, and the ridge-lines seemed to be deep impenetrable hedges of jungle-bush.

On our right heaved the deep far-stretching Atlantic Ocean, waiting calmly for the fierce tropical tornado to stir its waves into action and fury, and submerge the saucy little 'Dauntless' and her crew in its soundless depths. But the Atlantic waves, though placid enough in our vicinity, expend their strength and mighty energy to our left, where they thump the shore with tireless fury and noise.

The afternoon of the second day the 'Dauntless' was racing past a lengthy stretch of glistening beach dotted here and there with palm-groves, each of which embowered a village. This palm-dotted beach separated a lagoon of the Volta from the sea. It was but a narrow spit of land thirteen miles in length, on the eastern termination of which was Addah Forh, where Glover's force was reported to be encamped. At the end of a couple of hours we saw a break in the sandy beach, deeper groves of palm trees, a village with glistening white houses, a number of surf-boats and native canoes drawn upon the beach, and in the foreground of all a raging surf, a rolling line of tall waves and white foam.

When we had arrived opposite the white houses we stopped the steamer and blew shrill signals to those ashore to bring boats, which were unheeded for a long time ; but persistency finally was rewarded, and a long surf-boat, manned by wildly gesticulating paddlers, commanded by a tall brawny fellow, who stood upright on the fore thwart and who gesticulated like a madman, was seen now on a lofty wave, anon hidden from our sight, labouring towards us. While witnessing the frantic antics of the commander, the gesticulations of the paddlers, the uneasy throes of the boat, we anticipated but little enjoyment when it would be our turn to experience the same dangers.

The steamer was hauled up astern of a brig lying in the road-stead, we entered the surf-boat, and after an exciting race with a huge wave, which bore us on its white crest, we were finally dashed, boat and all, high on the sandy beach of Addah Forh, amid the strange welcoming cries of a couple of hundred sable savages, who were grouped ashore to watch our perilous but

triumphant progress through the surf. So much of our journey
was ended safely, and we breathed free.

Then we made inquiries of a trader as to the whereabouts
of Commissioner Glover, and were answered that he had gone
up the Volta in his steamer, the ' Lady of the Lake,' but would
be down again next day to the camp at the mouth of the river
Volta. We were told also that the camp was barely a mile
from Addah Forh, and that the river was only about 500 yards
beyond the camp. The trader was a Scotchman, who offered us
the usual hospitable invitation among men on the West Coast
of Africa to refresh ourselves with anything in the shape of liquor.
Did we fancy brandy, English and German beer, claret, hock,
sherry ? ' Name anything you like, gentlemen.' The sultriness of
the atmosphere, the buffeting by the waves, the subsequent hot
flushing of the face, the parched tongues, demanded German
beer, with which we were immediately gratified. Bidding our
new-formed friend a good evening, we strolled away across
marshy ground in the direction of the camp, through scenes
which deserve from their novelty and picturesqueness to be
immortalised in oil on canvas.

Addah Forh is embowered by palms. Under the grateful
shade of the cocoa-nut trees it is twilight even in mid-day ;
imagine the deeper gloom which prevails just after the sun
has plunged down beyond the horizon ; people this gloom with
two or three thousand half-dressed and over-dressed black
savages in all attitudes, with hundreds of camp fires lighting
up hundreds of queer black faces ; imagine you hear a noisy
hubbub of human voices articulating languages you have no
conception of, shouts to comrades with strange names, and now
and then a startling report of a gun, with flashes of red fire,
and you have a picture that an artist with a true artist's
soul might fervently desire to paint.

Through the palmy colonnades, uprearing above all this
scene the graceful pendant leafage, we marched a short distance,
and gained comparatively open ground seamed with irregular
depressions half filled with stagnant water, whence issued a
dolorous chorus from the *anourae* with which the muddy
water swarmed. When there was no swamp the ground was
half covered with dwarf clumps of bush, from the depths of
which cricket and mantis grieved the ears with their stridulous

notes. Soon we beheld the bell-tents and marquees of the camp rising through the fast deepening darkness, a multitude of dark forms gliding through uncertain flickerings of soft fire-light and several crouching men, and distinguished shortly after the low humming sound that indicates the neighbourhood of a camp.

After inquiring of a coloured man in a cap and European clothes, whom we were certain from his costume could speak English, the whereabouts of the head-quarters, we soon arrived before a large hospital tent, wherein we introduced ourselves to Captain Goldsworthy, who is second in command of the Volta expedition.

Unlike the reserve and diplomatic dumbness which char acterised our intercourse with the authorities at Cape Coast Castle, we met a gentleman in Captain Goldsworthy who, while tendering the most gratifying hospitality, omitted not to impart the information we had come so far to obtain. He furnished us with ample materials of a letter, and I soon found myself compelled to have recourse to my notebook to treasure manifold items.

One singular thing I noticed in Captain Goldsworthy, and that was the enthusiasm which marked his conversation when Commissioner Glover's name was the topic. He never cared to restrain the impulse to endow his chief with all the high qualities which a great chief should possess. It was singular that Captain Goldsworthy should share in the enthusiasm for Glover which I had perceived strongly imbued the remarks made on him by every merchant and trader on the Gold Coast. I was compelled to reflect that when a man receives the high encomiums of which Commissioner Glover has been the recipient, from both civilians and soldiers, that there must be some, if not much, good in him. As there was no equivocation whatever in the praise bestowed on Glover, I was prepared to see and converse with an extraordinary man in some respects. Before introducing him I might give in brief some account of him.

Commissioner J. H. Glover was born in the year 1829, and consequently is in his forty-fourth year. He entered the Royal Navy when very young, and rose rapidly to the rank of Commander. He was specially detailed for service on the West

Coast of Africa, where he has undergone some remarkable adventures. After cruising about several years in the waters, he left the Naval service and became attached to the Colonial service as Governor of Lagos, which is an island and port on the Guinea Coast. His career while in charge of this African colony in the Benin Bight, has established his fame as one of the wisest administrators the British Government ever exiled to Africa ; he has endeared himself to the whole native population of Lagos and its dependencies ; his name is as universally known among the tribes of the interior bordering the West Coast as Livingstone's is known in Central Africa. He has enriched the country beyond belief, until from a miserable settlement, with an unenviable notoriety for fevers, its annual imports and exports amounted in 1872 to 5,432,310 dollars and the port of Lagos is considered as healthy a place to live in as any town in West Africa. Its wharves, piers, drains, houses, public buildings, order, systematical government, its local police, and its volunteers, mark the port of Lagos as having been under the fostering care of a man who took an interest in his work, and was possessed of the requisite energy and will to carry out his projects of reform and improvement. One of the most important acts of his administratorship at Lagos was the raising of a force of Mohammedan soldiers, or armed policemen, emigrants from the country of Houssa, and a faithful loyal band of Mussulmen from the Yoruba country, with which he contrived to repress the turbulent and contumacious tribes in the neighbourhood, and to inspire them with respect for the British Queen's Government. When this able man resigned his administratorship of Lagos last year, there was a general feeling of sorrow among all West African natives, as well as white, and it was the universal belief that no man, however great he might be, could ever replace the great Governor ' Golibar,' as he was called by the Yorubas and Houssas.

The military camp at Addah Forh was the nearest thing I had seen, since I joined the proposed expedition to Coomassie, approaching a warlike aspect. Cape Coast Castle, with its commissariat surroundings, the Government House, and its resident, Sir Garnet Wolseley, surrounded by a brilliant staff, the medical hospital and control huts, marked and numbered and guarded, were all well enough ; but they did not strike the eye and mind accustomed to a long series of military scenes so effectually as did one glance at the trimly-set bell-tents, the well ordered arrays of cloth houses, the lines of sable soldiery at drill, the outdoor piles of Government stores, the row of cannon open-mouthed and ready pointed towards the distant enemy, and the constant outgoing and incoming of military orderlies.

We were awakened very early at Commissioner Glover's camp. Soldiers may not sleep late, newspaper correspondents in soldiers' camps must wake early whether they will it or not. At 5 A.M. we struggled hard to sleep, despite the premonitory rolls of a large drum, and had in a measure succeeded, or perhaps would finally have succeeded, had we not ten minutes later been suddenly startled with a full-blown bugle blast close to our ears, which drove away all desire to sleep. The Governor's voice was heard calling for his coffee; his gallant subordinate Goldsworthy halloed with lungful force for his unusually dilatory servant; the commissary screamed for his chocolate, and the hum of a stirring camp was wafted into our ears by the morning air. Before I, to speak personally, had quite recovered the full use of my faculties, Governor Glover had dressed, drank his coffee, and had gone. A desire, growing stronger every minute, to watch the strange man, had more effect in rousing me than my coffee had.

'Which way has the Governor gone, boy?' I asked of a coloured menial adjunct of the Governor's staff, named Sam.

'Ah, massa, the Gobner, sir? He gone; praps ribber, praps town, praps Yoruba, praps Houssa palaver, praps Accra; no tell, no savey.'

It was not yet daylight, but I sauntered towards the river, whose noble breadth was overhung by a thick haze, and whose further bank was a long black line of mangroves. The stream was but 300 yards from the camp. The landing-place was about 500 yards from head-quarters.

I soon discerned the sturdy form of Governor Glover striding hither and thither, and recognised his cool calm voice giving orders. He was superintending personally the loading of the 'Lady of the Lake' for an up-river trip with ammunition; he was giving orders to a blacksmith; he was showing a carpenter what his day's duties were to be; he was speaking to the engineer about his boilers; he was telling the coloured captain at what hour to be ready and what sand bars to avoid; he was assisting a man to lift a box of ammunition on his shoulders; he was listening to a Yoruba's complaint about some unfairness in the distribution of accoutrements; he was inspecting the crews of the steam launches; he was directing some of the steamboat men how to treat the wild bullocks; he was questioning the commissariat officer about his supplies; he was rebuking the Accra king, Tarkey, for the dilatoriness of his men; he was specifying the day's duties to a Houssa sergeant; he was here, there, and everywhere; alert, active, prompt, industrious. He was general-in-chief, quartermaster-general, commissariat officer, military secretary, pilot, captain, engineer, general supervisor of all things, overseer over all men, conductor of great and small things, a most remarkable man, and in short, the impellent force of his army.

I was led to think by the varied scenes in which I saw him engaged within an hour, ' What on earth would become of the expedition if this ready-handed, ready-witted, energetic commander and supervisor, were to fall sick and be incapacitated from duty?' He has good, able, clever men, a few, under him; but capable as they may be, there is only one Glover. The Houssas, the Yorubas, the Accras, and Aquapims will tell you that there is but one ' Golibar' whom they will obey. ' English officers very good, sir, but they no Captain Golibar.'

I don't suppose they would actually disobey Captain Goldsworthy or Captain Sartorius, they have too much respect for them; but they have not become so intimately acquainted with these officers, they have not been objects of such solicitude with them as they have been with their dear old ' Golibar.'

' Golibar' brought them out of the jungle and marsh of the Niger-country, treated them kindly, fed and dressed them, made them soldiers, and endeared himself to them. He has led them to victories against their perverse brothers of the savage bush, counselled them, given them self-respect, made them feel they were men and soldiers of the British Queen; but Glover's subs have not as yet. This is why I was led to think of the consequences resulting from Glover's sickness.

I am glad I undertook to discover Glover's whereabouts. It has been a pleasure to me to have seen this great man in camp, where a great man may best be seen. I shall ever think of the time when my eyes first lit on the kindly face, the massive features with the genius of commanding men lighting up every lineament; on the sturdy form so full of energy, as an event. Glover, in my opinion, is really a great man; he has the elements of greatness in him. Perhaps he would not shine in the small details of an army, such as a man well educated for the pure military line would; but I doubt that I shall soon see his equal on the West Coast of Africa in the purely administrative and the organising line. Sir Garnet J. Wolseley would, perhaps, strike an observer as being the most brilliant of the two, but Governor Glover's greater power and solidity would be acknowledged as being very marked.

Glover compels men to love him. There is something specially attractive in him; all who come in contact with him thaw before his general kindliness of manner. There is no affectation about him; he is a plain, hearty, genuine man.

He is not one to endeavour to make men believe that his mission in Africa is one of universal importance; that a word of explanation respecting his intentions and purposes would destroy the whole subtle machinery of this campaign. Oh, no! he speaks without reserve, answers all questions about his mission as if his mission were really a public one, to be canvassed, talked of, criticised by the public which sent him out. On an enterprise like this he does not think it would injure his reputation or his mission if the British or American public were told something about it, were to be informed of all and everything concerning it.

The 'Lady of the Lake' was about to ascend the Volta river to the second post occupied by a portion of Governor Glover's expedition. The Governor kindly asked if we should like to visit the first and second posts, for, if we did, we were welcome to proceed in company with Captain Goldsworthy. Of course we instantly accepted. Captain Goldsworthy conveyed some mess-dishes and what provisions and wines his mess could furnish. I saw the gallant soldier was fidgetting about, desirous of making the trip enjoyable to his visitors.

Before we started, some amusement was afforded by the antics and pugnacity of the wild cattle Commissary Blissett had procured from tribes to leeward of the river Volta—the largest of which, though full-grown, was not as large as a year old heifer. The Houssas who drove them were butted rearward and frontward, were dashed at on the flanks and tumbled over to the intense joy of the crowds drawn together by the vociferous merriment which first signalled the onset of the cattle and the defeat of the Houssas. Reinforcements of the uniformed warriors arriving with long coils of rope, however, turned the tables on the valorous beeves. They were, after a few ineffectual struggles, finally captured, bound fore and hind legs, and dragged ignominiously along the gangway planks to the lower decks ; and the steamer, swinging to midstream, soon revolved her paddles and shot down stream, until nearing Dolben's Point and the surfy bar, she rounded the extremity of Kennedy's Island, and, with her stem pointing up river, breasted the noble current of the Volta.

The Volta, near the mouth, is from two to three miles wide ; unless, indeed, you dare call the lagoon, formed by the confluence of the Volta and the Troyeny rivers a mouth, when you may say it is near thirty miles wide. From Dolben's Point, on the Addah side of the Volta, to Richards Point, on the opposite side, is but about a mile and a half. These two points, however, are the eastern and western extremities of narrow sand spits, which, with the exception of a mile and a half of river exit to the sea, enclose a large area of water, in which are many islands overgrown with mangroves and bush, the resort of numerous aquatic birds and crocodiles.

The Volta needs exploring, and as we ascend it from the sea it must be described, because we are ignorant of its source and course above Pong : otherwise, we might take readers from its cavernous birthplace, or its natal dell in some far-away mountain range in the interior, and glide downwards singing of its charms, as the current, now rapid, now slow, bore us on. We might give them glimpses of the picturesque and inspiring scenes through which it rushed, of new countries as yet untouched by a traveller's pen, of strange animals and vegetables. But above Pong we know nought of this river, and very few ordinary stay-at-home people know anything of its mouth.

All can imagine what a river like the Volta must be near its mouth. It is an African tropical river of some magnitude,

H

not as large as the Nile, the Congo, or the Niger. It is as wide, when from bank to bank of mainland you see its breadth, as the Missouri is at Omaha, as the Hudson is at Sing Sing, or as the Thames is at Greenwich. I should say it was altogether as large a stream, as many miles in length, as the Hudson, though not so deep nor so useful for navigation.

The Volta has all the peculiarities of a tropical river. Its course is wayward, running at random around numerous low islets in midstream, at one time giving you a view of a long oblique stretch of water, a mile wide, and three or four or five miles in length, with a shallow depth averaging five feet, because a vast quantity of water has been absorbed by the creek-like channels which separate the islands from each other, and into the innumerable small arteries which lead to the broad lagoon.

The banks of the mainland are fringed with the indispensable and everlasting mangrove, while the low islands appear to be dark, gloomy, forbidding masses of the same bush, fit lairs for the amphibious monsters with which the African rivers of the tropics swarm.

Were you able to so crane your neck as to look over the dark mangrove lines on either side, you would see a vast extent of morass, half shaded by dense overhanging clumps of bush, vivid green thick bulbous-leafed plants, only inhabited by crocodiles, divers, and other aquatic birds. No sign of man or vestige of cultivation would be seen in all the wilderness, because Nature's laws will not permit that man shall breathe the fatal air of those marshes and live. To live there would be like eating of the forbidden fruit. The black vegetable mould of the morass is still in a state of chaos, the air and light which belong to the habitable portion of the globe may not enter until the morass shall have been transformed by the subtle, slow proceeding of nature into healthy productive earth. At present man flies the vicinity, and wisely leaves it until time shall have prepared it for him.

Though we on board the steamer hurry by the fatal district, the wind wafts the deadly exhalation towards us, and in its unrelenting embrace we are involved for a time, where particles are deposited in the lungs and the stomach to incubate and torment the frame with the rackings, the torments, and agonies of perhaps a remittent fever which will last four or five days, and show the verge of the dark pit so many poor bodies have already barely escaped on this Ashantee campaign.

Even while we hurry by we feel the influence of the forbidden region, we feel unusual lassitude of frame, are conscious that a veil of death hangs over us, and sensible that for breathing that noxious air we shall suffer. It is low tide, the river has decreased in height, and the mangrove roots stand revealed in all their hideous deformity. The bushes seem as if they grew on stilts, for so the mud-painted stems appear, entangled amongst one another, and disgustingly naked ; and below is the rich deep alluvial deposit of the river, fermenting in the fervid sun, and adding still more to the disease which has begun to permeate through our frames.

Despite, however, the sad reality of the picture I have attempted to give you, any one fond of ' gunning' and solitude would here find ample opportunity for the exercise of his skill with his weapon, and ample space in the broad stream, in the mazy, sluggish creeks and lagoons, to indulge in the luxury of thorough solitariness. Here are crocodiles, and hippopotami, divers, cranes, pelicans, storks, whydahs, paddy birds, and ibis by thousands. With enough powder and ammunition, a man in time, might slaughter a pile of birds as high as Cheops' Pyramid, and still leave a vast number for his successor. From the upper point of Kennedy's Island, where a good view of the main river is to be seen, I can see that the banks on both sides are as thick with divers and whydahs as the mangrove branches are able to bear. On a single branch I see as many as ten divers perching, and above their heads, resting on higher branches, are so many white paddy birds, that it seems as though they have been reared thereon, and the two lines of mangroves, stretch I should say, three miles up on either side.

About 2 P.M. the steamer arrived opposite Gravie Point, the site of the first camp beyond Addah. The intelligent coloured officer, Mr. R. Bannerman, whose presence with Commissioner Glover is a great satisfaction to the Accras, came aboard and saluted Captain Goldsworthy, and informed him in answer to a question proposed to him by his superior, that there were 2,000 natives, armed, equipped, and organised, at the camp. The camp was in a satisfactory state ; it lacked nothing, but the men were somewhat impatient to be on the march against their enemies.

Gravie Point is to be held until Glover's force of Accras, Houssas, and Yorubas arrive; when, after garrisoning it with a couple of hundred men, the Point will be abandoned by the main body, which intends to cross to the opposite side, and strike north-easterly to hem in the enemy, which the Commissioner thinks will be a very easy matter.

About 5 P.M. we arrive within view of Blappah, the second post on the Volta, chosen by Glover. The aspect of the country changes for the better about Blappah. The mangroves have been gradually thinning and have now finally disappeared. Stately silk cotton trees, doum and Guinea palms lord it over fields of grain, undisturbed by the aggressions of the mangroves and the shrubs of the morass. Little hills begin to show their crest above the river bank, and villages are more frequent. The atmosphere is more salubrious, the river current is swifter and browner, the banks are higher, and the red clayey earth begins to show itself.

The debarkation of our party, and the landing of the pugnacious bullocks, furnished us with humorous scenes.

Captain Reginald Sartorius, of the 6th Bengal Cavalry, is Commandant at Blappah.

Sartorius is hankering for the time when Glover will tell him, ' Cross the river, dear Sartorius, and slay them.' It did not need him to tell us how tired he was of staying at Blappah, how he was suffering from the inactivity which he could not help. We had seen his impatience in his face. Sartorius reminded me by his physique of Guy Livingstone, and that class of large-limbed, resolute, undaunted heroes Laurence delights to describe. When the fight with the Awoonahs, and Aquamoos begins, I am sure they will be as much surprised by the vigour of this English officer as they will be by the spiteful alarming hiss of Hale's rockets.

I have described Captain Sartorius first, because he is the chief of the forces at Blappah; but his brother officers are no meanly formed men either. Especially one, Lieutenant Moore, of the Royal Navy, a tall, long-limbed, energetic fellow, with the lives and vivacity of a hundred common men in him. His long boots flitted with his body from one end of the camp to the other in a surprising way. The man seemed to be everywhere at one time. Surely though Glover has but ten officers with him, he cannot say but what he has selected the very fittest men he could wish to assist him in his enterprise, for ten such men as I have seen with him are worth a hundred ordinary men, and so long as they retain their health, I have no doubt Glover will effect, with their aid, what he has determined upon. Sartorius thought, with Captain Goldsworthy, that the enemy should not be permitted to stand quietly and undisturbed on the other side of the river, to survey the actions of his men, and accordingly sent a nine-pound shell, and a Hale's rocket, one after another, at the curious and over-much unconcerned group, which had the effect of dispersing the insolents in a very precipitate manner, in all directions. Sartorius said after the excitement was over, he wished he could show us better fun, but if we only stayed a week with him, he would enable us to see how his men would punish that party across.

The exact number in his camp at Blappah, as made out by the sergeants, was 2,948, consisting of Aquapims, Crobboes, a few Crepees, Houssas, and Yorubas, all of which Captain Sartorius assured us that night would be drawn out in number for our special information the next morning. Unfortunately, however, the next morning a tornado set in, accompanied by a heavy shower of rain, which entirely precluded the possibility of muster or manœuvre.

Captain Sartorius, under directions from Glover, while his chief was compelled to be inactive at Accra, travelled through the country lying between Accra and Pesse on the Volta above the rapids, and his map of his route furnishes most invaluable information concerning unknown lands. He says the country lying between Accra and Accropong is an open country, interspersed with hills, until, on approaching Accropong, the hills become attached, and form a range which, from salubrity and fertility, has no equal on the West Coast of Africa. Beyond Accropong to Pong, or Amedika, lying near the rapids of the Volta, the land is exceedingly diversified by wooded plain, hill and dale. North of Pong the country becomes open again, entirely devoid of bush. Clumps of forest make travel agreeable and pleasant. The traveller is protected from the heat of the sun, while being free from the annoyances of the jungle. Captain Sartorius thinks that the steam launches may be easily drawn over the rocks of the rapids by which an advance of the expedition will be advantageously made by means of the river as far as Pesse. From Pesse three moderate marches will enable the expedition to reach the Prah, whence no difficulties interpose beyond to Coomassie.

After the most generous hospitality shown to us by the captain and his brother officers, we descended the river next morning, and arrived at Addah Forh, where the Commissioner and chief of the expedition was ready to welcome us with the news that nearly all the Accras had joined them, and to prophesy that in four days Addah Forh would be a deserted camp.

I hope that Governor Glover will be able to make his prophesy good; but after some experience of the native African, I am not prepared to be so sanguine as to say that on the fifth day from this date the Governor will have departed to the war. I incline to the opinion that perhaps within eight days he will be ready, and once having started, the great obstacle to rapid travel and movement of forces has been removed; and he may be able, if he will be stern and uncompromising with the dilatory chiefs and permit of no delays, to be on the Prah even before the end of this month.

The entire strength of Glover's expedition now assembled on the Volta numbers about 23,000 men. 11,000 are at Yankitty, consisting of Akims and Aquapims; 3,000 are at Blappah; 2,000 are at Gravie, or Agravie Point; 550 are Houssas, 600 are Yorubas, encamped at Addah Forh. 5,000 Accras are advancing to assist him, under kings Solomon and Tarkey.

The assembling of this powerful force proves the qualities of Glover and his subordinates. If the commission had about 100 British officers, all capable men, the expedition under Sir Garnet would be unnecessary.

To-morrow I return by the 'Dauntless' to Cape Coast Castle, at which point, perhaps, there may be by this time something interesting to describe. So far this Ashantee campaign has been devoid of interesting incidents, which may be ascribed to the fact that the campaign has not as yet begun.

CHAPTER VII.

THE 'BLACK WATCH' AND ITS OFFICERS—TEETOTALISM—THE RAGE AMONG THEM—STEWARDS DISGUSTED—CHRISTMAS—COMMISSARY O'CONNOR—THE QUALITY OF THE CONTROL OFFICERS ON THE GOLD COAST.

December 16, 1873.

ARRIVING from Addah Forh, I found that the white troops, consisting of the Rifle Brigade, the 23rd Fusiliers, and the 42nd Highlanders—the Black Watch—had at last come, to the immense satisfaction of Sir Garnet Wolseley and his staff, and the press correspondents. The troop-ships 'Tamar,' 'Himalaya,' and the 'Sarmatian' conveyed the soldiers from their natal shores to the African coast.

I had always a strong desire to see the Highlanders in their kilts, and from my knowledge of Africans, I was pretty certain that the sight of such a splendid body of kilted men would have a fine effect on the native mind, but on visiting the 'Sarmatian' I heard to my regret that they had left their kilts behind them, and that they were breeched, like ordinary British soldiers, in the unpicturesque Norfolk grey. The cause was, I am told, the kilts were not fit for the bush-work; that the authorities feared that thorns and cacti would have so wounded the epidermis of the soldiers, that they would in a short time be disabled from active service.

To me, who am an old campaigner on the West Coast of Africa, having been here now seven weeks, it was amusing to watch the officers' faces. They are mostly all young men of very good families, with cream-coloured complexions, light hair and whiskers, as if they had all been turned out of one mould—they appear to be so very much alike.

They all seem weighed down by the one pervading thought, 'How shall I stand the climate?' and their features conse-

quently are cast in an interesting pensive melancholy.   Not
that they are a bit afraid of the campaign or of however adverse a future, but they have heard so much about the disagreeable climate, that they naturally feel anxious about it.
Each has taken great efforts to study the very latest suggested precautionary measures by those who fortunately or
unfortunately have been to Africa and still survive, and whose
effusive correspondence on the African climate has done more
harm than good.

   With a naïveté quite interesting, these young Highland
officers tell you of the plans they have adopted to guard against
the fever ; of the good advice which some friends have been
kind enough to give them, which they think the best thing
in the world.   One beardless boy, whose soft face and pure
complexion neither sun has tanned nor wind embrowned, tells us
confidentially that he has adopted the mosquito bar plan : ' They
tell me it is a sure protection against fever ; what do you think
of it ?'   Another, tells us he means to stick to quinine : ' Two
grains in a glass of sherry taken each morning before coffee or
chocolate, will do the job sure, won't it.   Don't you think
so ?' he asks anxiously.   Another, a fleshy stalwart young man,
whose muscles have found ere this ample exercise over Scotch
moors, is determined to eschew all liquors, ' None of your
brandies, soda, sherry, gin, or anything of that sort for me.   My
opinion is that people wouldn't die if it were not that they kill
themselves drinking.'   This he delivered with an air of one
who knew all about it long before Adam was created, and to
prove how brave his resolutions are, he bawls out to the steward
to bring him an ice lemonade mixture.   The stewards of the
' Sarmatian ' complained to me that during the voyage the
officers drank nothing but lemonade.   ' Hi never seed such a
teetotal set in my life,' uttered a cockney waiter, as he passed
by me, bearing a tray loaded with lemonade mixtures.

   I never saw such a number of youthful looking officers belonging to any regiment.

   How these young fellows will stand the climate, and how
they bear up under privations in the African tropics, must be
left for future chapters to tell.   They will do their duty, no
doubt, and up to the moment of debarkation and the day of
departure for the interior, the British Government will assist

them with prudent wisdom and care to retain what energy they possess for the fulfilment of that duty. The white regiments will remain quartered on board the transports; nay, they have already departed, seaward for a cruise which is to last ten days.

Meantime, during this cruise, Sir Garnet Wolseley will hurry the final preparations for the great march. The road has already been cleared of bush as far as the Prah river. Stores in large quantities are being conveyed to Prahsu, the frontier post of the Protectorate situated on the other side of the Prah. There are probably some 4,000 carriers altogether engaged in the work of transport, and each station is amply furnished with a supply of stores. The arrangements of hospitals for the sick are superb, and a strong force of medical men has been placed at each camp.

The British Government has been lavish in the outfit for this campaign; every imaginable thing requisite for a soldier in health and in sickness is to be found here in abundance. As yet I have not heard of a single deficient article, though I have heard of the arrival of many unnecessary articles, but that is a small fault.

Sir Garnet Wolseley leaves Cape Coast for the Prah on December 27th; the staff leave on the 26th. I also intend to depart for the front on the 26th, and will push on as far as the Prah river, hoping to meet with scenes and incidents worth describing. Excepting a description of the country and the camp, I do not expect that there will be much of interest attached to an eighty-mile march. The excitement of the campaign will begin when the white troops shall have crossed the Prah, and shall begin to struggle through the bushes of Ashantee-land, peppered by iron slugs. That is the lively time to which I am looking forward.

On December 23 Sir Garnet proceeded to Chamah at the mouth of the Prah, in the flagship of the African squadron, the 'Active.' He expected to have to rebombard the village which behaved so treacherously to Commodore Commerell, but he found no inhabitant on the beach, though he landed with a force of marines to search for the natives who have been so vicious.

We hear that the African mail steamer, the 'Soudan,' has been lost between Madeira and Sierra Leone. She had on board 15,000l. in specie for the use of Sir Garnet, who is without any money at present, having sent his cash lately down to Glover.

Glover is to be at the Prah by January 15, and as this river, or a branch of it, runs diagonally across the country from north-east to south-west, he will be able to reach the Prah within three days after leaving the Volta, and both forces will then continue their march to Coomassie along two lines of a triangle, the apex being the capital of Ashantee.

Christmas of 1873 is over. We commemorated the great Christian day with as much goodwill towards one another as good Christian men appear to possess elsewhere. To temper the comfortless sterility of such a day, passed alone in our own houses, on such cheer as tinned Australian meat, which is an ununctuous heap of coarse fibre, we met by invitation in social circles at the houses of our friends.

Sir Garnet Wolseley entertained his friends at Government House, with what means of hospitality his caterer permitted him. There were any number of convivial tables scattered among the dwellings of officers at the Fort and other places. The Press Corps met at the residence of Mr. William Cleaver, the agent of the great house of the Swanzy Brothers, where we dined on turkey and other edible substances, and drank hock and champagne, to the regeneration of the physical man and the improvement of our minds.

We have been strongly tempted to criticise Sir Garnet's apparent inattention to transport matters. Some men will perhaps think that Sir Garnet should escape censure, and that the officer of the 'Control' in charge should be blamed, but though this may be right in a measure, still Sir Garnet is chief of the expedition, and any failure in the undertaking must rest on Sir Garnet's shoulders, as well as the principal merit of success. Sir Garnet determines the number of officers required, appoints the official in charge of the whole, or at least, his name is submitted to him for approval. If he leaves the whole charge to this official, without supervision, and if this official is unqualified by experience, or want of tact, or forethought to provide for his necessities in the commissariat and transport line, he should be justly held responsible for this disqualification.

Please bear in mind that I, for my part, augur nothing

but completest success eventually in the conduct of this expedition, but it is an anxious question just now. ' Shall we be able to march this distance of 140 miles to Coomassie and be back to Cape Coast before the rainy season commences?' when we evidently are much behindhand already. If not the English people will have ample cause for dissatisfaction; if we do, they shall have as ample reason to bless the Government which selected Sir Garnet Wolseley for the command. We all like Sir Garnet so much that we would wish to see him return with honour and glory to himself.

There is no doubt of the capability of the British troops, now at sea off Cape Coast, to force their way to Coomassie against all odds. They consist of picked regiments—the famous ' Black Watch,' the 42nd Highlanders, the Royal Welsh Fusiliers, and the Rifle Brigade, with a naval battalion of picked men. We may confidently back these white men against any number of Ashantees. Even the tribes of the entire West Coast of Africa combined could not oppose them successfully, bush or no bush. Their numbers are comparatively few, but their discipline is perfect, their pluck is traditional, their heroic endurance of evils undoubted, their chief is a soldier of reputation and great promise, their officers are superb men individually and collectively.

Give them food, look after their healths as best you can, do your duty to them, and they will walk to Coomassie, though every tree in the forest, between here and Coomassie, became transformed to armed men.

Well then, whence may we fear serious obstacles? I answer in the dilatoriness and inefficiency of the Control Department, which has the transporting of stores and provisioning of the troops in its charge. Hence the important necessity of unceasing watchfulness and supervision of this department by the General in command. He should seize upon every able-bodied man in the Protectorate for this duty, pay him well in sterling coin if he behaves as he should behave, punish him if he will not by forcing him to the work, even with a bayonet if necessary. For the General has the safety and welfare of the army under him at stake, and this is no time for feeble effort and windy orations about 'humanity.' It would be the humane thing to exercise the severest measures towards the deserters, the unwilling, and obstinate, for the sake of every white man at present in the country, and the future prosperity of themselves and their country.

I do not say that there are any inefficient officers in the Control Department ; far from it ; what I mean to say is that Sir Garnet Wolseley should be certain that he has a sufficient number of officers and carriers for the work on hand, and that he should exercise the greatest caution that he is not persuaded by any feeling of personal friendship or bias, to retain any man in charge of the Control, who may prove himself, by want of energy and general incapacity, to be unfit for his position. It is not a time for personal friendships, it is a time for serious work of body and mind, and if Sir Garnet recognizes the importance of the hour and this duty—the greater praise will be awarded him—that he sacrificed personal feeling for duty.

There is one man on the Control Staff whom I am compelled, as an impartial eye-witness of the man's vigour and energy during most trying weeks, to commend, and he is Commissary O'Connor. At all times and all hours this poor overworked officer has dauntlessly braved the fierce heat of this sea-coast town, has manfully stood at his post when associate after associate was stricken down, and sub after sub invalided. His constitution has enabled him to withstand the fever in a great measure, but even his iron frame has been racked with its agonies several times.

Indeed there are half-a-dozen sterling men of whom I might speak solely as an eye-witness, with unstinted praise : as I might of the zeal which has animated such officers as Messrs. Ravenscroft, Coates, young Hamilton lately from Sandhurst, and the untamable vigour with which they have performed the most arduous duties.

A peep within the Castle yard would give one a better idea of the labour of a Control officer than anything else.

Fancy a castle, a huge structure of stone and mortar, whitewashed a glaring white, with galleries and walls enclosing a spacious courtyard. Lounging in the galleries, out of the torrid heat, are the invalids, officers, who being unable from sickness to work, look down on their fellows out in the painful glare reflected from the snowy whitewashed walls. In the courtyard are immense piles of stores, being checked by a boyish youth who braves tropical heat for the first time. He is fresh from an English military school. His slender figure, dressed neatly in Norfolk grey, his head with short clustering curls of light brown hair escaping from under his helmet, his face almost transparent, and pallid complexion forms an agreeable picture in the midst of the nude athletes of Fantee land, who obey his slightest gesture with an affecting readiness. He is like a young Daniel in the midst of lions.

Yonder before a great heap of bread sacks just landed from the transport ship stands Commissary O'Connor. He is never tired, never complaining, his jacket is off, beads of perspiration stand out on his practical face. His hand is just now upraised directing a burly chief of a tribe how to pile the stores under another shed. His frank bearing and humorous language make him a general favourite with the Africans. His artless commands win for him respectful and ready obedience from all, though he has all sorts of irrepressible characters to deal with.

To my right is another officer, Mr. Ravenscroft, who has also a cheery, hearty way of ordering men about. He is also in his shirt sleeves. The hot sun has burnt his face to redness, and though he perspires, and is, I am sure, uncommonly thirsty, he strides from place to place, with eyes roving about to see that all goes on well.

Endless files of half-naked bearers come and go, go and come, with bread sacks, meat kegs, flour, and rice boxes, chests of tea and sugar, and an indescribable variety of things which troops on a campaign will require.

A large number of men and women are arranged in rows, ready for the porterage of supplies required immediately at the front. They will shortly take their leave in charge of an officer, who must give account to another officer at the front of what he brings.

Some sinewy, athletic fellows are preparing to move off to stations in the interior, with hammocks. Their beautiful bronze figures captivate the eye for an instant, until it is distracted by another fresh gang of men coming in with great water-casks on the hand-carts, for the use of the Fort, from the condenser at Salt Pond.

Over all shines the sun relentlessly with a fierceness which few who have not experienced its uncomfortable heat in the torrid region can understand.

Just outside the castle yard are other scenes, where the Control men are observed distinguishing themselves by their devotion to their hard duties. Perspiring faces, and coatless figures of Europeans are seen here also superintending the landing of stores amid the roar of surf which is almost stunning. Surf-boats go out and come in, but nothing is done unsystematically. Though the picture is one of a large concourse of men, white and black, in a confused mass while engaged in their various duties, and a wild, foaming, surfy sea dashing against rock and whelming boats and men, every article landed must be rigorously accounted for, every group has its overseer and director from the Control Corps.

These are a few of the labours which the men in this department are engaged in; to enumerate them all would be almost impossible; to dwell upon them longer is needless.

But a gentleman has lately landed at Cape Coast Castle, who is about to restore our confidence in the General commanding, and we anticipate a perfect triumph for him over all difficulties. This officer is Colonel Colley, reputed to be a first-class organiser. His duties will be to apportion the transport gangs along the road to Coomassie between the several stations, to procure as large a force of bearers as may be possible, to guard against the wholesale desertion of natives which has been going on, to the delay of the advance.

Strong contingents have been already procured. From 600 to 700 have been received from king Blay at Dix Cove, and from 500 to 600 from Elmina. But there will now be a constant accession of carriers to the already large number in active employment, and we may begin to congratulate Sir Garnet that the great difficulty of the hour has been tided over so rapidly.

The stations at which the European troops will rest on their journey to the Prah are as follows:—

		Miles.
1st day,	Inquibim from Cape Coast	6
2nd ,,	Accroful from Inquibim	6
3rd ,,	Yan Coomassie from Accroful	10
4th ,,	Mansu from Yan Coomassie	10
5th ,,	Sutah from Mansu	9
6th ,,	Yan Coomassie (Assin) from Sutah	12
7th ,,	Barracoe from Yan Coomassie	10
8th ,,	Prahsu, on the Prah from Barracoe	6

Total, 69

CHAPTER VIII.

TO THE FRONT!—THE JUNGLE—THE FOREST—THE CAMPS—ALONG THE ROAD
TO PRAHSU — THE OFFICERS COMMANDING DESCRIBED — MAJOR HOME,
FESTING, WEBBER, EVELYN WOOD—BAKER RUSSELL.

January 6, 1874.

At 3 p.m. we deserted the houses that had sheltered us
during the probationary term of our stay at Cape Coast Castle.
The sun was hot, the incandescence of the stone walls was
disagreeably warm, the streets were almost empty of white men
as we emerged out into such oppressive sunlight; but the
natives were out, that is the female portion, for the men were
' absent in the war.' The little children were grovelling on
the warm streets, and tumbling about in the gutters and on
the door-steps, recklessly unconscious that the ' Press Gang '
were moving for the front, the little savages !

The road to Coomassie led by the main drain of Cape Coast,
and as we looked at the green slime and lazy rillet, we congratulated ourselves that we should have little of its evil odour
any more. Then, skirting the base of Connor's Hill on one
side, and passing by Prospect Hill, we were soon shut off, by
low hills such as are found scattered along the coast line, from
all view of Cape Coast Castle, and finally from the sea, the last
impression of which was that of a blue expanse of water in
perfect tranquillity.

The road was about fifteen feet wide, with a rise along its
centre, and a furrow on each side to drain off the water. The
telegraph line, a very novel institution in Fantee land— a long
Fetich charm, as some of the natives believed it to be—was
laid along on bamboo poles, and as we looked at that the
barbarity of the country was somewhat relieved.

We had heard a great deal about the ' bush ' while at Cape

Coast; here it was, on either hand, covering hollows, hills, slopes, and summits, overwhelming the earth with its luxuriance and density.

Fancy an untrimmed English hedge, if you can, fifteen and twenty feet high, scores of miles in depth, where you see only an innumerable variety of brushwood and plants trying to overgrow one another, and you will form an approximate idea of the thickness and character of the bush. There are but few large trees to be seen the first few miles; bush has usurped every inch of ground: which proves that in former times—how long ago no one can tell—this tract of country between Cape Coast and Inquibim was pretty extensively cultivated. Adding to this impression are the frequent patches of the undying and ever-living plantain plant, whose broad fronds of a pale green colour are in agreeable contrast to the darker green of the ordinary bush.

After six miles of such scenery, the road every now and then dipping down into hollows and rising up short abrupt steeps and brush-covered knolls, rising to sufficient altitude to relieve the monotony, we arrived at Inquibim.

At Inquibim we have, for the first time, an idea how well the British Government can and does take care of its troops. Accommodations for sleeping, in the shape of immense sheds thickly thatched, and walled with plantain leaves arranged symmetrically, are constructed to relieve the soldiers from further labour upon arrival at camp. Under these sheds are long platforms of split bamboo, raised a couple of feet above the earth to prevent the tired men from throwing themselves upon the ground as they arrive hot and perspiring after their march.

Opposite these sheds are the spacious huts constructed for the reception of stores, for the accommodation of Control and regimental officers. In the centre of the camp is an open shelter, covering an enormous iron tank weighing over two tons, which contains a filter invented by Captain Crease of the Royal Marine Artillery. One of its qualities is that it filters beautifully clean water as fast as it may be required. The tank contains many hundred gallons of water. By this provident measure the thirsty troops are saved from the inconvenience and illness consequent upon drinking the unfiltered water of these parts, the properties of which have an exceedingly bad effect on the constitutions of Europeans.

Three or four bullocks imported from San Iago, of the Cape de Verd Isles, in the meanwhile feed on the sweetest plantain fronds around the camp, until they are converted into beef for the troops.

Between Inquibim and Accroful, where we stop on our first night, the scenery is of the same character as between Inquibim and Cape Coast, though it is noticeable that the large trees are more numerous. The bamboo, the most useful of all woods for

ACCROFUL.

an army, grows more frequent. We passed by two or three remnants of Fantee villages which were abandoned by the people when the news of the Ashantee invasion startled them from their homes.

Accroful was a pretty station occupying the crown of a low hill, in charge of a youth just from Sandhurst College. He was a rosy-cheeked, beardless lad when he arrived, but he has already been converted into something resembling a campaigner. The sun has tanned his face, and he has been a victim to two or three attacks of fever. But the

I

insouciant way in which he takes his punishment, shows, that though there is but gristle and aptitude in him as yet, there will be bone and muscle and value in him before many years. These young tough Britons show remarkable plasticity and adaptiveness in their characters. Old General Jackson, our ' Old Hickory,' would have at once embraced the young hero, and pronounced him to be of the ' right sort.'

Filters, iron tanks, sleeping accommodations, stores, &c., await the arrival of the troops at Accroful station, and I am told that these will be found at every station.

On the 28th we left Accroful and resumed our journey. About a couple of hundred yards beyond, we came to a still deep pool and creek, bridged over by Major R. Home and his engineers. In the pool sported several dusky mermaids in a state of nature. When the waters were undisturbed, the pool lay black and still, with here and there a water-lily resting its large leaves on the surface.

A mile or two beyond the creek we came to a perfect arbour, formed by the embracing stalks of two rows of bamboo clumps, where the shade was gratefully cool. At frequent intervals, after this experience, the over-sunny road became sheltered by gigantic trees, cotton-woods, and teak ; palms also cropped out above tall masses of brush, and by their graceful feathery leaves added a charm to an otherwise tame scenery.

The soil about this part is evidently different to that which sends out such luxuriant crops of bush near the coast. There the soil is composed of coarse sand, covered with a slight coat of vegetable mould ; here the vegetable deposit is of a greater depth, and has more consistency, hence the more vigorous character of the shrubs and trees and the greater variety of plants. Flowers were also more numerous ; in some parts, from their rich colouring, they were a perfect feast to the eye, and diffused a subtle, fragrant aroma. And as we passed over more than usually moist places, a singularly sweet fragrance, as of lotus and lilies, was borne on the gentle air, from the bosom of the forest, as if far in the dusky recesses there were deep and still expanses of water nourishing the beautiful plants.

About noon we came to Dunquah, commanded by Lieutenant-Colonel Festing. This post was first occupied by this brave soldier on October 11, 1873. Here were also Lieutenant

Filleter, of the 2nd West, Lieutenant Vandermeulen of the 50th Regiment, Doctor Gore, Mr. Irvine of the Scotch Fusiliers, Surgeon Major Fox, Messrs. Beardmore and Hamilton of the Control Department.

Whatever reputations may be made by officers for gallantry around Coomassie, it will never be forgotten that long before Sir Garnet Wolseley and his Special Service Corps arrived at Cape Coast Castle, Lieutenant-Colonel Festing, Royal Marine Artillery, had several times distinguished himself in actions with the Ashantees.   Until Sir Garnet Wolseley came to take command, it was the prominent figure of Festing which attracted men's attention.   He had met and defeated the enemy near Elmina in August, had driven them back from several points along the Prahsu and Cape Coast road ; had defeated them at Dunquah, and indeed contributed by his restless activity and intrepidity to the demoralisation of Amanquatia's army more than any other officer.   As he is very likely to be kept behind in charge of Prahsu, it will be well to remember these things, that they may not be obliterated by memories of later achievements.

Upon conversing with one of the Control officers, I find the British Government is evidently determined to win praise for the very liberal allowance of food distributed to the soldiers, as well as for many other prudential measures it has adopted for the welfare of the troops.

The following scale of rations for each man, is exceedingly liberal, and any soldier who can retain his appetite must survive all malarial influences on such fare :—

> 1½ pound of meat, salt or fresh.
> 1 pound of pressed meat.
> 1¼ pound of biscuit.
> 4 ounces of pressed vegetables.
> 2 ounces of rice, or 2 ounces of preserved peas.
> 3 ounces of sugar.
> ¾ ounce of tea.
> ½ ounce of salt.
> 1/36 ounce of pepper.

The meat is either to be tinned Australian beef or mutton, salt beef, or salt pork, which is carried in 50-lb kegs.   The fresh meat is procured from the bullocks imported from San Iago.

Up to this time there are 60 officers on service in the Fantee country belonging to the Medical and Control Departments, exclusive of the Line officers.   There are also new arrivals on each steamer.

I sometimes think how well Sir Garnet Wolseley could spare a few of these officers for the service of Captain Glover.   Altogether, Sir Garnet must have over 250 officers in his command, though his force numbers only 3,000 men.   Poor Captain Glover has only 10!   Yet Glover had a force eight times numerically stronger when I visited him.   I have been fearfully chaffed for my championship of him.   Many officers ask me when I mean to start an expedition in search of Glover the second time.   But I still maintain my faith that Coomassie will not fall without the assistance or the influence of his force being recognised, and in that faith I mean to remain until the war is over.

Dunquah as it appears now, under the able and efficient management of Colonel Festing, has a very neat and symmetric appearance.   Rows of well-built huts, sheds, store-rooms, meet the eye everywhere, and on the northern side of the camp is a redoubt fenced with abattis.   It is also considered a tolerably healthy station.   Certainly the large clearing about it tends to create that belief; besides, Dunquah being placed on a low hill, is fortunate in having sufficient steepness of slope on each side.

Out of the sunlight, garish and dazzling, into the gloom again, as we resume our journey to Mansu, the next station. The forest grew denser, loftier; there were cotton-woods of extraordinary height, and girth, and lengthy prolongations of buttresses to support the colossal stems; there were sturdy teaks with lower crowns of foliage, and tamarinds spreading out in broad terraces of lighter green, gum-trees upreared their dark foliage, and slender palm-trees shot their graceful heads to view, all of which become more frequent, assume taller proportions and greater girth, as we move on.

The road is all a glade; an impervious depth of leafage formed by the commingling branches overhead, shelter us from the sun.   The engineers have done their work so well that it becomes a pleasure to walk on the smooth, broad road, though there are parts where the corduroys make it unpleasant for a short time.   Steeps and descents, up and down hills, vary the uniformity of the level forest land and jungle; a few rippling streams which we have to cross during the day are not unpicturesque reliefs to eyes jaded with the flanking walls of vegetation.

Mansu is a camp of which Lieut.-Colonel Webber, of the 2nd West Indians, is Commandant. It is to be the great central depôt between Prahsu and Cape Coast.

Report had conveyed to us an impression that this camp was most salubrious, but at a glance we perceived its incorrectness. As if any one spot in this great tract of jungle-forest could be more salubrious than another! Why the very sight of the dull skies and moisture-dripping woods and the wet tangle, made darker, more melancholy by the murky atmosphere, disabused us of the impression that we had found a sanitorium.

Report had also erred and gone widely astray from the truth when it stated that Mansu was situated on a plateau. I have been compelled to state that certain stations are situated on hills, but unless the statement is qualified by the remark that these hills are mere crests of forested land-waves, people may have an idea that this is a hilly country, and therefore healthy. Neither is the location of Mansu on a plateau.

Perhaps a truer idea of this country may be gained if I state that from Cape Coast to Mansu the land, stripped of its thick garniture of forest and jungle, would appear to be undulating, some land-waves higher than others, some troughs, or hollows, or dells, deeper than others; but there is a general uniformity of undulation, all of which is covered with forest, undergrowth of all tropical trees, and plants. This intense block-like mass of vegetation spreads out on each side of the road, and the road, shaded by the commingling and embracing branches, is more like a tunnel than a highway.

There were about 1,500 natives at work at Mansu, when we passed through, engaged under various officers in the different services. Colonel Webber, of the 2nd West India Regiment, despite the frequent invaliding of his subordinates, has contrived by attention to sanitary rules to make Mansu as healthy as its melancholy surroundings will permit. A pool near by, to which some youths have given the term 'lake,' affords the advantages of bathing, which if visited with moderation conduces to health.

Two of the 2nd West India Regiment were the first to cross the Prah River.   Colonel Webber had invited volunteers to step forward for the very dangerous task of following the retreating Ashantee army, and making their way to the river, to discover whether any of the Ashantees were still on the southern side.   Two brave fellows volunteered, one of whom is Robert Fagan, B, 905, 2nd West India Regiment.   I regret that I do not remember his comrade's name.   These two plunged into the bush after the enemy, traced them to the river, then crossed the river, and writing their names on a piece of paper, nailed it to a tree as a witness that they had performed their mission faithfully. .

These West Indians are undoubtedly fine men.   I do not bestow on them this praise being affected by the incident just related, but from impartial observation of their conduct and discipline.   For size they may well be called Her Britannic Majesty's West Indian Grenadiers, to which title may be attached, for their long and trying service, the yet more honourable title of ' MOST FAITHFUL.'

As an evidence of their fidelity may be cited the readiness of these soldiers to carry loads, when the Fantees, by their desertion, seemed to have demoralised all arrangements of transport.   It must not be said that the West Indians were more fitted for the work than European soldiers ; the statistics of mortality amongst them during their period of service on the Gold Coast and Sierra Leone prove that their constitutions are as frail before malarial diseases as those of Europeans.   Had they been so disposed they need not have volunteered as porters, for all medical men would have refused to countenance arbitrary orders.

A great deal of the merit—though it does not detract from that belonging to Colonel Webber, of having brought this regiment, by kindly treatment and constant personal supervision, to the present high standard of excellence—is due to Major Bravo.   It is well-known that up to a few years ago British officers had the reputation of being very stern, and displaying too much of the zeal of martinets to make themselves beloved by the coloured troops in Her Majesty's regiments, but this belief is utterly at variance with the attachment the 2nd West India Regiment bears towards Major Bravo, who has evinced his interest in his profession and care for the welfare of his command, through long months of most arduous service on the West Coast of Africa.

Captain Grant, also of the 2nd West India Regiment, is another officer who has distinguished himself on this campaign. The memorable exploit of his, in precipitating his company against the besiegers of Abracrampa, will not be readily forgotten. Acts of great gallantry and intrepidity serve to give

SUTAH.

an historical interest to a regiment, and when the history of this campaign is written, it will be found that the 2nd West India Regiment has a conspicuous name in its pages for gallantry and devotion to the cause of their Sovereign Lady the Queen.

The march to Sutah, and thence to Yan Coomassie (Assin), was through a forest, where powers of description fail. In density and wild luxuriance it eclipsed everything I have ever seen in Africa. I do not suppose that anything in Brazil even can give one a more thorough realisation of tropical luxuriance than a march through the district that intervenes between Yan Coomassie (Assin) and Mansu.

Yet even through this forbidding region of swamps and streams, shadowy forest and tangled jungle, Home's Engineers have laboured triumphantly to our comfort, convenience and health.

Aye, what languishing heaviness of soul fills a man, as he, a mere mite in comparison, travels through the lofty and fearful forest aisle? If alone there is an almost palpable silence, and his own heart-pulsations seem noisy. A night darkness en-

YAN COOMASSIE (ASSIN).

velopes him, and from above but the faintest gleams of daylight can be seen. A brooding melancholy seems to rest on the face of nature, and the traveller, be he ever so prosaic, is filled with a vague indefinable sense of foreboding, and becomes affected and morbid in like manner as the forest. When he emerges out of the depths of gloom and awesome shades, how like a burst of dazzling glory appears the glaring daylight !

At Sutah there were vigorous efforts being made to prepare the camp for the reception of the troops, and Dr. Thornton was

in hopes of being able to ameliorate the sanitary conditions of
the place within a very brief period.   At Yan Coomassie, Com-
missaries Marsh and Rainsford were bestirring themselves for
the same purpose, and sheds fifty yards in length were in process of construction.

Between Sutah and Barracoe, the forest retained the same
density and luxuriance.   At Barricoe, as I rode up on my mule,
the first of the species seen in the front, my appearance was
greeted by a broad smile from each of the blue-jackets and
marines encamped here.

BARRACOE.

Query. Were the sailors amused at something odd in my
appearance, or at the solemn, serious face of the long-eared
animal I rode ?   I incline to the belief, though I should feel
gratified at the thought that it was I who afforded the amusement, that the sailors were amused by the sight of the mule.
But what is it, can anybody tell me, that lies in a mule's
appearance that a sailor must always laugh and joke when he
sees one ?   I have noticed in many other lands that the sight
of a poor donkey, or of his cousin, the mule, provokes a smile

in the face of Jack. It is an exceedingly interesting psychological fact which deserves attention. I should be sorry indeed if there was any relation, even in the remotest degree, between a sailor and a mule. I have always been of the opinion that there was not the slightest kinship, because a sailor's sense of the ludicrous is very keen, while a mule is an uncommonly grave animal, seldom, if ever, disposed to wear a smile. There is a stately gravity in him, even under the most ludicrous circumstances. A mule might be carrying a monkey on his back, but he would be just as serious and solemn as though he carried the great Lord Mayor of London, or the President of the United States.

The sailors are certainly lively fellows, and do their duties manfully whether on ship-board or ashore. Though there is as much discipline in their camp as in any military force, yet their awe of their superior is not so great as to repress the high animal spirits which they bring with them on the campaign. They are like great school-boys, fond of sport and levity, and as they swing their lithe bodies round they seem to enjoy the roving life ashore. As their hearty laughter is heard echoing through the woods, I can fancy the wan face of some invalid in the hospital hard by lighting up pleasurably at the sound, as though it were delicious music. Their songs at night tend to enliven both officers and men; and though the fare is hard and the work is plenty, and the campaign is not yet very interesting, their hilarious liveliness may be taken as evidence of the spirit with which the sailors of the Naval Brigade will perform their allotted duties when called upon.

The march from Barracoe to Prahsu was a short one, along an excellent road. The labour performed by the Engineers was illustrated by the sight of the immense logs of mahogany and hard African teak, which lay on each side of the road. Some of these logs were four feet in diameter, and the way in which they were chopped through, was a witness with what raw, unskilful material Major Home and his subordinates Buckle and Bell effected a road through the mammoth timber of this region.

As we arrived at the outskirts of the camp, we realised what mighty changes had been wrought since December 14. A space of about thirty acres was here cleared of the great lords of the

THE CAMP AT PRAH-SU.—NATIVE ARTILLERY CROSSING THE RIVER.

forest, and the tangle of bush that formed the undergrowth: lengthy huts arranged on each side for the different regimental quarters, for the Control Department, the hospitals, and Staff met the eye in orderly array.

Travelling through the vast camp, which echoed to the sound of axes and forges, and vocal commands, we came to the river bank, where Rait's Houssa Artillery were encamped with their tiny cannon and Gatling menacingly pointing across the river towards the enemy's land.

Captain A. Rait, Lieutenants Knox and Saunders of the Artillery, had constructed a redoubt at the landing, close to the old redoubt of 1863, with the aid of the fine force of Houssas, which, by constant drill and persevering and patient attention to discipline, has been brought to a high state of efficiency. The Houssas, as I have said before, are Mohammedans from the interior of West Africa, beyond the Benin Bight.

I witnessed the evolutions of the Houssa force yesterday, as conducted by Captain Rait, and I must candidly confess that few European artillerymen could have gone through them in better style; they certainly could not have done so well under the hot sun in which the Houssas worked.

And just here I must digress to give a warning note to Great Britain. Though I do not really think she needs it, those of England who hear it may rest assured that it is sound advice. It is said, though I do not know how true it is, that the British Government intend after the campaign is over to surrender the Protectorate to natives and the Fantees, to which I say to English people—Do no such thing!

In this people, specimens of which I have just seen in the Houssa Artillery, in Wood's and Russell's native regiments, you have the elements of unlimited power and usefulness. You may do with them what you desire; subject them to discipline, and you have allies of no contemptible order to labour for your interests in this land. You see in them, poor and ignorant though they be, the nucleus of a great empire; if your ambition be to establish one, an empire that in process of time will rival that of India. The soil of the land is as rich, the forests are as useful, as those of India. The natives of the soil you may convert into allies as faithful as your most loyal Indian sepoys. Raise a fort on the banks of this river, leave a garrison of

Houssas here, let the telegraph stand between Prahsu and Cape Coast, extend the clearing all about the fort, invite natives to settle in the settlement, keep your road to Cape Coast open, and the peace of the Protectorate is assured for ever. This present expedition has begun the work of utility and civilisation, has opened the path for progress and beneficent influence. Why should you not reap what you have sown? Continue the good work, and set at nought all doubts about the future of the land and the capability of the people. You have sufficiently demonstrated the capabilities of the people as to the land, no one can deny its superabundant richness.

By keeping a garrison at Prahsu, your timber merchants, who buy the valuable teak of India at 3s. 6d. the cubic foot, may obtain the material they need at one half the cost. The Prah river will float anything needed on barges to the coast. When this clearing is extended around the fort, cotton planters may be able to export a large cargo of cotton annually to England, for if the seed is properly planted, I am certain that the capacity of the soil is so great as to yield a bale of six hundred pounds to the acre. If you prefer to cultivate sugar cane, what part of the world is better for the purpose than this extent of alluvial deposit, which lies for miles between here and the sea? Many other staple products, such as rice, and other things might be grown to the enrichment of men disposed to work under the auspices of the protection which the little fort of Prahsu would guarantee to them.

The Commanding officer of Prahsu camp is Colonel Evelyn Wood of the 90th Regiment.

Here are also Major Baker Russell's Regiment, Major R. Home's Engineers and Rait's Artillery. Sir Garnet will arrive to-morrow or the next day, and the European battalions will be rolling in day after day.

Major Home, upon arrival of Sir Garnet, will begin to bridge the Prah river. The pontoons and materials for the bridge have all been brought up. There is a faint rumour flying about, that the general advance across the Prah will be made on January 15. Captain Glover must be ready by that date, Butler with the Akims, and Messrs. Moore and Dalrymple, if successful in their mission, will have the honour of leading the Wassaws to the dread Ashantee land.

CHAPTER IX.

AN ASHANTEE EMBASSY—CAPTAIN ARTHUR RAIT EXHIBITS THE GATLING GUN
—ARRIVAL OF THE NAVAL BRIGADE—THE 2ND WEST INDIANS—BOW-
DITCH'S BOOK—COOMASSIE AS IT APPEARED IN 1817—QUESTIONS.

Prahsu: January 8, 1874.

SIR GARNET WOLSELEY, the young chief of the Anglo-Ashantee
expedition, arriving here on the morning of the 2nd inst., was
well received.   He was mounted in solitary state on the top of
a light buggy, which had been drawn all the way from Cape
Coast Castle by six strong Fantees, assisted now and then over
the deeper sloughs by the police bodyguard.   Colonel Wood,
Majors Russell and Home were in the great square of the head-
quarters ready to receive their chief, and when the news passed
through camp that the general had arrived, the enthusiasm of
the native troops manifested itself in a hoarse murmur and
deep hum of joy.

Sir Garnet, having reported himself, in excellent health
and spirits, to the numerous queries propounded to him,
expressed himself very well satisfied with the vast clearing, the
construction of the spacious huts, the ample accommodation
provided for the white troops, and with the prospective sanitary
condition of the camp on the Prah.   The officers of his staff
corroborated to a man the descriptions I have already given
of the country between the sea and the Prah, and gave the
palm for general discomfort, muddy cataclysms, desponding
sloughs, dreary, depressive, melancholic appearance to that
part which lies between Sutah and Yan Coomassie (Assin.)

Early the same morning Lieutenant Grant, in charge of
the pickets on the other side of the Prah, was informed by his
people that there were some strange people, probably Ashantees, skulking behind trees at some distance beyond.   When
Lieutenant Grant arrived upon the scene, he found both pickets

and strangers gesticulating violently towards each other, ducking their heads and dodging behind any defensive tree that offered itself, each evidently desirous to impress the other party with the feelings of amity which possessed them ; but, in the words of a British officer who criticised the performance, 'both parties were in a beastly funk.' The Lieutenant, however, contrived to calm the gesticulative pantomine, and to induce the small party of Ashantees to come forward, when he was told by a small, stoutly formed Ashantee, wearing a large square gold-plated badge on his breast, that he was the town crier of the Ashantee capital, come upon an errand from the king to the English chief; that his companions, at least six of them, were sent with him to see that he did his duty.

'But who are the others? There are eleven of you altogether,' demanded the Lieutenant.

'Four of them,' replied the messenger, 'are Ashantees, who overtook us on the road from Amanquatiah's camp.'

'Well, what do they want here if they are not of your party?'

'They came to find out where the white men are.'

'Oh, indeed! they are scouts, then,' said the Lieutenant. 'If they will accompany me they shall see where the white men are.'

'Could you not let them return to their chief, who is waiting to hear from them?' demanded the messenger.

'No, I could not do such a thing. They came to look for the white men. The white men are on the other side of the Prah ; but they must come and see them for themselves, that they may be better able to report about the white men.'

The eleven were accordingly brought across the Prah, and, as the General had arrived, were conducted to him ; but as he was too fatigued to see them just then they were taken to a hut guarded by a detachment of the Second West India Regiment.

The gold-badged ambassador had brought a letter to Sir Garnet Wolseley from the Ashantee king, the contents of which, when read, created no small astonishment among the staff.

The King of Ashantee, in his customary, ambiguous way, which may be seen by a glance at his former letters published

ASHANTEE AMBASSADORS CROSSING THE PRAH.

in the Blue Book, had written to say that he was grieved to hear from his messengers the losses his troops had incurred from an attack made on them at Fasua by the white people. The attack, he said, was perfectly unjustifiable. His young men were returning to their own country from a visit which one of his chiefs had made to the king of Denkerah, when, after the Denkerah king's death—which had occurred during the visit — his men were set upon by the white people, who slew many hundreds. He had always loved the white people, had ever been friendlily disposed toward them; had always wished to cultivate nearer and more amicable relations with them; but this wanton attack on his people in the Forest of Fasua he could not understand. Would the English chief condescend to let him know what this hostility meant, and send his messenger, who was in his confidence, back to him in safety?

The feelings of all who heard the contents of this strange letter may be better imagined than described. Either the Ashantee king is a profound diplomatist or he is a poor tool of the Ashantee chiefs, who have made war upon the Fantee protectorate without informing him of the enterprise in which they have been engaged since February, last year.

Either he is an unmitigated rogue or a fool, who is not aware of the valuable time he is losing by writing such puerile letters, when he still might save his kingdom, and perhaps his crown and life, by asking on what terms can peace be secured at this late hour.

This morning the Ashantee ambassadors had the gratification of witnessing the mechanical powers of the Gatling gun. The gun was constructed by Sir William Armstrong, and cost the sum of 245*l.* It weighs, without the caisson, nearly 400 pounds. The caisson contains four drums, each drum containing 240 cartridges. By the time the gun was wheeled behind Sir Garnet's house, with its ominous muzzle pointing up a snaggy-bosomed reach of the Prah, it was apparent that the Gatling's harsh thunder was to have a larger audience than had been anticipated. Every officer to whom its wild cracking, ripping notes were unknown had gathered about the Ashantee ambassadors, in the rear of the monster, and away rearward was a vast concourse of Fantee labourers, who had crowded every available spot to witness the effect of the Gatling upon the unconscious Prah.

Captain Rait and Lieutenant Knox, of the Artillery, who superintended the arrangement, found some slight hitch in the mechanism of the Gatling, for, though the preliminaries were begun in a calm way, the Gatling came near being pronounced a failure by some of the strong conservative gentlemen. One gentleman had already given his verdict and said that he never saw such ' a rotten arrangement; that wherever he saw it tried, no matter where, there was always sure to be some stoppage; that so many men as were engaged on the Gatling, armed with Sniders, might have made a far more effective impression on the minds of the ambassadors from Ashantee.' The gentleman had hardly done speaking when a new drum, loaded with shot, was placed on the top of the gun, and, the handle being turned, the Gatling began to speak with startling emphasis. That part of the river at which it was directed began to shoot up tall columns of water and spray, until it appeared as if the Prah was about to form itself into so many grey columns of liquid and to join in a dance. The contents of the drum were expended without a halt, and the effect of the exhibition was hailed with boisterous applause by the Fantee spectators and by the Ashantees with low remarks and expressive looks towards one another. The officers were also well pleased with the effect, and the Gatling,[1] which had its reputation previously endangered, had redeemed its fame and become more highly appreciated than ever.

Yesterday morning the camp of Prahsu, which is rapidly assuming vast proportions, was stirred to its centre by the arrival of the Naval Brigade, consisting of 250 picked blue-jackets from the Cape Coast squadron. These brave fellows had marched from Barracoe, seven miles off. They advanced in perfect order along the road, one half singing the well known song, ' When Johnny comes marching home,' the other half keeping step and chorus to ' John Brown's knapsack is number ninety-two.' If any set of men ever looked adapted for hard work the lithe-bodied, soft-paced men of the Naval Brigade did. Their frames looked green, sapful, and their faces so cheery and healthy that one could scarcely believe they had marched through the eighty miles of irreclaimable forest and

---

[1] The Gatling was not taken to Coomassie on account of its weight. Something lighter should be invented for such campaigns as that in Ashantee.

swamp between Cape Coast Castle and the Prah river. Sailors always march as if marching were natural to them, as if they were animated men, of joints and muscles; while English soldiers appear stiff and rigid, more like walking machines in comparison.

The uniform of the sailors is the naval blue shirt and wide trousers, which they use on shipboard, while they appeared somewhat jauntier in their broad-brimmed straw hats, covered with a canvas cape fastened around the hat by a brown muslin veil. The Naval Brigade is armed with Sniders.

To-day 300 of the 2nd West India Regiment arrived, under the command of Lieut.-Col. Webber, who has been rusticating so long with his people at Mansu. The 2nd West India Regiment consists, as most persons are aware, of coloured soldiers enlisted from the negro populations of Jamaica, Nassau, and other British settlements. Physically they are fine, huge fellows, some of them giants in frame, but as soldiers they don't strike me favourably. They are faithful, loyal men though, and I suppose as ardent in Queen Victoria's cause as the blue-jackets are, and no one can rob them of the laurels they have already acquired on this campaign in the bush skirmishes of Abracrampa, Dunquah and Fasua. But their appearance contrasted unfavourably with that of the Naval Brigade. Their marching was a mere slouch, a heavy, careless, leaden walk. There was no dignified military pace among them; they jerked their heavy feet forward, after which they permitted them to fall, without order, unison or emphasis, on the ground; and, of course, their bodies were as ill-regulated as their pace. Their arms were swung about as if they were using sledge hammers; their heads bobbed up and down irregularly, and their bodies swayed indolently from side to side, with a downward droop distressing to witness.

One need only have glanced at either of the battalions whose *entrée* to Prahsu I have described to know at once where must lie Sir Garnet's chief anxiety. The Naval Brigade was accompanied by a force of carriers laden with baggage and hammocks, the sight of which was almost disheartening, and the battalion of the 2nd West India was almost as bad. There were probably 800 carriers for 550 soldiers. How many carriers will at this rate be required for the 42nd Highlanders, the 23rd Fusiliers, and the Rifle Brigade? At the least, I should say, 2,500. Where are the control carriers to come from? For a body of 2,500 white men, 300 West Indians, 1,000 native allies, 500 Haussas and 500 irregulars, how many carriers will be required to provision them on the march to Coomassie? A movable force of nearly 5,000 soldiers, exclusive of their auxiliaries, in the shape of burden-bearing men, requires no small amount of provisions for thirty days. The intention is to march by battalions on the 15th inst. across the Prah to the next camp, ten miles distant. The bridge has not been completed as yet, and the advance force, which was expected to leave on the 3rd, has not begun its forward movement. I have been at this camp now four days, and not an ounce of provisions have I seen arrive as yet, although every new arrival must pass before my tent door, as the tent is planted within a few feet of the road. If we are to leave this on the 15th inst. and the expedition is destined to have an early success, this transport and control matter is really all-important. The road must, in some measure, be widened to admit of the artillery and rapid progress of the troops, a strong detachment must always be engaged some distance ahead, and supplies for such a concourse of men as will follow ought to be constantly pushed forward. Unremitting attention and energy paid to these matters will far more insure the success of the expedition than the most consummate skill in war tactics, because Sir Garnet is not about to measure his military knowledge with that of the Ashantee generals; he has only the task of showing the Ashantees to his gallant Scotch Highlanders, Fusiliers and Riflemen, who will know how to deal with them, without any generalship. This is as certain as though it had been an already accomplished fact.

We have come so far fairly enough; it would be a thousand pities if any laggardness were shown now, and a fine body of white men compelled to wait at Prahsu until they perished by hundreds from the fatal unhealthiness of the climate. I do not suppose that they will be permitted to remain thus exposed; still, it is a matter of anxiety that four days have elapsed and we are not a foot advanced forward. I have confidence in Sir Garnet that he has in him all the elements of a good leader— that he is a dashing, intrepid chief—and if I hesitate to give him the full measure of praise that more enthusiastic and less prudent admirers have already accorded him, it is that I but wait

SHARPENING CUTLASSES IN THE CAMP AT PRAH-SU.

to see those elements of fitness for his work developed. In what has been done heretofore Sir Garnet has evinced considerable ability, but it is in that which lies immediately before him that he must show the full extent of his genius.

I have just finished reading Bowditch's 'Mission to Ashantee,' and a more interesting book of travels I have seldom read. The only fault to be found in it is his exaggerated notion of such scenery as we have passed through on our journey from Cape Coast. I would hardly take it upon me to criticise at this late hour a book so long ago obscured by later volumes on the same subject, except that it has just emerged from obscurity, and is so valuable a guide to the country which we are now bound for. As it is so generally commended, and as I can heartily join in the commendation, it only remains to be said, as a guide to those who may entertain wrong ideas of the country from its reading, that its scenic descriptions must have been written by a very young man in the flush of ardent youth, gazing for the first time upon a wilderness of tropical vegetation. Such expressions as 'enchanting,' 'grand,' 'magnificent,' 'beautiful,' 'unrivalled,' 'extensive area of summits,' are entirely out of place in connection with what we have seen. To be literally true, he should have used such strongly contrasting words as 'disagreeable,' 'monotonous,' 'melancholy,' 'depressing,' 'gloomy,' 'fetid,' &c. We have seen nothing, as yet, that could be called grand unless you were to call the lofty cotton-woods by that title ; and the most extensive areas have been ancient clearings not more than twenty acres in extent. As for anything being magnificent in this densely-forested land, where your vision is limited by tall, thick brush and colossal cotton-woods, I can hardly conceive the meaning of such a word in this country. 'Beautiful' is evidently a misnomer, for nothing can be regarded as beautiful that inspires disgust and nausea; and the expression 'unrivalled' could only be used by one who had never seen much of tropical vegetation, though the density of forest and brush is really unrivalled. Apart from the enthusiastic freshness of language kindled by tropical novelties, there is a wonderful amount of industry manifested in Mr. Bowditch's book, and it has been with the utmost pleasure I have taken his narrative to glean some idea of the country still before us, for our march

Crossing the Boosumprah, according to Bowditch, we come to Prahsu, three quarters of a mile from the river. (This has no existence in 1874, and its site can barely be recognised.)

Kickiwhirri, beyond Prahsu one mile and a half—' a larger town, not so regular '—is a collection of old huts, untenanted in 1874.

Atobiassee, four miles from Kickiwhirri, a ruined village, insignificant and uninhabited, but now occupied by us as a post.

Asharaman, or, as we call it, Essiaman, nine miles from Atobiassee, a small village on an eminence, was occupied by a few Ashantee families January 6, 1874, who, excepting two women, deserted the place on the appearance of Gifford's scouts. The two women, who expected death or slavery, were very much astonished when Lord Gifford informed them that if they did not leave the camp he would compel them to do so.

Akrofroom, twelve miles beyond Essiaman. Path thither, swampy ; forest thick.

Moissee, the furthest town in Assin, eleven miles from Akrofroom, situate at base of three high hills (Adansi Hills), which are the southern barriers of the Ashantee kingdom.

Ascending a steep a mile and a half in length, which is very rocky, thence descending and crossing a small river called the Bokmen, noted for the virtue of eloquence which its water is supposed to give, and passing through the first Ashantee village, called Quissa-Quisha or Kevisa, we come to an important town called Doompassy, after a march of six miles.

Two miles beyond Doompassy is Tioboso, situate on a high hill ; thence to Datiassoo is seven miles, and one mile further is Dadawassi.

Descending a hill and a march of eight miles brings us to Assiminia.

Proceeding through Boposo, situate on another high hill, and passing through several unimportant villages, we come to Sarrasoo, at a distance of eleven miles from Assiminia.

Crossing the river Dah, a stream sixteen yards wide and four feet deep, a march of six and a half miles brings us to Coomassie, or Kumassi, the capital of Ashantee.

The distance, then, we have to march before reaching Coomassie is :—

	Miles.
Prah River to Atobiassee	6
Atobiassee to Essiaman	9
Essiaman to Akrofroom	12
Akrofroom to Moissee (Adansi Hills)	11
Moissee to Doompassy	6
Doompassy to Dadawassi	9
Dadawassi to Assiminia	8
Assiminia to Sarrasoo	11
Sarrasoo to Coomassie	6½

Total, 78½

According to Bowditch, who was there on a political mission in 1817, Coomassie is built upon the side of a large rocky hill of iron-stone. It is insulated by a marsh close to the town northwards, and by a narrow stream, half a mile distant from it northwest, and sixty yards broad; close to it northeast, east, southeast, and south, and about a hundred, twenty, seventy, and fifty yards broad at these points. In many parts the depth, after heavy rains, was five feet, and commonly two. The marsh contains many springs and supplies the town with water, but the exhalation covers the city with a thick fog morning and evening, and engenders dysentery. It is a little extraordinary that we never saw a mosquito in Ashantee. I could find none but bird's-eye views of the city, which were uninteresting, presenting nothing but the thatch of the house. It was encircled by a beautiful forest, which required more time than I could spare, and a more expressive pencil to portray. Coomassie is an oblong, four miles in circumference. Four of the principal streets are half a mile long and from fifty to a hundred yards wide. The streets were all named, and a superior captain in charge of each. The palace was situated in a long and wide street running through the middle of the town, from which it was shut out by a high wall, terminating at each end at the marsh, when it was discontinued, that being a sufficient boundary. I reckoned twenty-seven streets in all, which I have laid in a ground plan of the town. The small grove at the back of the large market-place was called Sammonpone, or the Spirit House, because the trunks of all the human victims were thrown into it. The bloody tracks, daily renewed, showed the various directions they had been dragged from, and the number of vultures on the trees indicated the extent of the recent sacrifice. The stench was insupportable, and the visits of panthers nightly.

The Ashantees persisted that the population of Coomassie, when collected, was upwards of 100,000. I say when collected, because the higher class could not support their numerous followers, or the lowest their large families in the city, and therefore employed them in plantations, generally within two or three miles of the capital, where their labours not only feed themselves, but supply the wants of their chief, his family, and more immediate suite. Perhaps the average resident population of Coomassie is not more than from 12,000 to 15,000.

'The markets were held daily from about 8 A.M. to sunset. The larger contains about sixty stalls or sheds, a small square frame covered with cotton cloth. Among the articles for sale, were beef, mutton, wild hog, deer, monkey's flesh, fowls, pelts of skin, yams, plantains, corn, sugar-cane, rice, peppers, vegetable butter, oranges, papaws, pineapples, bananas, salt and dried fish, large snails smoke-dried and stuck in rows of sticks in the form of a herring-bone, eggs for fetich, palm-wine, rum, pipes, beads, looking-glasses, sandals, silk, cotton cloth, powder, small pillows, white and blue cotton thread, calabashes, &c.'

(Bowditch's account of Coomassie, as it appeared to him in 1817, will stand very well for the Coomassie of 1874.)

The Rifle Brigade has arrived at Barracoe, the first station between here and Cape Coast, and, to quote an officer's words, 'is in a great rage because of detention.' The battalion will be kept there for a week longer, owing to failure of transport.

The 1st West Indians are said to have been also added to the list of carriers between Cape Coast and Dunquah. You must remember, as you pass your judgment upon this glaring mistake of the authorities here, that the detention of the white regiments at the unhealthy stations on the road means much more than delay—it means sickness and mortality—and also remember that to make carriers of West Indian negroes, who are as little capable of standing this climate as white troops are, is to make them nearly all unfit for the active prosecution of the war.

There is another mistake which Sir Garnet is committing, and that is to pay such attention to the road in front after the white regiments have been landed and are beginning to be exposed to the malaria of the climate. He has actually despatched black native regiments in advance, beyond the Prah, while the work which an English regiment is naturally capable of performing they have been withdrawn from. In such a climate as this, the least the ardent Englishmen are fit for is to be detained inactively at an unhealthy post.

Why were not the Rifle Brigade and the 42nd Highlanders sent across the Prah at once to occupy the Adansi Hills, accompanied by Colonel Wood's and Major Russell's regiments of Fantees, Bonnymen, Kossoes and Sierra Leone natives, as carriers of supplies for them? The natives might return and employ their whole strength to moving as much eatables to the Adansi Hills as they could carry, while the native carriers, reorganised under efficient non-commissioned officers, might continue the transport work between Prahsu and Cape Coast Castle. In the meanwhile a small force only under the Royal Engineers might improve the path into a proper road between the Adansi Hills and the Prah river.

There are enough black men at Prahsu Camp, engaged in constructing huts and other improvements, to form a very respectable force of carriers. Major Home informed me he had about 1,500 men under him for engineering work, which, in such a crisis as the present, is a number exceeding the necessity of such work by exactly 1,000.

Why Major Russell's black warriors should have the honour of occupying advance posts, while the white regiments are detained at the lower stations to sicken and die of disgust and inactivity is incomprehensible?

Major Russell's regiment is expected to arrive at Accrofoomu to-night.

CHAPTER X.

SCENES IN CAMP AT PRAHSU—THE PRESS CORPS—THE SAILORS AND
MARINES IN CAMP—A SULLEN YOUNGSTER—CATCHING DESERTERS—HOW
A SERGEANT OF THE RIFLES WAS FRIGHTENED—RUNNING AWAY FROM
A MULE.

January 14, 1874.

DESPITE the frigid reticence and chilling reserve maintained
by Sir Garnet Wolseley and his staff toward newspaper correspondents respecting future movements, the past week has furnished us with sufficient food for comment. We have seen a
young, ambitious General, who at the War Office drew up an
elaborate plan of the Ashantee campaign, and subsequently, on
his arrival at Cape Coast, proclaimed the date on which his
little army should cross the Prah, as confidently as if his army
had already crossed that river, struggling to keep his word intact; though most men, seeing the lack of material at Prahsu,
believed the thing to be impossible. Sir Garnet has gone to
great lengths to secure the fulfilment of his promise; he has
employed Colonel Wood's men, the 1st and 2nd West India
Regiments, as carriers to bring up supplies from the down
stations, and finally the famous Black Watch, the 42nd
Highlanders, have volunteered to carry provisions, which in
such an enervating climate as this is an act deserving of as
much praise as the most conspicuous bravery on the field of
battle.

While waiting for the advance, let me tell you something of
ourselves. We are now told that certain regiments shall move
on to-morrow. These will comprise Colonel Wood's regiment
of mixed natives, and the 2nd West India, and perhaps the
Naval Brigade, though the latter name has not transpired as
yet. They will proceed to Essiaman, and the next day to
Accrofoomu. We are not told, however, how these troops are

to be fed, what supplies there are in front, what arrangements
have been made for the regular transport of provisions from
Prahsu to the advanced posts. There cannot be a large supply
at the front, as Major Russell's force has been engaged in improving the road beyond the Prah. It is very probable, however, that each regiment will convey a few days' provisions
with it, which will allow for certain arrangements to be made
after their departure. When a staff officer was asked the other
day, whether the white troops would depart on the 15th, he
replied that when the white troops would march they would
proceed from the down stations directly for the front. This
reply is in the usual vague way the staff employ, according to
the instructions of Sir Garnet.

The 23rd Fusiliers have been wondering all this time what
is to become of them. The day has arrived for their debarkation, after a weary cruise some distance out to sea; but after
one wing had debarked, an order arrived at Cape Coast that
they should stay on board. Indignant at this treatment, the
first wing continued its march through Cape Coast, driving its
ancient goat ahead, which presently died from sheer old age
and change of climate. On arriving at Accroful this wing was
halted until further orders. We are informed that only 100
men of this regiment will be permitted to enter Coomassie
with the European brigade under Sir Archibald Alison.

The Royal Engineers under Captain Buckle, have left
Prahsu for road-making and otherwise improving transit.
Major Home, the indefatigable chief of the Engineers, some
time ago proceeded to Essiaman, where he has prepared abundant work for his people. Major Russell has advanced eight
miles beyond Essiaman, and will immediately prepare a camp
for the oncoming troops. The Naval Brigade has been ordered
to provide its own transport, which it is very well qualified to
do, having a whole fleet whence to draw carriers from the Kru
boys on board. Four hundred and twenty of these faithful,
hard-working Kru boys will accompany the Naval Brigade in
that capacity.

For the last few days there have not been heard the joyous
chorus and the roaring songs which made the evenings lively
and pleasant at the camp of the Naval Brigade. A profound
silence has brooded over that locality lately. This was so

unusual that inquiries were made as to the reason, and we were told by a weather-beaten blue-jacket, in most respectful but melancholy language, that 'Sailors can't sing, sir, leastways it oughtn't to be expected, when they have no rum. It makes us boys think better of this beastly country while we have rum, or we forget it when we sit and sing round a blazing fire; but when there is no rum, sir, ah; we begin to tell one another what people say of the fever of this country; how such and such a mate died or how some shipmate was sent to the hospital very much gone, poor fellow, and such stories, sir, carry one far into the night, and finally to bed with the feeling that something isn't as it ought to be.'

Last night, however, the songs and chorus startled every one into smiles and social chattiness, and we all knew that a fresh supply of rum had come up, and the sturdy blue-jackets had been fired into blissful unconsciousness of the mortalities and unhappy memories of an African climate.

Very few amusements suffice to make life enjoyable in a camp. The Press Corps have been very fortunate in their mess and surroundings, and they eke out what is pleasurable in such a country as this, and depend upon themselves for amusement and social joys. Our first week on the Prah was passed in constructing a comfortable hut, out of which we contrived to make two apartments, a dining room, a bath room and store room. Our tents, of course, sheltered us at night. Having possessed ourselves of a hut, we sought to increase our comfort with the adventitious aid of architectural improvements; we railed off our quarters by a lattice-like rail, and within the enclosure still more embellished our quarters by making a flower-garden, which elicited the admiration of the entire camp. An abundant supply of creature comforts enables us of an evening to contemplate a longer stay in this camp with philosophic resignation.

From our hut door we can see across the square to headquarters, which have a very deserted aspect during the day; but towards evening things begin to look livelier. An officer lounges out, gives a yawn or two and casts an eye about for a human face; presently another emerges out of his wattled den, and the two engage in talk and form a nucleus, which attracts a group. By-and-by Sir Garnet appears, bright and genial looking, and a promenade is begun which probably ends across the river. At the late hour of 7 P.M. the headquarters mess dines; at 9 P.M. the symposium breaks up, and each member betakes himself to rest. At decline of day others of the camp betake themselves to the river for a bath.

It is singular, considering that all African rivers are peopled with amphibious monsters, what a complete immunity the bathers of the Prah have enjoyed from the attacks of the crocodiles. That these animals have their haunts in the river there can be no doubt, for a party of officers captured a sullen youngster of the deep the day before yesterday. This little fellow measures about three feet in length, and affords wonder and amusement to all who look upon him. Wonder, at the peculiar way in which nature has fitted him for both land and water, at the strong, flexible, ridgy tail, at the scaly armour on his back and neck and the horny shield which covers his eyes; amusement, at the way in which he takes his imprisonment, his unconquerable sullenness and the way in which he affects death. There can be no doubt that there are crocodiles in the Prah, but so far we have escaped an accident, for which, in my opinion, we may congratulate ourselves.

We are collecting cattle here to an important number, but the mismanagement of the Superintendent, in surrendering them solely to the charge of natives, has caused the loss of some of the finest of them. Even the very light work of tending cattle requires to be superintended by a European, otherwise the Fantee permits his charge to wander astray where nothing but the most meagre vegetation can be procured for their subsistence, and where they drop down quietly to the earth from sheer starvation. Apropos of the Fantees, the Kossoes, a tribe of hunters from Sierra Leone, who have the faculty of scouting and searching out paths and traces of human footsteps almost as much as the North American Indians, have discovered the hiding-place of the Fantees.

Away, deep in tangled woods, surrounded by a fetid morass, an uninhabitable spot, but one of the best hiding-places that could be imagined, these deserters had sought and found a locality which Ashantee scouts could not find, into which no Fantee policeman could penetrate unless guided thither by the unerring instinct of a Kossoe man. Hither they had invited a few trustworthy women, who were made to purvey food for them, while they enjoyed themselves serenely after their brief period of hard toil and high wages.   A few bruised twigs, a few crushed blades of grass, were sufficient for the Kossoes, and the trail was followed swiftly until the deserters' lair was discovered and the recreants were brought to camp for punishment.   In these fastnesses of nature many such lairs might be found, if it were worth while to discover them ; but time presses.   The Kossoes, as well as other men, are needed for other pressing services, so that the majority of the deserters will never be apprehended.

But justice is to be meted out to the rogues who have thus placed obstacles to our progress, by not only refusing to despatch their men to aid the expedition, but digressing from their way to preach sedition and mutiny among the disaffected still in the employ of the British General.   A chief of Abra was tried by a court and sentenced to be shot.   When Sir Garnet heard of the sentence, it is said, he at once ordered that the chief should be hanged, and not shot.   The chief was a traitor, and did his best by precept and example to disaffect others.   As regards the natives, the English Government cannot be accused of cruelty or injustice, but certainly of gross leniency.   The British have treated Africans as we treat dainty children. They have petted them and spoiled them, but they have never punished them.   Sir Garnet seems to have discovered this fact.   Leniency is weakness with the vulgar and ignorant, and in future dealings with the native Africans England will do well to remember this.   In a great measure this excessive leniency aided to produce this campaign.   My references for this statement are the Blue Books.   Those who care to read them will blush at the depth of self-degradation the Colonial Office and its employés have been guilty of in their written intercourse with the sable potentate of Ashantee in past times. I cannot say worse of England and her colonial government than that in her past intercourse with the king of Ashantee she has deeply humiliated herself.

But passing by such topics for the present, I will attempt to describe the vast difference that lies between our life on the Prah and life on the Hudson.   I have the advantage over folks at home for, by the slightest effort of fancy, I can sketch New York and its teaming population dwelling under the rule of a hard frost; the filthy snow heaps ranged on each side of Broadway; the long, long lines of animals struggling amid the débris of winter more than knee-deep; the cheerless faces of the citizens, made more cheerless by the nipping cold, and many more uncomfortable and dismal things attending a residence in New York during winter. Could they imagine in their turn our life in camp on the Prah? The terrible sun which has death in its fierce, merciless heat, apopletic strokes in its beams; the dark earth steaming hot, emitting its baleful vapour; the dense forest and denser undergrowth whence issues the fever; the long continuous lines of naked blacks bearing burdens of rice, flour, sugar, tea, beef, &c., to the control sheds; the unjacketed officers who pass by us complaining of the heat and the lassitude it engenders; the crowds in the evening who hasten to cool their parboiled bodies in the waters of the Prah; the perspiring people who brave the heat and the glare of the tropics and sigh for one drop of cool water to cool their parched tongues.

The newly arrived soldiers give us rich fun. Sailors have always carried off the palm for ingenuous sayings, but in this campaign I am certain the soldiers will win the day. What they have done has convulsed the camp with laughter. The Rifle Brigade, or, at least, a portion of it, is encamped at a place called Barracoe. It may be remembered that Prince Arthur is a captain in this corps, and that the regiment has a very high reputation for gallantry. Soon after its arrival at Barracoe one of the number was placed at night as sentry at a place looking toward the deepest part of the forest, whence issued during the darkest hours of night a chorus of unmelodious sounds which were sometimes followed by a series of harsh, unearthly screams. The sounds caused by a laughing hyæna were as music compared to them; the shrill cry of the fish eagle is not half as alarming. These startling screams were uttered at regular intervals and seemingly in close proximity to the bewildered sentry. He dared not fire, for to alarm the camp by a shot fired at random into the black forest with the hope that the beast or whatever it was might be frightened into silence was not to be thought of. But the sentry determined that by preparing himself against attack by fixing his sword-bayonet, he would be guarding himself in some measure against any manifestations of hostility the beast might exhibit. Thus he stood constantly peering into the impenetrable darkness, vainly seeking to pierce the almost palpable blackness of the night until the relief party came round.   The sentry had no time to recover himself and sheath his sword before the party in charge of the officer was close upon him, so he thrust his sword under his arm, and, after challenging, brought his weapon to a ' port arms ' to receive the relief.   Seeing his evident confusion and the sword under his arm, the officer demanded the cause, and was told by the sentry, ' If you plaze, zur, there's some snake of a wild baste a constantly screaming close by here. Divil a bit has he stopped since oi have been standing here, and oi'm thinking the crathur can't be far off.   Sure the divil must be in him.   So oi just fixed my swurd for him to give him some cold steel.'   ' Oh, you mean that lemur which cries out so.   Why, my dear fellow, that wild beast is not as large as a rabbit.   You need never fear anything from him.' The poor sentry, it is reported, has been fearfully chaffed by the Rifles for his fright at an animal not bigger than a rabbit.

A soldiers' camp-fire is a cheery sight.   Each face as you see it by the firelight seems tinted with a warm, rosy hue, and there is very much of genial life and kindly spirit around it. Fires are not always kindled for the warmth they give, and even a winter fire is as welcome for its blaze and cheeriness as for its warmth ; but in the tropics, the military camp fire is kindled for the magnetic geniality its bright blaze evokes. With the Rifles as with the sailors the camp fire is a nightly institution, and around these the bonny faces are gathered and become wreathed in smiles as some highly-coloured story is being told for their delectation.   Every imaginable thing affords a subject for graphic illustration and embellishment, and the Rifles enjoy the tales with a zest which speaks well for the good fellowship which distinguishes them.   But the stories generally treat of wild animals and their savagery, of thrilling adventures with beasts of prey.   One night the fire circle had sat up rather late, the anecdotes had been frequent and various, and more than usually illustrative of the dangers to be encountered in an African forest.   The weird sounds which issued from the deep, dark forest surrounding the camp at Barracoe, and stole through the darkness on their quickened ears, lent somewhat of an alarming reality to the themes which had engaged their ears that night, and if the brave Rifles had been subjected to the confessional there is no doubt but that many of them would have confessed they would much prefer encountering the dangers of the forest during the daylight than at night time. It chanced that a sergeant was compelled to proceed alone to a distant quarter of the camp, after a whole répertoire of alarming stories had been exhausted. The night was pitch dark, the tall cotton-woods seemed to rise to infinity in the darkness, while the noises that issued from them were extremely depressive to the soul. The sergeant thought he might as well prepare against contingencies ; for, ' who knows ? something might happen to a fellow in such an unchristian country.' He unbuttoned his holster, and kept his hand on the stock of his revolver. He stopped once or twice to listen, as he imagined he heard footsteps. Again he continued his fearful way, and again and again he halted to listen. Through the darkness he could faintly distinguish the outlines of a huge beast ; but, whatever it was, the beast remained motionless when he stopped. Did the thing mean to spring upon him ? lions and tigers and leopards generally did on a benighted man ; that was the usual mode of attack. Arguing thus, the sergeant drew out his revolver and kept it directed at the monstrous thing, which still persisted in following him. As he drew near his own quarters the sergeant, whose fears had risen to the highest pitch, shouted out to his comrades in most agonising accents, ' I say, Bill, Jack, Tom, hurry up, for God's sake, and show us a light ! quick, for mercy's sake ; here is a wild beast going to attack me !' Responsive to the poor sergeant's cries for assistance came the cheery ' Aye, aye ' of his messmates, who sprang out of their tents with lights and Sniders and swords to the rescue. But imagine the astonishment of all when the lanterns disclosed the form of—a mule !

The sergeant, however, is not the only man who has been frightened nearly out of his wits by such a domestic animal, for it was but the other day when a native, being requested to lead a donkey to water, started with him gaily enough, until the animal began to bray, when the native ran away from him with the speed of lightning, and never halted until he found himself a good half mile away from the strange brute, which could excel even the African lion in noise.

CHAPTER XI.

January 15, 1874.

THE answer to Sir Garnet's third ultimatum has been received
from Coomassie. It speaks of the king's desire for peace with
the white men. The envoy who brought it and his suite of
fifteen men were halted at Essiaman. Yesterday, also, the apparition of a pale-faced captive—a shadow of a man suffering
from pulmonary consumption, one who had lingered in com-
pulsory detention at the capital of Ashantee since June 12,
1869—startled the entire camp. This sick captive is the Rev.
Mr. Kuhne, of the Basle Mission, a zealous missionary, a native
of Neufchatel, who, after five years' residence at Coomassie, was
suddenly permitted to leave Ashantee and proceed to the camp
of Sir Garnet Wolseley.

The poor man is in such a wretched state of bad health that
I could not find it in my heart to subject him to a prolonged
interview; but in the following remarks you will find inter-
woven such information of interest as I have been able to
glean :—

The king of Ashantee is a young man thirty-five or thirty-
six years old, strongly pitted with traces of small-pox, possessing
some natural ability, which however has been warped by the
assiduous and constant flattery of his parasitical and trembling
subjects.

The adulatory cries of the town crier of the capital are
more than sufficient to transform any man, white or black, into
a vain, incarnate despot. They are generally in the strain of—

' O King, thou art the king above all kings !

' Thou art great!

' Thou art mighty !

' Thou art strong!

' Thou hast done enough !

' The princes of the earth bow down to thee, and humble themselves in the dust before thy stool !

' Who is like unto the King of all the Ashantees ? '

King Coffee, though arrogant and vain, and cruel beyond measure, has the eye of a king, which means that it is the eye of one possessing unlimited power over life and death. So little does he value human life that he frequently says, ' By the slaughter of one hundred shall I be able to produce a thousand.' The king is placed on the stool by the united voice of the chiefs ; but immediately he is seated, in him becomes vested the supreme power over life and death. If the council of chiefs and captains propose that he shall engage in war, he dare not absolutely refuse to make war, lest some ask him, with a sneer, if he is afraid, when, if he replies in the affirmative, the monarch's prestige departs from him. When the chiefs propose that he shall make peace, after engaging in a war which turns out to be ruinous, he may then proclaim that peace shall be declared without incurring disrespect.

A great power in Ashantee is the king's mother, who often with her advice has plunged Ashantee into war or hastened a peace.

The king's palace consists of many houses connected together, after the native style of architecture, and adjoining it is another residence, the two-storeyed stone house constructed for him by Dupuis, a traveller who visited Ashantee about 1820. The missionaries have lately constructed another house for him, 44 feet long, 25 feet wide, and two storeys high, of brick which they themselves manufactured for the king.

The king possesses about three hundred wives, though they do not all live in the palace, but are scattered among the suburban crooms, or villages. It is death for a man to look at the female possessions of the monarch. When any of the harem ventures out, the criers hasten ahead to clear the way, and warn the citizens of the advance of the fatal procession with cries of ' Ecow! ecow ! eh! eh ! eh !' the first two words being drawn out like our ' halloo,' while the last three are delivered in quick succession and in higher tones. As they approach, the passengers turn their backs to the road and cover their faces with their hands, so they may not be thought to take any advantage of the confidence of the king in thus exposing his human treasures to the light of day.

Bowditch's graphic descriptions of the stately processions and the wealth of barbaric gold dispensed at Coomassie on grand occasions turn out to be correct. This wealth principally consists of golden heirlooms, ornaments of massy gold, which for generation after generation the natives have treasured in secret places. If a hostile army approached the capital this gold would be hidden, and the death of all slaves would be almost certain, lest they might betray the localities where the treasure was concealed.

The permanent population of Coomassie is I find estimated at 15,000 souls—men, women, and children.

At the beginning of the war the army of Ashantee mustered 48,000. This force was collected from the combination of nations known as Ashantee and the tributary countries adjoining the kingdom, which, according to all accounts, extend far into the interior, one of the allied chiefs having volunteered from the borders of the Sahara Desert. The army that returned to Coomassie lately under the famous Amanquatiah numbered only 28,000, the large number of 20,000 having fallen victims to disease. With this sadly reduced army were the remains of 279 officers or chiefs who had fallen in battle or died of disease, enclosed in so many boxes. This army was received in the great market of Coomassie by the king, and was thanked and rewarded with presents of gold, 100 sheep, and forty loads of salt ; but there was no ovation paid to it. The reception was singularly silent, utterly unlike the indescribable triumphal fanfaronade formerly made on such occasions. Then the people dispersed to their crooms, their plantations, and houses. Fifty-five slaves were slaughtered to the manes of the departed chiefs on the custom day, January 1 of this year.

Amanquatiah and the chiefs who have been engaged in battle have sought to dispel depression from the mind of the king by boasting of his performances in the war, and declaring to him that there are enough men still left to defend the capital from any attack. It appears that the British letter created astonishment, alarm, and yet unmitigated contempt in the mind of the king. The first feeling was caused by the stern purpose which his words implied; alarm, by the amount of compensation to be exacted for the injury done to the Assins and Fantees; and unmitigated contempt at the long explanation which the General had seen fit to enter into concerning the cause which led to the death of one of his envoy's suite.

It is known now why the unhappy Ashantee ambassador committed suicide. It seems that it was proposed to send a white man to Coomassie with Sir Garnet's terms to the king, and that this man ventured a remark to the effect that if any white man went to Coomassie the king would kill him; whereupon the envoy, who was his brother, censured him for his unjust suspicion, and promised to report him to his Majesty. Fully satisfied of the extent of the king's anger on hearing the calumny, he put an end to his own life.

When the English General's letter had been read, the chiefs who composed the council began to extol themselves, and the tributary king of Adansi sprung to his feet and shouted, 'Behold! I am the king of the Adansis. Who can cross my country without my permission?'

Attifowah got up and said, 'Lo! I am strong, my warriors are brave, and none can overcome me.' But Amanquatiah, who had just returned from the war, said, 'He who has seen no war cannot show a scar.'

Presently cooler counsels and more prudent advice prevailed; hence the letter which Sir Garnet Wolseley received the day before yésterday, treating of peace. The king's first intention after his general's return to Coomassie was to permit his warriors a week's rest and occupy the Adansi Hills, and when the white people should appear to attack them with vigour. If he were successful in driving the English back, he would slaughter every prisoner, white or black, he had taken in this war; if he were defeated, he intended to await at Coomassie the arrival of his foe, reserving his prisoners to plead for peace for him.

The reputation of the Gatling is now spread throughout Ashantee. 'It is a terrible gun, which shoots all day. Nothing could stand before it; the water of the Prah ran back affrighted.' The effect of this, combined with many other things, has been to induce the king and his council to deliberate and reflect on the possibility of peace.

Of himself Mr. Kuhne said he had been captured on June 12, 1869, by the General Addo Buffo or Adu Bufu ; that a valuation of 6,000*l.* sterling was placed by his captor on Mr. Kuhne, Mr. Ramseyer and wife, and M. Bonat, a French trader, which was finally reduced to 1,000*l.* He and his brothers were treated as captives who might probably be of value eventually or not. They were granted a sufficient amount for subsistence, but nothing more. What money was despatched to them was retained by the king, who doled out the sum of four and a half dollars per period of forty days for the subsistence of each white man and his servants. They have preached a few times in the streets to the Ashantees of the capital, but the king never gave permission to his subjects to become converts. The populace of Ashantee treated them at first with contumely; but, when the king gave permission to them to punish insolence with the whip, some more consideration was shown to them.

It was amusing the way in which some of the people endeavoured to exempt themselves from being subject to the privilege granted to the captives by crying out, 'Know ye what ye do ? I am the king's shirt-washer ! I wash the king's robes ! I wash the king's feet !' While another would venture to cry out, ' My sister is one of the king's wives !'

The Fantee prisoners were exceedingly insolent to the white captives. They were pleased to utterly disbelieve that the English General would ever approach Coomassie, and omitted no pains to imbue the Ashantees with that idea. Of the Fantees Mr. Kuhne and his missionary brethren entertain the very lowest opinion.

On October 18, Mr. Kuhne wrote to the king requesting leave to return, as he was suffering from a disease which, unless he was able to obtain good, nourishing food, must soon kill him. Last Friday night his second request was granted, and he left Coomassie at once by torchlight, and arrived in Karsi, a place situate about three miles from the capital. During Saturday he travelled for eight hours, and reached Akankwassi. On Sunday he made a short march to Quisa, on the other side of the Adansi Hills.

On Monday he travelled to Essiaman, a village occupied by

the British advance, and by 10 A.M. on Tuesday he reached this camp, a subject of compassion to all.

Mr. Kuhne's opinion is that the king will not fight; that his fears have been so worked upon by the reports of his chiefs, who have exaggerated the power of the white men and their guns, that he has no combativeness left in him. He does not suppose, either, that King Coffee would perform the harikari upon himself, as his illustrious prototype did in Abyssinia, or that he has the moral courage to do anything uncommon. The flow of spirits which caused him to dance in the streets of Coomassie before his subjects when he heard that his army had arrived at Dunquah, in the early part of the war, and prompted him to order the remorseless butchery of the white men on the coast, has evaporated and left him like one who has just passed through a long debauch—nerveless, fearful, and an object of contemptuous pity.

As the advance of the British army, under the impetuous Russell, has already scaled the Adansi Hills and made good its position at Quisa, no one can conceive the possibility of defence left to the Ashantee king. The very fastnesses he boasted of— the frontier of his kingdom—have been unlocked, and between Quisa and Coomassie is but twelve hours' march. From which we may accept it as a fact that there is no fighting spirit left in the king of Ashantee, and that the British campaign of 1873–74 is destined to end in a peaceful parade at Coomassie. Sir Garnet Wolseley has sent his fourth letter to the king, and entrusted it to Captain Buller, chief of the Intelligence Department, who will deliver it to the envoy, who, with his suite of fifteen men, is awaiting it at Essiaman. It is rather disheartening to officers who have dared the fatigues and the climatic dangers of Africa to be told that peace will be made. It is more disheartening for an expedition which numbers so many Europeans to have to go through the form of marching to Coomassie only to sign a peace—such a peace as has been made already too frequently. It must be annoying to Great Britain, after such an expense—after cherishing the delusion that now she had the opportunity to crush the insolent Power which has insulted her representatives during the past century—to be told she shall not have the power to punish or avenge her wrongs; but she shall have her expenditure, the losses of life she has incurred, the sickness and fever which her sons have endured, all for nothing.

That is of course certain ; a score of valuable lives may be lost on the journey, half a million of pounds sterling will be added to the expenditure, but England's honour demands that her army shall enter Coomassie and go through the form of taking possession. Hostages will be demanded, and probably a much larger number of ounces of gold than the king can ever pay will be imposed as a fine, and the army will return to re-embark for home, having accomplished absolutely nothing, for in ten years or so the Ashantees will reinvade the Protectorate, and the same anger and impotent wrath will be roused just as fruitlessly as before. The desire of the king is for peace. Both the answer to Sir Garnet's letter and the release of the missionary confirm it. The missionary says that the Ashantees have lost so many chiefs that they have become disheartened. They have not lost so many in skirmishes, but by sickness. The warriors are returning to their villages ; and so satisfied is the king with the result of his letter that he makes no effort to keep them together. There is only one hope remaining, after this probably tame ending of the expedition, that, as the honour of England demands that the troops shall enter Coomassie, this event may not be accepted by the Ashantees as a sign of amity. The king may be very willing to declare his readiness to sign a treaty, to send any number of hostages to Cape Coast Castle, to pay any reasonable sum as an indemnity to the Assins and Fantees for the losses they have sustained ; but will he be willing to allow an invading army to occupy his capital and dictate terms of peace to him from an apartment in his own palace? That he will fear treachery on the part of the white men, some deep scheme of revenge, some bloody massacre, some wholesale spoliation, is the only hope left to the English that they will not be permitted to enter Coomassie bloodlessly or without some opportunity of inflicting a punishment on the savages, that it may become a tradition among them of what may be expected should they venture to attack any territory under the protection of white men again.

It is decided that the white troops shall not move from here before the 23rd of this month, because of the non-arrival of sufficient stores. The 2nd West Indians left here last night for

Essiaman. Colonel Wood's regiment left this morning, taking with it provisions for six days.

Sir Garnet Wolseley has just informed me personally that we shall move on the 23rd to occupy Essiaman; that on the 24th we shall reach Accrofoomu; on the 25th, Quisa, or Fomannah, at which place we may probably stay until the 29th, and that by the 31st we shall probably be in Coomassie.

CHAPTER XII.

January 22, 1874.

NAPOLEON III. writing from Compiègne, October 22, 1856,
to Marshal Vaillant, Minister of War, said that the able and
indefatigable minister who day and night occupies himself in
organising a large force, and in insuring to it at a distance of
800 leagues from his country the means of living, fighting, and
conquering in an almost barren spot, has merit fully equal to
that of a general who triumphs on the field of battle, and that
the nation should unite in its thanks him who prepares for
victory by combining the necessary means as well as him who
achieves it by plans well conceived and grounded.

If the Emperor could so highly recommend the provider of
the material of war, what shall be said of the young General in
command of this expedition to Coomassie, in whom is found
united not only the prudent qualities of a War Minister, but
the energy and will, tactical skill and forethought of a Commander of an army in the field!

I admit that I sometimes think that there are things to
which, in my opinion, more attention should have been and
should be paid by Sir Garnet Wolseley ; but while any man of
mortal mould is constantly, as he is, under the ever-searching
and critical eyes of a number of journalists, no man can escape
entirely from blame.

In his conduct of this expedition so far, I impute no blame
to him ; on the contrary, I give him large credit.   He has done
his best, and his best has been a mixture of untiring energy and
determination—youthful ardour, toned down by the sense of his

grave responsibilities; exuberant good nature, which nothing seems to damp; excessive amiability, by which we are all benefited; discreet judgment, which will surely meet its reward; wise forethought, which, assisted by his devotion to his work, prove that the trust reposed in him by the British Government will not be betrayed.

Looking back upon the great tract of forest country we have left behind, which stretches between the Prah river and the sea, and in imagination following the 74 miles of road already traversed, we are tempted to ask what value may be placed upon our progress to this date.

It is useless to recapitulate what is already contained in previous chapters; but it may be as well, before crossing the Prah to the real Ashantee land, to be sure that we properly appreciate what has already been accomplished.

When Sir Garnet Wolseley landed at Cape Coast Castle, with a retinue of about thirty special service officers, which seemed to be a kind of a gallant Forlorn Hope, he found the enemy swarming over the forest tract lying between the Prah river and the sea. In his front they were within a few miles; to his right and left they were masters of every inch of ground outside the forts. Even between Elmina and Cape Coast the road was considered so dangerous that communication was precarious.

With his Special Service Corps he began work, first, to create a semblance of order in the disorganised masses of Fantees; second, to induce the several chiefs of the friendly natives to a more ready submission to his dictates; third, to organise new regiments under European officers; fourth, to discover the enemy and his intentions; fifth, to measure the scant material he could obtain against the enemy's numerous horde; sixth, to worry and pound the enemy unrestingly, to attack them front and rear from all points, proving to them in this energetic manner that a new and strange force, of which the Ashantees had never before known, was now opposed to them to the death.

Sir Garnet's Special Service Corps, though much too inadequate for his wants in numbers, was composed of select men.

Such men as Lieut.-Col. M'Neil, wounded near Elmina, are

few. His efficiency as chief of staff met warm commendation from the young General in command.

Major Baker, 18th Royal Irish, a quiet and amiable character apparently, was a man of trust, discretion, and tact, who seconded Sir Garnet's efforts to the utmost of his power. When Colonel M'Neil was invalided home, Major Baker became temporarily chief of staff.

Captain Brackenbury, whose reputation as a writer on military matters is established, was Sir Garnet's military secretary.

Captain Buller was at the head of the Intelligence Department, in which capacity he exhibited traits of character which, on a more intimate acquaintance with them, will prove him not unworthy of filling posts of the most onerous and most responsible nature.

Captain Lanyon, of the 2nd West India Regiment, officiated as colonial secretary for a time, where his fitness for the post showed what an invaluable aid the new Governor and General had secured. He brought with him long experience in a similar office in the West Indies. Captain Lanyon is an eminently shrewd and diplomatic officer.

Captain Huyshe, who is at present suffering under severe illness, is another man of Sir Garnet's choice and liking; always efficient and alive to his duties, wonderfully hard-working, and faithful.

Captain W. F. Butler, author of the ' Great Lone Land,' needs no recommendation from me. He is a favourite officer of Sir Garnet, and occupies an enviable post. He has proceeded to Akim to raise an auxiliary force of natives. He is one of those energetic men who, by their own example, infuse a vital force in others, whence we may expect that important allies will be raised.

Lieutenant Maurice is Sir Garnet's private secretary. He has not had the opportunity to distinguish himself before the world ; but we may rest assured that he is a credit to Sir Garnet's choice, otherwise we should have heard something to the contrary ; and besides, he is the winner of the ' Wellington Prize,' a prize for which his own chief competed.

The Hon. Evelyn Wood, Lieut.-Colonel of the 90th Regiment, to whom has been awarded the Victoria Cross, a high distinction only conferred for undoubted bravery, was selected, at first, to fill the post of Commandant of Elmina, where he raised a force of natives; but he was soon sent to distinguish himself in the field, where his indefatigable zeal, and his intelligent performance of the commands of his chief, elicited high encomiums from those who understood the spirit which animated him, and the measure of his services.

Major Baker Russell of the 13th Hussars, who commands a native regiment, is one of those invaluable characters which a general must have—a reliable, efficient officer; a resolute, hard-handed man, with vim and nerve in him, a powerful sharp sword for an enemy, and a soft tongue for a friend. Lord Napier of Magdala had such a man in Abyssinia in Colonel Penn of the Artillery. Commissioner Glover has now with him a similar friend in Captain Sartorius of the Bengal Cavalry, who would be in his glory and proper place at the head of a Balaklava charge. Such men as he the eye of the special correspondent loves to dwell upon.

Major Home of the Royal Engineers has been before the public so often, and ever with praise, that it is almost unnecessary to do more than mention his name. He has chopped his way through over a hundred miles of forest and jungle tangle, made our road smooth for us, constructed an infinite number of bridges, miles of corduroy road across swamps, erected over a thousand huts, every stick and thatch of which he had to procure from the forest, constructed miles of telegraph lines; and while a portion of his men is now building the pontoon across the Prah, another portion is miles ahead cutting bush, widening roads, erecting more huts; and another is miles in our rear putting up the telegraph line, widening and improving the road; and Major Home has to employ mind and body in attending to these several onerous duties. Words fail to express our sense of the obligation every soul of this expedition is under to him. He has made the path through the wilderness smooth and pleasant, and all but bedded and boarded us. May his Government honour him as he deserves!

With most of the above heroes Sir Garnet set to work: the how and why have already been explained in previous chapters; but it has had this extraordinary result, that without aid of others than marines, blue-jackets, West Indians, and two native regiments, he has driven a force of (variously estimated) from 20,000 to 40,000 Ashantees in hurried flight across the Prah. This work occupied Sir Garnet Wolseley from October 1, 1873, to December 13, 1873. So much being understood, let us cross the Prah into Ashantee land.

Major Russell's native regiment, preceded by Lord Gifford's native scouts, were the first to cross the Prah river in force, and since their departure to the front we have received news from day to day of the rapid advance into the enemy's country made by the impetuous Russell and the gallant Gifford.

Close behind Russell, picking his camps up as he deserts them for other encampments further forward, are Colonel Wood and his regiment, and behind Wood's Regiment are Colonel Webber and two companies of the 2nd West India Regiment.

Major Home, Captain Buckle, and Lieutenant Bell of the Royal Engineers are in the advance with Russell, engaged in the usual arduous task of cutting bush, and making the road available for the European force and the native porters of the transport service.

We—that is, the members of the Press—left Prahsu on the 19th ; Sir Garnet and staff left the next day with the Naval Brigade and Rifles.

On crossing the Prah, we set foot on the soil of what is still an integral portion of Assin. But we searched in vain for a change in the landscape. No sooner had we departed from under the sunlight, which poured hot and strong upon the cleared camp of Prahsu, than we discovered ourselves to be involved in the impenetrable shade of cotton-woods, teak, and mighty tamarinds, with the insatiable jungle clustering thickly around their girths, and impeding the keenest eyesight, such as we experienced between Dunquah and Prahsu. Though we saw nought below the dense shadows which the thick foliage of the forest formed, above our heads, far, far above, the colours and tints of the leaves were discovered to be of an agreeable variety as the twigs and branches bent and rustled before the gales. Intermittent gusts of wind sometimes swept down on us, and caused us to gratefully doff our hats to enjoy the cooling draughts ; but Nature, ever grave and sombre in the forest shade, relaxed nothing of its sullen, lifeless aspect.

Mile after mile was passed of this forest. Sometimes the

eye caught glimpses of broad-fronded plantain stalks, or the tall feathery palm, or the slender parasitical rattan, or a huge length—like a monster serpent—of a gigantic lliane, swinging in mid-air, from tree to tree, which served to relieve somewhat the dead monotony of the march.

Four miles from Prahsu we came across the encampment of Attobiassi, constructed as a half-way station between Essiaman and Prahsu.

Beyond the camp the ground sloped abruptly a few yards to a narrow, sinuous stream, which in the rainy season overflows its banks, and forms a broad extent of water shallows and muddy sloughs in the tangle of bush on the river banks. The consequence is that in these depressions the over-fruitful soil shoots out with renewed vigour a denser crop of bush, plantain, palm, and hosts of parasites.

We had reason to be grateful to the genius of Major Home, who had laid down entire furlongs of corduroy while we traversed these miry troughs, for we were permitted to travel without losing our boots and stockings, and perhaps our feet, in the mud.

A short distance beyond Attobiassi we came across a third village of dismantled huts, and wrecks of Assin habitations; and a charred human skeleton, lying close to the road, reminded us of the purpose of the English expedition to Coomassie.

Another hour brought us to where a great tract of tall timber was deadened, and glints of daylight ahead announced our approach to the camp of Essiaman.

Essiaman is in charge of Captain Frank Russell, who has been displaying his knowledge of engineering, and making the camp habitable. An injury to his foot incapacitated him for active service in the front, to which he aspired like most others of the gallant fellows who have braved the heat of the day in this apparently irreclaimable and uncivilisable land.

Essiaman at present consists of the body of the old village, half-a-dozen huts massed together, and encompassed with strong timbers placed upright to the height of eight feet, until it bears the resemblance of a fortlet. Lengthy huts, which stand outside the fort in a row, and thatched with palm, bamboo, and plantain leaves, are an agreeable shelter for a night for even European troops.

A clear space of about 500 feet all around leaves room for light and air to penetrate into the camp, and an excellent chance for Snider bullets to hit straight at any foe having the temerity to advance from under cover.

At Essiaman we heard of the death of Captain Huyshe of the staff, and all who knew him sorrowed over the fate of the brave, faithful, and willing officer, and amiable gentleman.

The next day we left Essiaman, and resumed our march, in order to avoid the disagreeableness of travel in rear of a long, straggling line of people.

On leaving Essiaman we were tempted to congratulate ourselves that the country was improving. The bush was not nearly so dense, and something very like the ghosts of hills looming through the forest kindled a hope that the bush would gradually thin, until long views might be had. The sound of living waters, brawling over stones and down rock steeps, strengthened the hope; but for a long time we doubted the good prospects.

An hour's march brought us to the stream Fumasu, a muddied torrent; lumps of hematite iron ore, breaking up from the depths, impeded its flow, and made it gurgle, and murmur, and chatter most pleasantly to the ears. Rocky lumps appeared on the other side of the stream, and limited vistas through the trees gladdened us.

A deposit of sand with glistening yellow particles in it caused me to dismount from my mule to examine it; but, though I had flattered myself for a moment that the El-Dorado of the Ashantees had been reached, a very short glance at the glistening particles sufficed to inform me that they were only pyrites of iron.

I find I am not the only one who has been deluded by the deceptive shine and glitter of the sands in this stream. Captain Nicol and others of Russell's Regiment have been actually panning at the same place.

My mule seemed to understand my disappointment; for, as I resumed my saddle, the sage animal, before breaking into his accustomed jog-trot, wagged his head as if to say, 'No use, my dear master; the 'Shantees find the precious dust further north a good bit.'

At noon we reached the second station in the trans-Prah

land, Akrofoomu, a camp at present in process of construction. Half-naked black wood-cutters make the place resound with blows of axe and hatchet. Others are swarming on the tops of sheds at thatch-work ; others, with spade, shovel, and pick, are clearing weeds, grass, and dead vegetation away. Busy activity is exemplified by the aspect of labour which the hurrying forms of helmeted officers, and men black and white, give the scene. In a few days the scene will be that of an orderly, well-conducted camp, under rigid discipline, for so quick do ready hands and skilled minds change and subdue the virgin forest.

On the 21st, hearing that Sir Garnet would be at Essiaman about noon, we deserted our quarters at Akrofoomu, and plunged again into the forest shades, as we began our march to Fomannah.

Again spectres of hills or wooded heights were seen through the forest ; again were heard the pleasant sounds of running water, and again we anticipated an agreeable change. But looked we ever so long, looked we ever so far, waited we ever so expectantly, no natural clearing met the eye. It was all forest: forest here, forest there, forest monopolising every fathom of land, save the road.

An hour and a half brought us to the camp of Parakomi, in command of Lieut.-Colonel Webber and his West Indian coloured troops. Again was I compelled to wonder at the amount of labour an intelligent English officer can get out of such raw material as he finds at hand. Colonel Webber had but occupied this place two or three days before, yet in this brief period a large camp had been cleared and made agreeable to the eye. The camp was situated on a ' hog's back,' a ridgy eminence with an abrupt fall on either side for drainage.

The Parokomi stream ran south-west towards the Prah, under arbours of weed, grass, and brush. The bush brake was very dense along the banks, and for a few yards beyond the forest seemed to vie in density with the jungle.

After leaving Colonel Webber's camp, a quarter of an hour's brisk trot brought us to the junction of the ' King's road ' with the Army road—a locality about which Sir Garnet Wolseley seemed to be anxious, as he had despatched an urgent command to Colonel Webber to occupy the position with a portion of his force.

The King's road runs obliquely from the left of the road, and skirts one of the Adansi peaks which we shall have to surmount. The road then runs north-west into the Elmina, Borborassi, Becqua, and Coomassie road. That which the army of invasion must follow runs parallel with the King's road, from three to ten miles off, by Fomannah, Doompoassi, Amoaful, and Agimamu, into Coomassie.

Five miles distant from the Parakomi stream lay Moinsee station at the base of one of the towering hill peaks of Adansi, commanded by Lieut.-Colonel the Hon. Evelyn Wood. The troops occupying the post consisted of Wood's Native Regiment and Rait's Artillery.

Colonel Wood and Captain Furze, with their usual hospitality to wayfarers in search of the wherewithal to satisfy hunger, offered us a substantial repast ; and just as I was about to seat myself, Colonel M'Leod of the 42nd Highlanders appeared. I had heard that this gentleman, so highly spoken of, was a veteran, and something of a hero. The Colonel in appearance is a tall man with bowed shoulders, hazel eyes, bright with fires that need but a breath to kindle to a blaze, and a powerful frame. But I have seldom seen so quiet and modest a hero. I found him a very affable gentleman, disposed to wise observations, not unapproachable by reason of his position or power, nor ungenial.

Colonel Wood is fortunate in having splendid officers as his associates. Captain Furze is one of these geniuses, who, like the natives of Connecticut, are so ready-handed and ready-knacked. He takes photographs, improves tents d'abri to wall tents ; he constructs huts with a view to comfort and convenience ; and Wood's neat encampments have always been touched up by Captain Furze.

Lieutenant Eyre, son of Sir William Eyre, is another whom Colonel Wood has secured as an associate. Though heir to a large fortune, the young fellow has volunteered for this campaign, and roughs it with as cheery a face as though he were surrounded by every home comfort. He is one of the most industrious of young men, and always assiduous in his duties.

Leaving our hospitable hosts sighing for the time they

should be permitted to surmount the Adansi or Moinsey hills and view the promised land, we turned our faces up the great looming mass in our front, and were shortly engaged in panting and perspiring, as our failing breath became spent in climbing the steep slope.

Major Home's engineers had endeavoured to reduce the difficulty of climbing by making ever so many zigzags through the tall woods that clothe the hill from base to crown ; our progress, nevertheless, induced considerable exhaustion. But the very tallest of hills has a top, as all roads have an end, and so in time we reached it completely fagged out.

Our first impulse was to ask for water to cool our parched throats and furred tongues ; but before we had framed the request, the young commandants of the post, Lieutenant Richmond and Lieutenant Woodgate, offered a cup of warm tea, which in this country is much superior to cold water as a quencher of thirst.

There was a delightful sense of freedom, revivification, and elasticity pervading one on the summit of this hill. An oppression and vague feeling of dissatisfaction and biliousness disappeared. A man felt permeated through and through with pure, healthy air ; he became glad, and less austere to himself and fellows.

From the summit of the Adansi hills we obtained a better idea of the nature of the country which we had left behind than we had while traversing it, and the nature of that portion which we are destined to travel through in a few days if we proceed to Coomassie was revealed to us.

The Adansi hills appear to be a line of truncated hills, rising to an equal altitude, and spreading out crescent-shaped eastward and westward, inclining northerly, as if to hem in that land in which we are informed Coomassie lies. Each hill is wooded, each dip and depression of land is wooded, high land and low land ; far and near, north, south, east and west, every-where the forest land of Ashantee north of the Adansi hills heaves and rolls, wave after wave, varicoloured and uneven, now a ridge, then a hollow. South it bears the same aspect ; west it is a line of peaks ; east it is alike.

At our feet, as we gaze down from a height of 1,500 feet above sea level, that contiguous to us presents us with a beautiful play and mixture of colours, the mixture of the sere with

M

the green, and the rich autumn hue with the spring.   There
are terraces of tamarinds, great wide-spreading branches like
parachutes, globes of silk cotton with pale green leaves, round
uprising towers of teak leafage, flat extents of ordinary vegetation, deep hollows, which plantain and palm fronds combine to
fill, until the eye tracing the variety and form of the foliage
insensibly is carried away to where the colour of the vegetation
is lost amidst purple haze and blue ether.

This scene is similar to that which might be obtained by a
sail in a balloon from Cape Coast to Coomassie.   The eye could
not pierce through the heavy drifts of foliage heaped up in
ridges, globes, pyramids, or the thick carpet which covers the
hollows.   Indeed, looking down from this height on such a
scene, the power or force of the eyesight is compelled to give
way to that of the mind ; for, after an ineffectual effort of the
eyesight, 'tis the mind which observes here.   It is the unbroken
expanse of foliage that baffles the eye, subdues its power and
vitality, and compels the mind to an indolent and careless
observation of the whole.

I have noticed this peculiar power of African scenery elsewhere.   It is derived from its unimpassioned aspect, its want
of tremor or motion, the drowsy calm or lifeless phase which
a warm, hazy atmosphere imparts to the ocean of foliage which
such a large and wide extent of forest presents.

Leaving the peak, we were soon concerned in the necessity
of restraining the impetuosity which the steepness of the
descent induced.   It was almost a continual slide down hill ;
our strides were lengthened to an incredible distance.

As we descended we were sensible that the delicious freshness of the air we breathed on the summit of the hill, laden
with the fragrance of tender leaflets and buds, and sapful twigs,
and young verdure of the topmost forest boughs, had left us,
and that we were breathing the heavy, infragrant air circulating
languidly between the massive stems and forest colonnades.

We passed several small streams, which raced over pebble
and rock merrily, along many a bend and curve, down towards
the dark dells which surround the village of Quisah.

As we emerged into the light of the clearing around the
village, we saw a circle of officers seated before a pretentious
house in the shade, and I soon recognised Major Russell and his

subs, and the gallant boy Lord Gifford, who has handled his scouts so well as to merit unqualified praise from his superiors.

Quisah is the first village of Ashantee proper which we have seen. It is situated in the kingdom of Adansi, whose king holds court at Fomannah, at which place we shall arrive to-morrow.

CHAPTER XIII.

GIFFORD'S ADVANCE—FOMANNAH DESCRIBED—PALACE OF THE KING OF
ADANSI—NATIVE ARCHITECTURE AND CUSTOMS—THE FETICH PLACE—
RELEASE OF THE GERMAN CAPTIVES—CAPTAIN BUTLER.

Fomannah : January 24, 1874.

QUISAH, the frontier village of the Adansi kingdom, was
vacated by Russell's regiment and Lord Gifford's scouts the
second day of our arrival in Ashantee proper.   The advance of
the army and the staff were at Moinsey, on the other side of
the Adansi Hills, and as room would be needed for their lodge-
ment, Russell was compelled to move on.

The position of Quisah was felicitous in the extreme ;
abundance of fresh clear water was provided by the numerous
streams pouring down the Adansi slopes, and quantities of fruit
(plantain) were discovered by the Fantees in their ramblings
through the woods.   Our more esthetic senses were refreshed by
the views of bold hills looming upward in a crescent shape
towards the south, with their heads buried amid wreaths of
gray cloud.

Young Lord Gifford, who has, throughout the march from
the Prah to Quisah, exhibited great tact and energy, informed
us how, when struggling up the stiff steep of Adansi, he was
warned not to make further advance into the country by a
detachment of the enemy under the king's head scout.

' We have no palaver with white men,' said the scout. ' Go
back ; we were not sent to fight you. We may not fight until
the king tells us.'

But Gifford pressed on up the hill, and the Ashantees retreated before him with reversed muskets, showing by this that
they were not prepared to resist. When he arrived on the
crest he found two white kids impaled alive, and the head of a
simiad on a stake. This was the first specimen of the super-

stition which enslaves the Ashantee warriors. On the next day Gifford pushed his scouting parties forward as far as Quisah, where he encountered the still palpitating body of a woman, just beheaded in the name of fetichism, that most monstrous faith which insatiably demands human blood as a propitiatory offering to its deities.

After securing himself at Quisah, Gifford scouted the country round in search of paths and villages, of the former of which he found a great many, and having ascertained a knowledge of his surroundings, he pushed forward to Fomannah the capital of the Adansi kingdom, whence I date this chapter. The Ashantees on discovering him and his party, again warned him to desist from further advancing, as the vengeance of the king would be most awful if the white men angered him; but they still held their muskets reversed as a token of amity to the white men. Gifford, having received orders not to fire, retired upon Quisah for further orders, but visited the town again on the 21st and discovered it to be deserted, upon which his native scouts precipitated themselves upon the 'loot,' consisting of handsomely carved stools, short ivory tusks, belts, guns, hides, skins, earthenware and empty jars.

On the 22nd, after due permission from Sir Garnet, Major Baker Russell with his regiment moved upon Fomannah.

The distance from Quisah to Fomannah was traversed within an hour, so that it was not very great. Along the road we saw the strange delusion under which the Ashantees laboured. Some of their scouts having very probably seen the telegraph line, imagined doubtless that it was some powerful fetich, and imitated it by stretching along the path a cotton string extending from Quisah to Fomannah. Besides this they had piled up heaps of black crockery, strung an endless number of charms on the bushes, and trees. But little recked Russell's Regiment while the men had Sniders in their hands for the false fetiches of Ashantees, and no doubt before this campaign is over the Ashantees themselves will have lost much of their former faith in their abominations and frivolities.

As we entered the capital of Adansi we discovered that it was not by any means a despicable village. A broad avenue, flanked by family residences one story high, with very steep

roofs, led through the town, and when we had arrived in the
central part, where the king's house was located, we saw that
another broad street ran at right angles from it.

There was an instant rush to occupy good houses, but good
houses happened to be plentiful in the town, and I should say
there were quarters enough for a force of 5,000 white troops in
Fomannah.

You will naturally ask what an Ashantee house is like. I
mean to answer the question to the best of my ability, and I
take the house wherein my press colleagues and myself find
ourselves lodged as a fair specimen of Ashantee architecture.
Externally this house presents us with four houses, each of which
is about ten feet long by six wide, standing corner to corner
and enclosing a quadrangle, or court. Abutting against two
of these houses are two others of similar size joined together by
a wall of mud, and pierced by a doorway which admits you into
the outer court, or court of receptions, for the inner court is
evidently devoted to family uses, for retirement and privacy.
Each house, though joined by its corners, seems to be independent, for as many houses which go to form this family residence
have independent roofs, which consist of layers of palm leaves
neatly laid over one another.

The exterior of the house is plain and unpretending. We
see mere expanses of walls roughly plastered over with mud,
but when we cross the threshold and enter the first courtyard
we are at once struck with the fact that a strangely original
people have been found in the Ashantees, infinitely superior to
the Fantees and Assins, whose countries we have just traversed.

The courtyard is of course littered with the pottery of the
inhabitants, which but proves how hasty was the departure. We
lift our eyes to the walls and elevated alcove, and are astonished
at the immaculate cleanliness of it, and the elaborate ideas of
ornamentation which they possessed.

For a height of three feet above the ground the walls are
painted an ochrish red colour, and so is the floor, but above this
they are of a waxen white, covered their entire length and
breadth with designs in alto-relievo, half an inch thick, cornices
are set off with many grooves, friezes with singularly bold
diamond-shaped designs with embossed centres, pediments are

something of the Ionic order, severely plain and square, the walls with intricate scroll-work relieved by corollas in alternate squares.

The alcoves are open, raised to the height of the ochrish-coloured portion of the walls, and are six feet in depth, by about eight feet in length. In the inner courtyard, the front of the alcoves are divided by one or two round columns supporting an open lattice work in waxen-hued plaster obtained from white clay. You at once understand the divisions of the family, and their respective apartments, for the largest of them, which has a drop curtain of woven, or plaited straw, is the one evidently used by the lord of the house and his favourite wife, another is that given to the other wives, another is devoted to the use of the children, and another must be the store-room, for there are numbers of black earthenware jars, wooden chairs, and knick-knacks of wood-work, ladles, spoons, basins and platters of wood. In the centre of the inner courtyard is a little tree which we have already found to afford shade, as we sit at our table for our meals.

Many little things which we see about us evince the taste and industry of the Ashantees. Take one of their stools for instance and examine it. Formerly it was one square block of white wood, very like sycamore. An artisan has chiselled and shaped a beautiful stool which any drawing-room might possess for its unique shape, design and perfection of workmanship. The seat is crescent shaped, the ends of which when we are seated in it come up half-way to the hips; a central column beautifully carved, resting on a flat board twelve or fifteen inches in length by about eight in breadth supports it, on each side of the column, is a side support chipped until it resembles a lace-work pattern. There is art in this stool, and whether it is original with the Ashantees, or borrowed from strangers, it is certainly a most interesting specimen of woodwork, the whole of which is cut from a solid block of wood. I think I have seen the shape of a stool similar to it painted on one of the tombs of Thebes.

Take again this soup, or water ladle, and regard the designs which ornament the handle. A European turner would be proud of the work, yet all that finical carving and cutting was done with an iron knife of native manufacture. Yonder are some specimens of earthenware, the figures on which are very

praiseworthy, and here at my feet are sandals excellently done;
nay, you might almost declare upon oath that they are as good
as anything worn by a middle-class Turk or Egyptian. Sandals!
At the mere repetition of the word one's thoughts revert to the
inhabitants of Egypt, Syria and Asia Minor. Why not to
Tunis, Tripoli, the Touaregs of the desert, to the Moors of

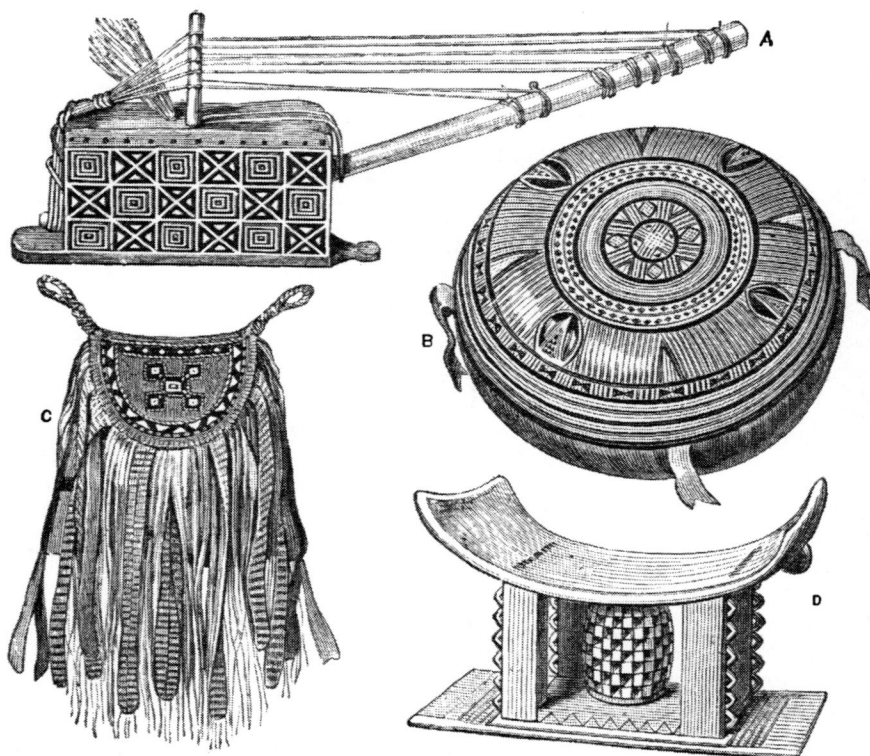

A. Sanko, musical instrument. B. Cushion. c. Ashantee Minister's state bag.
D. Fantee stool, cut out of one solid block of cotton-wood.

Timbuctoo? Then, if these sandals of such shape are one of
the necessities which intimacy with Moorish visitors taught the
Ashantees, why may not this elaborate scroll and flower designs
on the walls have been an esthetic taste imparted to them by
the Moors? Is it only an ingenious deduction, or may it not
be likely that this semi-civilized taste has been imparted to
them by long contact with Moors?

Bowditch in 1817 mentions that there were Moorish visitors in Coomassie when he was a resident in that city. It is likely enough the Moorish merchants were there before his time, for the fame of Ashantee prowess has been spreading amongst the tribes of the interior ever since 1740 when Sy Tutu founded the reigning dynasty.

If we only visit the king's palace—which is reserved for Sir Garnet and his staff—we shall find this ornamentation and elaborate tracery in *alto relievo* on a larger scale.   Court after court opens to the view, expanses of ochre-washed walls, and lengths of white plastered walls and columns, arcades, deep alcoves, lattice work, and great catharine-wheel shaped circlets in plaster.

It is with some feeling akin to a compassionate half-regret for the fugitive lord of Adansi, who but three or four days ago administered sternest justice within the precincts of this palace, that we wander through this many-courted residence. But he must have been a polygamous and jealously uxorious chief; all these inner courts so screened from peeping Toms were never intended for men-servants, unless they were eunuched to guard the peopled harems.

No two of the courts are ornamented alike.   The prevailing design of one court is an involved and complicated coil pattern, while another affects the square within square work.   One has numbers of circles with depressed centres, another has circles with globular bosses.   One alcove is so large that it becomes a good sized room partitioned with three columns supporting open lattice-work : another alcove is divided by square pillars severely simple, and another is rounded, with a raised band circling around and curving upwards.

Nor are the courts of uniform length and breadth.   One large court seemed to be that for receptions and audiences, and others are for domestic purposes.   There are only two double storey buildings in the king's house, but I do not think the upper storeys are for any other purpose than store-lofts.

Almost every house is decorated, more or less, but as we have not seen other villages or towns in Ashantee, except the small frontier village of Quisah, it is difficult as yet to know what importance as a town we shall ascribe to Fomannah the Adansi capital.   If Fomannah presents such extraordinary evidences of semi-civilised culture, what must the capital of Ashantee, to which this town is tributary, be like? I anticipate many curious revelations when we reach there.

Close by our house is a shady spot formed by two banian trees, a gum tree, and a palm tree railed around, and enclosing a circular space which I have no doubt is devoted to fetich uses. The altar is composed of poles lashed together resting on forked uprights. The sacred vessels consist of the bottom of a glass bottle, an earless mug, and a very ancient china tea cup, relics of a visit paid to the shops of Cape Coast Castle in byegone peaceful times. Bits of cloth, rag, cotton, twine, form the only things approaching to hangings we see; splashes of cornmeal water mark the trees over which the fetich priest flung his consecrated liquid. A large black earthenware pot, filled with some suspicious mixture, is dragged out to daylight from a recess in the circular space; but we cannot divine what its contents are.

An examination of the interior of the houses revealed to us much of the domestic life of the Ashantees. We saw the sleeping apartments, those set apart for reception, and those used for store rooms. These last were filled with a strange medley of articles, hide chairs, carved stools, hide bedsteads, wooden basins, pipes, drums, snuff-mills, ladles, platters, spoons, large earthenware crockery, bolsters stuffed with silk cotton, antelope skins, bags of leopard skins, rolls upon rolls of hard pressed tobacco, a heap of Indian corn on the cob, spiked walking staffs, iron knives and cleavers, curious teak boxes studded over with brass tacks, old locks, keys, gun flints, leathern pouches and belts, forsaken accoutrements and disused rusted guns, and many more things too numerous for detail.

Some of the houses contained long suspicious rolls covered with long grass and tightly bound, which, on being opened, were discovered to be dried corpses of chiefs or heads of influential families. There were six of these bodies found altogether, but as soon as they were discovered they were taken out of the houses and deposited in the jungle.

About a couple of hours after our arrival at Fomannah we heard that the remainder of the captives, Mr. and Mrs. Ramseyer, with two children, a boy and a girl, born in Coomassie, and M. Bonat, a French trader, had arrived at Major Russell's head-quarters with a considerable retinue of servants (Fantees) who had shared their captivity with them.

Mr. Ramseyer was a sturdy bearded German, with a true Teutonic cast of face, the French trader was one of our ideal Frenchmen, a lively vivacious man who enjoyed detailing his experiences to a knot of sympathizers which had gathered about him. These gentlemen confirmed the news that M. Kuhne brought to us at Prahsu. The king emphatically declares 'he

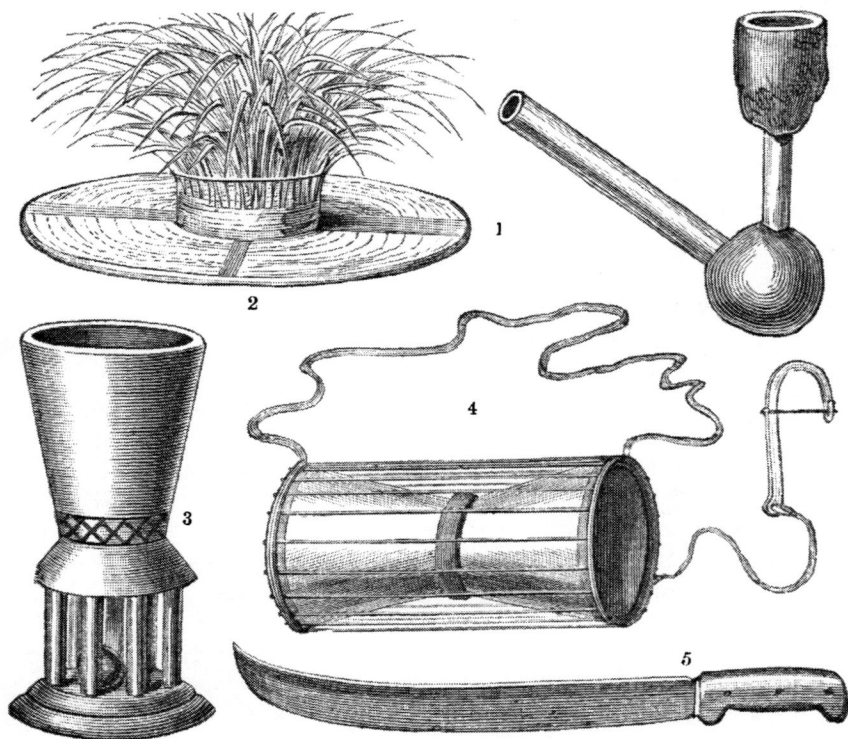

1. Pipe. 2. Hat. 3. Snuff mill. 4. War drum. 5. Sword.

has no palaver with white men,' and that his sole object in sending his army of invasion, February 1873, was to recover his provinces of Akim, Assin, and Denkerah—his right by conquest long ago—tributary to him until they rebelled. He wishes to be a friend to the white people, and to establish commerce with them.

M. Bonat informs us that King Coffee could never pay half a million pounds sterling indemnity, and that stories current in camp of his fabulous wealth are based upon the state pageantry, at which everything of value or ornament is exhibited, which taken at a lump valuation cannot exceed 50,000*l.* altogether.

These gentlemen brought a letter from the king to Sir Garnet, wherein he betrays the great anxiety he labours under and a most conciliatory spirit. The letter states that he is willing to pay the indemnity, that he can in time make Amanquatia pay the whole, but Sir Garnet must stay where he is and not advance further. This letter makes the third received from the king, and is in answer to the general's ultimatum despatched from Prahsu.

In the ultimatum Sir Garnet informed the king that he would wait a certain number of days at Prahsu for his answer ; if no answer complying with his demands for indemnification to the Fantees and Assins for the terrible desolation and misery inflicted upon the people of the Protectorate, he threatened to move upon Coomassie, attack it, and destroy it utterly.

Sir Garnet Wolseley of course cannot wait even to receive the indemnity, this would be out of the question. He has the welfare of the troops to provide for, as well as the realization of the punishment meditated on Ashantee, or the conclusion of a treaty of peace. Delays in the transport service may, however, interpose for a time, of which delays King Coffee Calcali may avail himself to his own peace of mind if he be so disposed, but if he neglects his opportunities he cannot stave off the entry into Coomassie of the European troops by ever so many protests that he has ' no palaver with white men,' or vain regrets that he did evil without knowledge of what he was incurring.

But apropos of this news from the king, Sir Garnet Wolseley has telegraphed the news of the king's promise to Cape Coast, with orders to the Senior Naval Officer of the fleet to despatch the ' Sarmatian,' one of the swiftest steamers now lying there, to convey the news to Gibraltar for the benefit of Parliament about to assemble, February 4. This excessive desire which Sir Garnet has manifested to inform Parliament of his progress will cost the British people the little sum of 7,000*l.*! Surely Sir Garnet lays too great stress upon the value of an Ashantee king's promise ; if it were a fact now that he received 200,000*l.* or 500,000*l.*, whichever the sum it is that he demands, the sum of 7,000*l.* were a small matter. But on a promise which may never be fulfilled, which the king, perhaps, never intends to perform, the sum of 7,000*l.* is surely too large to spend.

I have noticed another trait in our young General which is a novel one. He seems to be animated with a desire to compete with the natural purveyors of news, with us special correspondents. I do not blame him, for it may be, perhaps, a very laudable idea ; but it is none the less novel and singular. I never observed it in any other general before. We, of course, cannot compete with him if he refuses us the same facilities he employs himself, as he has certainly done in this instance. The 'Sarmatian' had already departed from Cape Coast Castle before any of the newspaper representatives knew of the General's strategical stroke. As Sir Garnet may be priding himself upon the successful 'beat' over four special correspondents, I may as well record the feat in his favour with a good grace.

On the 23rd the German missionary Ramseyer and his poor wife and children departed from Fomannah for Cape Coast, exhibiting considerable emotion and gratitude on finding himself and family in the bosom of such friends as all the English officers and men were disposed to be. M. Bonat, on the other hand, preferred to stay with the army, to give the staff the benefit of his knowledge of the country and language, as a slight return for the service the English had done him in procuring his release from a long and tedious captivity.

Sir Garnet arrived at Fomannah on the morning of the 23rd, looking as bright and cheery as he used to look during the first days of his arrival at Cape Coast, before he was stricken down with a recurrence of the Burmah fever which has tenaciously clung to his frame. As he came in, mounted on his magnificent mule, he was greeted by kindly faces, for he is universally beloved.

Commodore W. N. Hewett, V.C., who has come up to see his 'bould sailor bhoys' taking Coomassie by assault, was also warmly welcomed, for he has a very enviable reputation for being a first-class officer and genuine man in the Naval Brigade.

Captain Grubbe, of the troopship 'Tamar,' is the Commander of the Naval Brigade vice Captain Blake, R.N., lately invalided home for a serious attack of acute dysentery.

The Naval Brigade marched into Fomannah to a merry tune, which they sang with gusto, with a marino-militario gait which their loose pliable dress of blue shirt and pants admirably set off.

After the sailors and marines came the Rifle Brigade, in grey suits and sun-helmets, numerically pretty strong, though over seventy of their number were invalided at Prahsu. Thirty-two out of a force of 275 sailors were left behind at hospital, incapacitated by reason of intermittent fevers and dysenteric affections from continuing the march.

Captain Rait's splendid body of Houssa Artillery appeared behind the Rifles, officers and men looking remarkably well. The Highlanders remain at Moinsey until the day after to-morrow.

Since the European troops have arrived at Cape Coast, many a man has been stricken down with fever, some have died, many have been invalided home. The officers, from the nature of their duties, have been the worst sufferers, and the Special Service Corps, twenty-eight in number, who came out with Sir Garnet, show a gradual but certain reduction in number.

Up to date the list of casualties of the Special Service Corps reads thus :—

Lieutenant Eardley Wilmot, R.A.	Died in Action
Captain Huyshe, Staff . . .	Died of Dysentery
„ Townshend . . .	„ „
„ Charteris, Staff . .	„ „
Lieutenant Wells, R.N. . .	„ Fever
Colonel McNeil, Chief of Staff .	Invalided Home. Wounded in Action
Captain Godwin	Wounded. Invalided Home
„ Gordon	Invalided Home
„ McDermott . . .	„ „
„ McAlmont, Staff . .	„ „
Surgeon-General Home . . .	„ „
Captain Crease, R.M.A. . .	„ „
Major McLean, Rifles . . .	„ Cape Coast
Captain Lanyon, Staff . . .	„ Home

It will be thus seen by the list how serious a thing even a small expedition becomes in a malarious country like this, and what a waste of human life and wreck of brave men's constitutions have been incurred. It is in striking contrast to the comparative immunity from disease the British troops enjoyed in Abyssinia.

The arrangements entered into by Surgeon-General Home have proved admirable so far, in the transport of sick to the coast, and in the immediate replacement of the loaded hammocks by long relays of empty ones. There are about 300 hammocks, each of which is borne by four natives, thus giving a force of 1,200 men at the disposal of the Medical Department.

The Medical Staff is constantly receiving fresh accessions of army-surgeons, some of whom are most valuable and efficient. Among those who are very likely to distinguish themselves on the campaign, having shown by their assiduity to their profession and their fitness for their duties, are Surgeons McNalty, Fox, Turton—the energetic Sanitary Inspector—Jackson, of the 100th Foot, Mosse, and Waters. The task of controlling the Medical Staff now devolves upon Surgeon-Major MacKinnon since the retirement of the veteran Doctor Home.

The news from Captain W. F. Butler, formerly of the 68th Regiment, and the well-known author of the 'Great Lone Land,' is bad. Poor Butler, a man of undoubted zeal, not only as a devoted friend of Sir Garnet's, but as an officer who regards his profession with a noble pride, I fear will not be able to return from this campaign with the laurels which are bestowed upon those who prove successful. I do not think that his want of success can be attributed to lack of tact on his part, but rather to the natural stubbornness and obstinacy of the savages whom he was sent to organise into an auxiliary force on the right flank.

The report is, to-day, that he is on his way to camp with the same number of policemen he started with. Captain Brabazon, Lieutenant Paget, and Mr. Lowe, a medical officer, who composed Butler's staff, are, I hear, returning with him.

Apropos of this failure of Butler, I am reminded of a remark made to me by an officer a few days ago, when we spoke together about Glover and Butler's chances of success.

Said he : 'Captain Glover has done considerable injury to Sir Garnet by withdrawing from the Akim country all the best fighting men that were available. Now if we (Sir Garnet's officers) had had a special representative in the Akim country to enlist a force of Akims to co-operate with Sir Garnet's column, it would be impossible to estimate the value of such a force ; but what possible good can Glover do so far from our neighbourhood. However, Butler writes encouragingly from Akim ; he says he will be able to muster a force of 2,000 men to be ready to cross the Prah river by the 15th, but I do not think Glover will. Glover wrote in his last letter to Sir Garnet that if he had time he would be able to march with 12,000 men into Ashantee territory. Time, indeed ! how much time does he need ?—but that since his orders were to be on the Prah by the 15th he could muster barely 1,000 men, with which force he would start at once.'

Why this jealousy of poor Glover should exist I cannot understand. He has shown himself industrious, indefatigable, and capable, and willing to sacrifice everything of honour that he might have won with a strong force, by implicit obedience to Sir Garnet's commands. What could any commander desire more than a ready acquiescence to his wishes, allied with such efficiency as Glover brings ? I should hope that this feeling of depreciation of Glover's merits has not sprung from head-quarters, for Sir Garnet is generally so amiable, and so invariably disposed to be good-natured, that I should feel surprised if by any accidental expression of dissatisfaction he assented to the general feeling of jealousy of Glover which now unhappily, yet most undoubtedly exists.

If the news of Butler's failure be unfortunately true, it may serve to enhance Glover's efforts should he meet with success. I have not the least doubt myself but that Glover will come out of this campaign with more fame than he entered upon it ; and should he be successful, as I earnestly hope he will, Sir Garnet is that kind of magnanimous man that he will not be backward to officially recognise his merit.

I hope I do not bore you with this championship of Glover ; I cannot help it, since I have been a witness of the genius of the man in organising that large force he had on the Volta, and from what I see out here in the English correspondence, I think I am the only one who still retains faith in his ultimate success. The combined talent and energy of such a group of officers as Glover fortunately has to aid him, though by far too small in numbers for their health, must contribute to do something. Even Sir Garnet Wolseley can boast of no finer officers—though they are picked men—than Captain Sartorius and Goldsworthy. These last, when I visited their camps on the Volta were, to use a Western Americanism, 'just spiling for a fight,' which means that they were pining and fretting at the delay which the management of such a large force of savages as they had contributed to raise entailed on them. If they were pining and fretting at delays, they are more likely when the time comes to move at an astonishing rate, and to drive their men forward with astonishing vigour.

The left column, under Captains Dalrymple and Moore, do not seem to make any headway through the Denkera bush, though they were reported travelling at last accounts. It is very probable that these gentlemen encounter the same obstacles as Captain Butler, and that very little if any aid will be afforded by them.

Major Baker Russell (13th Hussars), with his native regiment, preceded as usual by Lord Gifford and his scouts, move forward to-morrow, and immediately after the advance guard will follow Colonel Evelyn Wood's Regiment and Captain Rait's artillery.

The following list of officers may be useful for a better comprehension of the status of affairs as they exist at present :—

MAIN COLUMN.

Sir Garnet Joseph Wolseley, C.B., K.C.M.G., Major-General Commanding.

Staff.

Colonel Greaves, Chief of Staff.
Major Baker, 18th Royal Irish, Assistant Adj.-General.
Captain Buller, Deputy-Asst.-Adj.-General and Chief of the Intelligence Department.
Captain Brackenbury, Military Secretary.
Lieutenant Maurice, Private Secretary.
Lieutenant the Hon. A. H. Wood, Aid-de-Camp.
Lieutenant Rolfe, R.N., Aid-de-Camp.

N

Surgeon McNalty, Medical Officer in charge of Staff.
Surgeon-Major MacKinnon, Principal Medical Officer.
Captain Baker, Chief and Inspector-General of Police.

Regimental Officers.

Lieut.-Col. the Hon. Evelyn Wood, V.C., 90th Regiment, commanding
Native Regiment of Apoboes, Kossos, and Bonnys, under Prince Pepple.
Captain Paul Methuen.
Captain Furze, 42nd Highlanders.
Lieutenant Douglas, 7th Fusiliers.
Lieutenant Barton, 7th Fusiliers.
Lieutenant Woodgate, 4th King's Own.
Major Baker Russell, 13th Hussars. Commanding 2nd Native Regiment.
Captain Bromhead.
Lieutenant Lord Gifford, in command of Scouts.

42nd Highlanders.

Colonel Mac Leod, Commanding ; Major Mac Pherson ; Lieut.-Major
W. Baird; Major A. Stephens, of 79th, with 42nd ; Captain Whitehead;
Captain A. M. Creigh ; Lieutenant I. B. Cumberland ; Lieutenant Souther;
Lieutenant Stevenson ; Lieutenant W. A. Berwick ; Lieutenant W. H. C.
Mowbray; Lieutenant R. C. Annesley, of 79th, with 42nd ; Lieutenant C.
D. Shoton, of 79th, with 42nd ; W. I. Smyth, of 79th, with 42nd.

The 42nd Highlanders will probably arrive to-morrow,
Fomannah being the general rendezvous of the army.

2nd Battalion Rifle Brigade.

Major Warren, Commanding ; Lieut.-Col. Warren ; Majors Stephens
and J. P. C. Glyn ; Brevet-Majors Nicholls and Sotheby ; Captains Slade,
Dugdale, Somerset, Cary, Lascelles, and Cope ; Lieutenants Hon. T. C.
Scott, Maberly, Sackville, Taylor, Hopwood, Thompson, Harrington, Smyth,
and Brune ; Sub-Lieutenants Hon. O. S. F. Cuffe, Noel, and Prittie;
Surgeon-Major Wiles, Surgeon Macrobin, Paymaster Harvey, and Quarter-
master Stanley; Colonel the Hon. S. Mostyn, Majors S. C. Millett and
T. B. Hackett, V.C. ; Brevet-Majors L. O'Connor, V.C., and J. de V. Tupper ;
Captains H. F. Hutton, E. M. Roe, and G. B. Luxford ; Lieutenants T. B.
Graves, G. H. Hutton, F. E. Shepherd, C. J. Gilbert, W. F. Cowan, Hon.
D. de Moleyns, R. H. Palk, H. R. Boyle, and A. F. Bernard ; Sub-Lieu-
tenants Ll. Griffiths and C. E. Clough.

2nd Battalion 23rd Royal Welsh Fusiliers.

Colonel the Hon. Savage Mostyn, and subs.

Native Houssa Artillery.

Captain Arthur Rait, R.A., Commanding; Lieutenant Knox, R.A.; Lieutenant Saunders, R.A.

Royal Engineers.

Major R. Home, Commanding; Captain R. N. Buckle; Lieutenant Bell.

Control and Transport.

Mr. Irvine, Chief of the Control Department; Lieut.-Col. Colley, 2nd Queens, in charge of Transport, Messrs. Ravenscroft, Coates, Hamilton, Felleter, 2nd West Indians, Lieutenant Graves, 18th Royal Irish, Assist.-Commissary Ward, Control Department.

Medical Department.

Surgeon-Major MacKinnon, P.M.O.; Surgeon MacNalty; Surgeon Major Jackson, 100th Regiment; Surgeon Fox; Surgeon Turton, Sanitary Inspector; Surgeon J. A. Murphy, R.E.; Surgeon S. Moore, 4th Dragoon Guards.

Royal Naval Brigade.

Commodore W. N. Hewitt, V.C., H.M.S. Active; Captain W. I. Hunt Grubbe, H.M.S. Tamar, Commanding N.B.; Commander Percy Luxmore, H.M.S. Argus, Second in Command; Lieutenant Gerard Noel, H.M.S. Active, Brigade Major; Lieutenant Edward S. Evans, Senior Lieutenant, H.M.S. Encounter, Commanding No. 4 Company; Lieutenant A. B. Pipon, H.M.S. Active, Commanding No. 1 Company; Lieutenant W. F. S. Mann, Senior Lieutenant, H.M.S. Amethyst, Adjutant; Lieutenant Angus, McLeod, H.M.S. Barracouta, in charge of Transports; Sub-Lieutenant Victor Bradshaw, H.M.S. Encounter, Sub-Lieutenant of Company; Sub-Lieutenant Charles Windham, H.M.S. Active, Sub-Lieutenant of Company; Sub-Lieutenant Gerald Maltby, H.M.S. Active, Commanding No. 3 Company; Sub-Lieutenant Tom Ficklin, H.M.S. Active, Sub-Lieutenant of Company; Sub-Lieutenant Wyatt Rawson, H.M.S. Active, Transports; Sub-Lieutenant Augustus Coker, H.M.S. Druid, Commanding No. 2 Company; Sub-Lieutenant Archibald Kennedy, H.M.S. Druid, Sub-Lieutenant of Company; Midshipman Charles Gladstone, H.M.S. Druid, Aide-de-Camp to Commanding Officer; Lieutenant Adolphus Crosbie, Royal Marine Light Infantry, H.M.S. Active, Commanding Royal Marines; Lieutenant Richard Deane, Royal Marine Light Infantry, Commanding No. 2 Company Royal Marine Light Infantry; Staff Surgeon Henry Fegan, M.D., H.M.S. Active, principal Medical Officer; Surgeon John Fisher, H.M.S. Decoy.

Special Service Officers.—Captain Hon. Paul Methuen, Scots Fusilier Guards; Captain Burnet, 15th Infantry; Lieutenant Pollock, 21st Fusiliers; Lieutenant Hare, 22nd Regiment; Captain Aldridge, 95th Regiment; Captain St. A. H. Phayre, late Inniskilling Dragoons.

The force now assembled between here and Prahsu may be roughly estimated at—

	Men.
Colonel Wood's Regiment	400
Major Russell's	400
42nd Highlanders	575
Rifle Brigade	650
23rd Fusiliers	75
Royal Naval Brigade	225
2nd West Indians	350
Royal Engineers	40
Rait's Artillery	50
Total . .	2,765

CHAPTER XIV.

ADVANCE OF THE ARMY—POSITION OF THE FOUR COLUMNS—ATTACK UPON
ADUBIASSEE—DEATH OF CAPTAIN NICOL—BATTLE OF BORBORASSI.

Quarman : January 30, 1874.

WE have tragic accounts of villages attacked and destroyed to write of now, and our interest in the campaign is daily increasing. The blood of the Britons is warming up to fighting heat, and, unless all auguries prove false, the great battle of the war will be fought to-morrow.

Our march hither from Fomannah has been through the villages of Doompoassie, Detchiasu, Akkankawassie, Insarfu, thence to Quarman.

Doompoassie was an important town about three miles from Fomannah. Detchiasu was a small village situated on the crown of a hill, with a well-watered country round about it. Akkankawassie was a large village occupying a terrace and slope of a prolonged hill, leading down to a small stream. Insarfu, a small village, occupied a clearing with very deep woods around it. Quarman is a still smaller village on the crown of a low hill from which the forest land slopes down to streams and sinuous water-courses, overhung by dense brake and bush.

While we have been agitated by doubts as to whether this campaign will furnish subjects interesting enough for historical record, and swayed by fears that the expedition is destined to terminate peacefully, the inhabitants of Coomassie, we are told, are troubled and disturbed by doubts, anxieties, and fears of other kinds.

Astonishing portentous omens have occurred at Coomassie. Braggart chieftains appear to regard the entry of the English into their capital as an impossible feat, and talk loud and windy. The queen-mother is said to have threatened to commit suicide if peace be not made immediately. The king perambulates through his capital in his usual state, and travels between Coomassie and the fetich places, the Bantammah and Manpontem, to consult his oracles on the probable issue of the war. The populace is subject to excitements and panics. Loud sounds are heard in the air, and strange noises are heard at a distance.

To add to this excitement and ferment, a heavy aerolite has fallen in the great market place. A male child on issuing from the womb is reported to have commenced to speak fluently. A house swept clean over night was found in the morning to be filled with bushes, and the wonderfully fluent child was dead. The great fetich tree of Coomassie has been destroyed, and many other things have presented such a category of omens, that the hearts of the stoutest braves seem paralysed. All these things may, or may not, be true ; busy rumour ever active, has given us the benefit of them. Virtue of fetichism seems to be on the wane ; the traditional terrors of Coomassie, the stronghold of the Ashantees, have lost their effect. Heaps of broken crockery, corn-meal flour splashed on trees, and stakes, and dwelling-houses, are of no avail ; human sacrifices, performed with all due ceremony, daunt not the palefaces ; lengthy lines of cotton thread strung across bush and branch impede not the advance ; all the howling and loud incantations of fetich priests are so much power of demented men's lungs spent in vain ; and vociferous vaunting of braggart chieftains deter not the Saxons' march. No, men with Sniders in their hands march on unrestingly !

A letter received from the king yesterday implores time to collect the indemnity, to which Sir Garnet replied that he would march on as long as the heir-apparent and queen-mother were not forthcoming as hostages, and the indemnity demanded to compensate the injuries committed upon the Fantees and Assins was not paid at once. Sir Garnet further charges the unfortunate king with having instigated the inhabitants of Borborassi to fire upon his troops, and he menaces him with his anger in consequence ; and declares that he will burn his capital to the ground if his terms are not acceded to at once.

The position of the various columns up to date are as follows :

Right column.—Captain Glover has reported up to the 17th instant. He had crossed the Prah on the 15th, with 750 Houssas and Yorubas, and had advanced to Obogo, whence he dates his despatch. He had taken this village by assault, his troops having been fired upon as they were marching towards it. He had seven Houssas and one Yoruba wounded ; the enemy's loss not being ascertained. Being in want of ammunition and supplies, he would be compelled to halt. His troops had behaved well, except in their ' wild and unnecessary firing.'

Second Column.—Captain Butler's last report is dated 24th instant. He writes from Yancoma, where he had arrived on the 22nd instant, about twenty miles north of the Prah. Hence two paths branch off, one towards Lake Boosum-Eching, the other towards Dadiasoo, in a westerly direction. He had proposed to follow the latter path, but the opposition of the kings was such that he was forced to abandon the movement. He hoped to get his whole force to move on the 25th ; but he thus describes his position—' I can scarcely convey to you a true conception of the position in which I find myself. All information is studiously withheld from me ; carriers are constantly refused ; distances are grossly exaggerated ; the most vexatious and trifling pretexts are put forward; anything and everything is done to effect the one great object,—delay.'

' *Third Main Column.*—Advanced posts are to-day on the River Dunsaboo, about two and a quarter miles beyond this place, and within three-quarters of a mile of the enemy's forces at Egginassie. The entire force is concentrated here and at Akkankawassie, two and three-quarter miles in rear, and it is probable that to-morrow we may have to fight whatever force the king has been able to collect to resist our advance. Further details of my position are given in the despatch which accompanies this.'

Fourth Column.—The last report from Captain Dalrymple is dated 25th instant from Kotakee ; he had arrived there on the 24th with King Apecoon about fifty men ; King Aquasi Bedoo was here with no men ; Kroo of Commendah had gone home ; Buarbin and Atrew were still at Damun on pretence of collecting their men.

Captain Dalrymple, though he had used every effort to influence the kings, was of opinion that he would be fortunate if he should succeed in crossing the frontier with 200 men.

The hill of Quarman slopes down to the River Dunsaboo, and at this little village are Colonel Wood's and Russell's regiments, and two rocket detachments. Headquarters, the Brigadier's staff, the 42nd Regiment, and 2nd Battalion Rifle Brigade, with Rait's Artillery, two 7-pounder rifled guns, and two rocket detachments, are at Insarfu.

Between Insarfu and Akkankawassie two and three-quarter miles in rear, are the head-quarters of the Naval Brigade, and detachments 2nd Battalion 23rd Regiment, headquarters and detachment 2nd West Indian Regiment, the small-arm ammunition reserve, and field-hospitals.

At Fomannah, to advance to-morrow is a detachment of the 1st West Indian Regiment, lately disembarked at Cape Coast Castle.

The following fortified posts have now been established between here and Prahsu :—

		Miles.
Prahsu to	Essiaman	10½
„	Accrofoomu	11
„	Cross Roads, near Parakoome River	6
„	Moinsey	5
„	Crest Adansi Hills	¾
„	Fomannah	3
„	Akkankawassie	8½
„	Insarfu	2¾
„	Quarman	2¾

Total, 50¼

The estimated distance between here and Coomassie according to guide-map, which has proved so far tolerably accurate is :—

		Miles.
Quarman to	Amoaful	3½
„	Agimamu	7
„	Ordahsu	7¾
„	Akkankawassie	1¾
„	Karsi	2¼
„	Coomassie	3

Total, 25¼

At the beginning of this letter I spoke of tragical events; these events were brought about by the keen intelligence and intrepidity of Lord Gifford. This young nobleman is not twenty-four years of age, has a fair, almost effeminate face, is a man slight of figure, and appears to be anything but robust in health. Yet his daring feats, his gallantry, have carried him over more ground afoot than any four officers have marched, and the intrepidity which has distinguished him, have been themes of praise with all on the Expedition, from the highest to the lowest.

Lord Gifford, leading his scouts westerly through the bush, came upon the king's road, which runs parallel with our road from the Adansi hills to Coomassie. Following the road, he discovered a village, inhabited, and after acquiring all the information he could of the place, the various routes that led to it from that traversed by the army, he reported his intelligence to head-quarters.

The intelligence amounted to this: that Adubiassee, a small village almost abreast of Doompoassie, was occupied by a force under the king of Adansi, and that further north, about three or four miles on the same road, was Borborassi, a more important village, and more strongly defended, situate almost abreast of Kiang Boassu, which is half-way between Doompoassie and Akkankawassie.

On the 25th instant, while at Fomannah, Sir Garnet warned the king of Adansi that he could not permit his people to assemble in the neighbourhood of his line of communication; that unless he moved away he would be compelled to attack him, a measure which was absolutely necessary.

On the 26th Major Russell's native regiment, with Lord Gifford's scouts, supported by a number of men from the Naval Brigade, were ordered to make a reconnaissance in the neighbourhood of Adubiassee, and if they found the enemy within the village, to attack it.

Major Russell approached the village from Doompoassie, and on discovering it to be inhabited by about two hundred of the enemy, made a dash upon it, and captured it without much trouble. The attack was made with spirit, and was singularly successful, without a single casualty, though the enemy opened a brisk fire as soon as the attacking party was discovered advancing.

On the 28th Lord Gifford, having approached the Ashantee village of Borborassi, and finding it defended by a strong force of the enemy, returned to Kiang Boassu, and imparted the intelligence to head-quarters. While lying in wait in the bushes between Borborassi and Becquah, he had also waylaid two men carrying with them some fowls and a keg of powder, of which articles he became possessor, after tumbling over one contumacious wretch who imagined he had a better right to the goods than Gifford with the stronger might.

While we halted the night of the 28th at Akkankawassie, an attack was planned upon Borborassi, and the charge of it was given to Lieutenant-Colonel McLeod, who acted as Briga-dier of the Advance Brigade.

The plan of attack was as follows : Colonel McLeod would start from Kiang Boassu with a detachment of the 2nd battalion 23rd Royal Welsh Fusiliers, consisting of Lieutenant-Colonel the Hon. Savage Mostyn and 79 non-commissioned officers and men.

The Naval Brigade, consisting of Captain Walter James Hunt Grubbe of H.M.S. ' Tamar ' 14 officers, 24 petty officers and 195 blue-jackets, 2 rocket detachments under Lieutenant Knox, R.A., of Rait's artillery, and one company of Annamaboes, 57 in number, under Captain James Nicol, formerly Adjutant of the Hampshire militia.

Lieutenant-Colonel the Hon. Evelyn Wood, was to start simultaneously with the advance from Kiang Boassu, with a portion of his regiment of native allies, and to penetrate to the road running between Borborassi and Becquah, along which he would deploy his men in the bush as an ambuscade to take the fugitives on the flank.

It is said that Sir Garnet gave orders to Colonel McLeod not to burn Borborassi, and he regretted that the village of Adubiassee had been destroyed. This has caused some talk among the officers, and a great many are wondering at this magnanimity in Sir Garnet. I presume that despite the presence

SETTING FIRE TO A VILLAGE.

of the large force in our front, the General still entertains a hope
that hostilities will be averted.

However, on the morning of the 29th instant the several
detachments took their respective roads, Colonel McLeod landing his party direct for Borborassi, Colonel Evelyn Wood leading
his force for the appointed ambuscade.

Within about two hours and a half, after a march through a
forest choked with a jungly under-bush, the force of Colonel
McLeod appeared before Borborassi, and as they emerged into
the clearing opened fire. The surprise was complete, and an
indiscriminate flying for shelter into the bush, and almost an
immediate desertion of the village was the consequence.

Gathering courage from the gloom and density of the bush,
the enemy, according to their usual tactics, attacked the Naval
Brigade on the left flank, but the sailors, delivering their fire
steadily into the bushes, soon overcame that force, and silenced
the fire of the Ashantees. Though a few sputtering shots were
heard at intervals, no damage was done.

The centre was carried in gallant style by the 23rd Welsh
Fusiliers; led by Colonel Mostyn, while the right flank consisted
of the Annamaboes of Russell's regiment, under Captain Nicol.
This last gentleman, while leading his men to the attack, on
seeing a group of the enemy, ran up to them and called on
them to surrender in the English language. The enemy of
course could not comprehend what the old man, who, animated
by genuine kindness of heart called out to them wanted, and a
savage deliberately shot him through the heart. The loss of
the fine old gentleman so infuriated his faithful Annamaboes,
that they opened a murderous fire on the wretches, stretching
several of them in death agonies on the ground, and pursuing
them with fatal shots far into the bush.

The village of Borborassi was soon in possession of the
English. Some fowls, a couple of goats running about the
village were especial objects of pursuit, over fifty Ashantee
guns and twelve kegs of powder were found and destroyed.
Several objects of curiosity and a small quantity of gold-dust
comprised the ' loot' the captors gained.

As the force returned from the scene of their success, the
Ashantees who had quietly concealed themselves in the bush,
rose from their coverts and attempted to fall upon the rear-

guard, but the Naval Brigade, to whom was entrusted the rear of the force coolly dropped on their knees and delivered such a scattering fire, that the enemy were but too glad to beat a precipitate retreat.   After this futile effort on the part of the enemy, the force were permitted to return in peace to Kiang Boassu, where they arrived at 4 P.M. after an absence of eight hours.

The ambuscade under Colonel Wood had no results save the killing of two Ashantees by the premature firing of the Apoboes. A party of the enemy was seen advancing along the path from Becquah, when the Annamaboes, unable to restrain themselves from firing at the unconscious and careless foe, delivered their fire with the above meagre results.

On this affair at Borborassi Colonel McLeod reported to Sir Garnet Wolseley as fellows :—

Camp Quarman, January 30, 1874.

Sir,—In compliance with the instructions received, I marched yesterday with a force to reconnoitre the village of Borborassi, situated on our left flank, and said to contain a large body of the enemy under Essamanquatia and other Ashantee chiefs.   Striking into the bush path at 8.15 A.M., we marched W.S.W., and at 11 A.M. arrived before the place, taking it quite by surprise.   The Ashantees, driven out of the village by our advance, took to the bush on all sides ; a party of them came back upon our left flank, and fired on the Naval Brigade.   This attack was brushed away with a few rounds of ammunition.   The village was immediately in our possession, but not without loss on our side.   List of casualties enclosed.

Captain Nicol, commanding the Annamaboe Company of Russell's Regiment, was killed leading his men, with the devotion of an English gentleman, round the right flank of the village.   I estimated the killed of the enemy at 50.   Fifty-three Ashantee muskets were collected. Twelve kegs of powder, and the umbrella of Essamanquatia were found in the village.   It seems this old chief ran away just before the attack, and so narrowly escaped capture.   The troops after their long march were halted in the village for one hour and refreshed themselves from their havre-sacks before commencing the return march.   On our return the blue-jackets formed the rear guard.   Before quitting the village they broke up the arms which were taken, and blew up the powder.   As soon as ever the Ashantees, lying perdu in the bush, discovered that we had gone, they returned to the village with shouting and blowing of horns.   Presently a body of them came down upon the rear guard and opened fire ; the blue-jackets calmly faced about, and poured amongst them such a fire of Snider bullets as sufficed to rid us of their presence for the rest of the day. The troops arrived at their camp at Kiang Booasu at 4 P.M. The discipline and conduct of the blue-jackets, under Captain Grubbe, R.N., who formed the advance and afterwards the rear guard, merited my special commendation. Captain Wood, A.D.C. to the Major-General, accompanied me and did good service ; after the loss of Captain Nicol, I placed him in command of the Annamaboes. I would desire to acknowledge to the Major-General the valuable services rendered by my staff officer, Captain Farquharson, V.C., in carrying out all the staff arrangements, which he did to my complete satisfaction. Lieutenant Wauchope, 42nd, was with me throughout the day as my orderly officer. My thanks are also due to Lieut.-Col. Mostyn, 23rd Royal Welsh Fusiliers, who was present with the men of his regiment engaged.—I have, &c.,

J. C. M'LEOD *Colonel,*
Commanding 42nd Royal Highlanders.

The Chief of the Staff, &c., &c., &c.

Camp, Royal Naval Brigade, Medowma, January 29, 1874.

LIST OF KILLED AND WOUNDED OF ROYAL NAVAL BRIGADE.

Killed.—One man, Fantee carrier—wound of chest. Wounded.— One petty officer, wound of head—very severe ; one petty officer, wound of chest—very severe ; one petty officer, wound of leg—slight ; one marine, burnt by explosion of gunpowder—severe.

ANNAMABOE COMPANY, MAJOR RUSSELL'S REGIMENT.

Killed.—Captain Nicol, wound of neck and chest ; one man, wound of head. Wounded.—One man, abdomen—dangerously ; one man arm—severe ; one man, back—severe.

H. FEGAN, M.D., *Staff-Surgeon, R.N.*

All the wounded are now under my care, and are comfortably provided for.

H. FEGAN, M.D.

To Brigadier-Colonel M'Leod, C.B., the Front.

The information received at headquarters to-day prepares us to meet a great struggle to-morrow. It is positively known that a numerous force is in front of us beyond Egginassie. Major Home's Engineers have prepared the road to within a couple of hundred yards or so of this village, or at least to such a close distance that they were enabled to hear the voices of the advance quartered within.

Lord Gifford, with an audacious temerity, not satisfied with ascertaining that the village of Egginassie was an outpost, made a reconnaissance to the left of the village close to the camp of the main body, and returned with the intelligence that the Ashantees are in force not more than three miles from Quarman, whence we may conclude that the decisive battle of the campaign will be fought to-morrow.

With the aid of the knowledge of the ground furnished to him by Lord Gifford Sir Garnet has been able to make a plan of attack which if successful, as no doubt it will, must lead to good results.

The very fact of the presence of this force in our front, is an evidence of the treachery of the king of Ashantee, and in divining the tactics of his wary enemy Sir Garnet has furnished further proofs, if any were wanted, of his capability to command this expedition.

The missionaries reported to us that the army of the king, disheartened at the results of their campaign beyond the Prah, had disbanded, that they saw no large bodies of the enemy as they were brought down to the Prah. This may have been true, so far as their own personal knowledge went, but it only tends to prove how successfully the king had deluded them, and how their ingenuous simplicity had contributed to conceal his designs. The skill with which Gifford, with his clever scouts, discovered them, has frustrated the king's plans, clever as they were. By to-morrow night he may know whether he has adopted the wiser course, or has more ensured the safety of his capital and the stability of his power by the system of duplicity and treachery he has pursued.

CHAPTER XV.

BEFORE THE BATTLE—ON TO THE BATTLE—THE BEGINNING OF THE END—
THE 'BLACK WATCH' IN ACTION—BATTLE OF AMOAFUL—VILLAGE OF
AMOAFUL.

February 1, 1874.

THE great battle of the campaign has been fought and won by
the English troops. The Ashantees were met and defeated
before Amoaful with immense loss, January 31. All skirmishes, and sieges hitherto fought in this land pale in interest
before this most decisive victory. The enemy's utmost strength
on their own selected position proved unable to withstand the
steady and determined advance of the British troops, and the
cunning designs he had laid, as well as the hopes he had cherished
of being able to drive back the invaders, have been frustrated
by the sagacity and bravery of his opponents.

We had retired to bed in the huts of Quarman, with a vague
foreshadowing of the great morrow, which dawned on January
31. We had conversed long together about the pros and cons
of what was fated to happen, around the cheery hospitable bonfire which Major Russell had kindled, to ward off the evils
arising from evening mist and hurtful night dew. We were
an uncommonly lively lot, being disposed to regard the event
of the following day only in the light of a precursor of a vic-
-torious and speedy return to the sea coast and home. And as
we regarded it, no doubt, many others in that little army
quartered within the huts of the Ashantees, regarded it also.
Some, perhaps devoted their time on the eve of the great battle
to writing their affecting adieus to their wives and children,
while others, perhaps ignorant of the exact character, number,
and position of the enemy, were sceptical that anything very
serious would happen.

But at about 7 A.M. as we had but barely finished our breakfasts in Quarman village, and buckled our belts on, our servants informed us that the white troops were close by. Hastening to the square, or plaza of the village, we were in time to witness the famous ' Black Watch ' come up, all primed and ready for action. This was our first view of the fighting 42nd Highlanders, and I must say I improved the occasion to get a good look at them as though I never had seen a British regiment in my life. I was so intent upon observing their march past which was done with an earnest, determined stride that promised well for their behaviour whatever might lie at the front, that I did not observe that one of my colleagues, through ignorance of the danger, no doubt, for he was a new hand in battle scenes, had foolishly followed their regiment. Gifford's Scouts were ahead of the 42nd waiting for them, before they began to move on to rouse up the enemy to action, and to develop his hiding places.

Close behind the 42nd Highlanders followed one hundred Fantee porters bearing the reserve ammunition, closely guarded by a strong detachment of soldiers scattered amongst them. Behind these went about thirty or forty hammocks each hammock borne by four men, and soldiers jealously guarding them also. Then went a detachment of the Houssa artillery, two 7-pound guns commanded by Captain Rait himself, and following these were a detachment of Royal Engineers under the immediate command of Major Home.

All the above force composed the front column, under the command of Brigadier General Sir Archibald Alison, Bart., C.B., son of the great Scotch historian of that name.

Now came the veteran Colonel J.C. McLeod, C.B., leading the left column, consisting of the right wing of the Naval Brigade, Captain Luxmoore, of H.M.S. ' Druid,' commanding. Major Baker Russell's Regiment of Native Allies, Rait's Artillery, two rocket detachments, and a detachment of Royal Engineers under Captain Buckle.

The blue-jackets still wore the careless gait which distinguished them on the road, as if they recked not what good or evil fortunes awaited them in the front, and Russell's native allies were brave in feathers, and horns, and ribbons, and had much singular garniture of the head.   The Houssas marched on with the serene appearance of veterans.

The reserve ammunition, and string of hammocks for the use of the left column followed behind carefully guarded, and watched by armed men who seemed determined that not one bearer of a timid heart should face about in a panic and demoralise his fellows.

Here comes Colonel the Hon. H. Evelyn Wood, a Victoria Cross officer, in the van of the right column which he is to command; and close to him is Captain Walter James Hunt Grubbe, of H.M.S. 'Tamar,' in command of the left wing of the Naval Brigade, Wood's Native Allies, a detachment of Rait's Houssa Artillery, two rockets, and a detachment of the Royal Engineers.

Following the right column comes Sir Garnet Joseph Wolseley, the young General who is about to manipulate this force of British troops in opposition to the Ashantee army.   He is smiling, cordial in his salutations, and in no way distrusts his fortune or his soldiers.   He is mounted on a Madeira cane chair elevated on the shoulders of four burly and semi-nude Fantees, a conspicuous object for a lurking enemy in the bush. His staff is about him, and in the beaming faces of the officers you see reflected the good hope and perfect contentment which possess their chief.

The rear column, consisting of the 2nd Battalion Rifle Brigade, is brought up by Lieut.-Col. A. F. Warren; and with the last man of the battalion has rolled by a perfectly equipped fighting force of over 2,500 men, Europeans and natives.

As the line extended itself into greater length along the road, the progress became very slow.   It wound in coils as it followed the sinuosity of the path, and uncoiled itself slowly as a straight stretch of the road permitted it.   We seemed to be marching to every alternate beat of the second hand of our watches, so slow the long procession moved on.   The road was but five feet wide, barely that, yet we were in double files.

The following represents the plan of the battle-field to which the tight little British army were now moving on, to occupy face to face with the Ashantee army.

AMOAFUL

AMANQUATIA	ASHANTEE ARMY	KING OF MAMPON

Front Column

42nd Highlanders

Naval Brigade   Rait's Artillery   Naval Brigade

STAFF

Rait's Artillery   *Village of Egginassie*   Rait's Artillery

23rd Royal Welsh Fusileers

Russell's Regiment   Russell's Regiment

Royal Engineers   Royal Engineers

Rifle Brigade

Left Column   *Right Column*

Rear Column

Road to Amoaful.

Quarman

At this slow pace we could but look listlessly on what we had often looked upon before. There were no strong contrasting colours to relieve the wearied eye. Grey coats and white sun-helmets in front of us, grey coats and white helmets in rear of us; dense dark bush to the right, dense dark bush to the left; a strip of blue-grey sky above our heads. Where is the grandeur and pomp of war here in this picture? Where the excitement and blood-heating scenes of battle? The quiet sober-coloured line has absolutely nothing in it to disquiet the glance or to attract a fixed gaze; on the contrary, the funereal movement of the procession has something somnific in its pace.

But suddenly—faint rumbling, detonating sounds pulsate

through the thick curtains of woods about us, and our eyes seek each other's faces with looks of inquiry.

'They have begun,' says one. 'Yes, the ball has opened,' answers another; and we press hard on each other's heels as though a simultaneous wish to be nearer to the sounds was transmitted through every mind.

The time of commencement was noted down, 8.5 A.M. Lord Gifford's scouts were feeling the enemy. Though the detonations were but faint they might have been counted by one interested in doing so.

By 8.15 A.M. we had either arrived much nearer the scene of contention and strife, or the firing was getting warm, for at this time we could hear long-continued rolls of musketry, and some of the staff remarked that the 42nd Highlanders must have become engaged.

We still continued to move on slowly, with the alarum of war constantly rising in volume, listening for want of something else to do—for we were imprisoned between lofty walls of vegetation—to the crackling, ripping sound of Snider rifles varied by the louder intonations of the over-loaded muskets of the Ashantees.

Five minutes thus, and nothing disturbed any man's thoughts, unless of those with responsibilities on their minds who had no time to think of the ideal, or the æsthetic aspect of the moment. An officer of the 42nd here rode up on a white mule, to communicate with Sir Garnet, who leaned eagerly forward on his chair to hear the news from the front. He was wounded, but his intelligence was to the effect that the left column had lost touch of the left wing of the 42nd Highlanders. The officer received orders to tell Colonel McLeod, who commanded, to cut diagonally to the right through the bushes until both columns should join.

At 9.15 our pace quickened somewhat, until suddenly light was exchanged for the gloom of the forest shade, and we entered the village of Egginassie, which the scouts of Lord Gifford had carried with a rush at 8.5 A.M.; and oh, what a scene! We travelling so soberly and unexcitedly through the narrow avenue leading through the forest had not thought of this— the possibility that while we listened to the rumbling and muttering of musketry, there were fellow-creatures being smitten to death in the dark recesses of the forest around Egginassie.

Right at the mouth of the umbrageous pass, the wounded, and dying, and dead, sat grouped or lay stretched on the ground, attended by the kind ministering hands of their white brothers. It may be that the readers as well as the writer of this have seen many a gory battle-field, but I must confess that I felt much affected at the sight of the first blood drawn.

The first eager enquiries as to who were wounded elicited the information that poor Captain Buckle, of the Royal Engineers, was shot through the heart and abdomen, and was dying. That three or four others were most seriously wounded, and seventeen or eighteen of the 42nd Highlanders, who were then bearing the brunt of the battle, were wounded more or less seriously.

During the interval that elapsed between 8.5 A.M., the time of the first shot, and that of our arrival within the village of Egginassie, much progress, considering the nature of the struggle, had been made by the front column, under Sir Archibald Alison; but the right column had remained almost stationary, having advanced but little. The left wing of the Naval Brigade was the only portion of the right column engaged in earnest.

The front column pushing on to occupy the village of Egginassie close after Gifford's scouts, had swept across the open ground of the clearing and deployed into position in the jungle. Reserving its fire until they encountered the enemy, the High-landers had continued advancing until they had penetrated about two hundred yards beyond the village, when the con-cealed enemy suddenly revealed himself by firing into their faces from cleverly contrived ambuscades. Henceforward the Highlanders continued to sweep the bush in front of them with steadily poured volleys, until they had silenced the enemy's fire, during which pause the Engineer labourers were pushed forward to cut the bush for a farther advance. When the labourers had succeeded in clearing a space of ground in front, the Highlanders moved forward until they discovered the enemy again. The road to Amoaful from Egginassie served as a guide to the wings spread out on each side of Rait's artillery, which continued to move down in line with the infantry. Whenever a favourable opportunity presented itself, Captain Arthur Rait with his brave Houssas sent telling shots. Thus Artillery and Highlanders slowly marched down the sloping ground, driving the foe steadily out of his numerous hiding-places which he had constructed of bush with a skill which almost defied detection by the eye.

The best means of discovering his whereabouts were found to be telling volleys from Sniders, and booming rounds from the tiny 7-pounders, which sent their shot with disastrous effect through the forest.

At the bottom of the slope ran a lazy stream, which coursed sluggishly through expanses of morass, and over depths of black slime. A hundred yards beyond this stream were seen the sylvan huts which the Ashantees had constructed out of tree boughs and plantain leaves. These huts numbered hundreds, spread out far on each side of the road.

Such was the place the Ashantees chose to defend, which they did with a pertinacity that won high praise and admiration from the Highlanders. The soldiers were put to their mettle, and the Houssas, as if catching the fierce enthusiasm which animated the Scotch Highlanders, laboured with a vigour and energy not eclipsed by any on the field. Captain Rait, halting at the same altitude above the stream below, as the Ashantee camp was on the other side of it, aimed his guns with such good effect at the huts, that on passing them, the ghastly heaps that met the sight, of rent bodies and disfigured dead bore a silent but significant testimony of the important service the Houssa artillery had contributed on this day towards crushing the pride of the enemy.

When the front column had dislodged the Ashantees from their several positions, and finally driven them with fearful loss from their camps, Sir Archibald pushed it forward; and while bagpipes blew their most strenuous notes, and the wild Highland cheers for victory pealed through the forest, the whole line surged across the stream, and swept up the opposite slope until the outskirts of Amoaful were reached.

Here Highlanders and Houssas, now animated to the highest pitch of valour, rushed forward at the top of their speed, to the entrance of the broad avenue which divides the town into two equal portions. As they appeared within the town at the foot of this avenue, they saw several excited groups of natives hurrying away from it, some bearing away wounded chiefs, others transporting their household property.

One group specially attracted the attention of Lieut. Saunders, R.A.—that of four slaves carrying on their shoulders the wounded body of their master, with two others following closely behind. Aiming a shell at them, the missile exploded but a few inches above the heads and in the centre of the group, killing every soul instantly. After a few more desultory shots the capture of the town of Amoaful was complete.

But the reader, hurried away with the front column by the narrator with such breathless impetuosity, must not forget that this is but one scene of the battle. We have noted the progress of the front column first, as until 11.30 A.M., the great interest of the battle-field concentrated on its movements, though it must not be supposed that the other columns were idle, or that we even who are with the staff, and are stationed in the very centre of the square formed by the columns, are disinterested spectators or listeners. We are with the pivot around which the sonorous sounds of the battle ring their loud echoes. We were better able to read, while grouped around head-quarters, the history of the advance of the ' Black Watch ' than we should have been were our attention diverted by immediate contact with the combatants.

At 11.30 we knew at Egginassie that the men of the glorious Scotch regiment had brushed the trucculent and contumacious foe from its front, and had elbowed him on against the right flank.

Wounded men of the 42nd ceased coming into the square of Egginassie ; they were the men of the right column that were now borne in to us. At 11.30 the right column was rudely awakened from apparent inactivity into a fierce blaze of excitement, and as the village of Egginassie was situated on the slope of the long low, forest-clad hill, the enemy when he crested it, and bore down on the right column in force, visited ourselves with a hail of slugs, which caused the trees around and the branches above us to shed their leaves over us as thick as flakes in a snow storm.

A few seconds after this, tremendous firing in our immediate vicinity began, Colonel Wood, commanding the right column, was brought in with an iron slug in his chest; then his aide-de-camp followed, disabled with a slug in his hip; then we find that in a short time fourteen blue-jackets have been assisted into the village, some of them grievously wounded.

The firing at such close quarters to us waxes terrific. The line of the fighting right column, now hotly engaged with a persisting foe, who crawls serpent-like closer and closer to them, is not fifty yards away from us, and we are plentifully touched and tapped, lightly it is true, by the hail of slugs. Men with whom I am conversing abruptly spin round as they feel the blows. Lieutenant Maurice sitting on a log, listening to the thunder of the unceasing, ear-splitting fusilade, is struck in the back. Doctor Fegan, of the 'Active,' while conversing for a moment with Commodore Hewett and myself, is violently struck on his scarf pin, and others have similar experiences to relate. Every man of the right column feels that this is a critical moment, and that he must roll back the tide of attack, or be driven himself in hot haste to infamous flight, and so he plies his faithful Snider with that nervous rapidity born of desperate necessity. Probably Sir Garnet feels that it is a critical moment also, considering that Wood's regiment of native allies only lies between his headquarters and the enemy, and he orders the 23rd Royal Welsh Fusiliers forward to the support, to advance in a north-easterly direction.

The firing rises to the deafening pitch, there is not a break or a pause in the thick volume of sound, lazy clouds of gunpowder smoke enwrap the forest tops as with a curtain. Things proceed at this rate for a short interval, wounded combatants drop in rapidly; there are about 100 wounded, dying, and dead in the village, though several of the wounded, having had their wounds dressed, have been borne to Quarman, when Sir Garnet orders up a second support of two companies of the Rifle Brigade, with emphatic orders to push on and drive the stubborn enemy from his coverts. We, waiting to hear this support of fresh men, can tell the very minute they commence firing, can mark the progress which they make through the thick jungle by the diminishing volume of the musketry, can almost reckon the rate at which they advance, and feel very much relieved when at 12.30 the wild cheers which the Apoboes utter tell us of the rapid retreat of the Ashantees.

But while we are congratulating ourselves that the important battle of Amoaful is ended—loud and continuous musketry is heard in our rear along the road to Quarman, whither the wounded have been taken for safety, and the entrance to the village from Quarman is choked by the forms of the frightened Fantee carriers. Sir Garnet thought of this possibility, and prepared his plan of battle for just such a contingency. The four companies of the Rifle Brigade, hitherto unemployed in the battle, are ordered to take the back track and defend the line of communication, and they are soon engaged with the Ashantees in vigorous earnestness, until 1.45 P.M., when a cessation of the musketry announces that the enemy, having attempted the power of the Europeans on the left, the front, the right and rear columns, is convinced that he has been defeated, and is unable to withstand the strange weapons which the white men use in war.

The battle, however, was not yet ended, for a reinforcement under the Ashantee General Essamanquatiah, who is considered the Von Moltke of the enemy, appeared to the right and rear of Quarman at about 3 P.M. Captain Burnet, of Russell's native regiment, having been left behind, severely ill of a fever, much to his annoyance, hearing the tremendous musketry in such close vicinity, rose from his sick bed, assumed command of the detachment of 33 of the 2nd West Indians, being senior officer at the post, and resisted them for several hours until the Rifles appeared from the direction of Egginassie, despite the confusion around him caused by the native carriers, whose first impulse had been to take to precipitate flight.

The provident sagacity of Sir Garnet saved him from even the shadow of a reverse even at this small village, for a wide clearing, extending a hundred yards around Quarman, and a redoubt of earth and bushes, enabled the detachment of the 2nd West to employ their Sniders to good effect.

Baffled at this point, the tireless and raging foe sought to make an impression at Insarfu, two and a half miles in rear of Quarman; but though they were engaged in it until midnight, they experienced nothing but repeated reverses.

At 3 A.M. of this morning (Feb. 1) they again attempted to assault Quarman with a diabolical vigour, which caused the harehearted Fantee carriers to rush into the jungle; but at dawn the Ashantee army retired from the long-protracted contest—a twenty hours' almost continuous battle.

The left column under Colonel McLeod, consisting of the right wing of the Naval Brigade under Captain Percy Luxmoore, Russell's Native Allies, Annamaboes, Mumfords, Cape Coast people, and Ja-Jas from the Bight of Benin, and detachments of the Houssa Artillery and Royal Engineers, were hotly engaged, as soon as they had taken their position on the left of the 42nd, for over an hour and a half. It was in this column that Captain Buckle was killed. But as they advanced into the jungle, endeavouring to keep pace with the front column, fighting their way now sharply for a breathless period, then subsiding into a few spluttering shots, they lost touch of the left wing of the 'Black Watch,' and had to cut their way diagonally by a road through the bush, arriving in the main road a little in rear of the front column.

But the historian of the battle, to be true to fact and to all scenes enacted in the stirring event, must not be beguiled from observing what transpired within the village of Egginassie, by the more absorbing scenes of the struggle.

Sir Garnet Wolseley, himself retaining the calm, proud air of the general who, wisely forecasting the issue, was prepared for the triumph, spared none of his willing and obedient staff. Captain Buller was kept trotting backwards and forwards with orders and reports until it was feared he would over-exert himself in the warm work. Captain Brackenbury was also an ardent co-operator in the designs of the general, and assisted him unremittingly by the sagacity of his suggestions, and intelligent reports of the progress of the battle.

Colonel Greaves, the chief of the staff, must of course be identified with Sir Garnet Wolseley, for on him devolved much of the responsibility and the faithful execution of his chief's plans.

Lieutenant the Honourable A. H. Wood, A.D.C., employed himself actively around headquarters, conveying orders to the front, and returning with hopeful reports, besides keeping a sharp eye upon all men emerging from the scene of the struggle.

Commodore Hewett, V.C., was beside Sir Garnet during the entire time, and throughout the various phases and fortunes was either engaged in assisting Sir Garnet with his advice, or superintending with interest the care and welfare of the wounded sailors.

And last but not least of the deserving, we must not forget the assiduous, indefatigable conduct of Surgeon Mackinnon, the principal medical officer, and Dr. Henry Fegan of the Royal Navy. Dr. Fegan simply behaved wonderfully. Soldier or sailor, white or black man, almost everyone had cause to bless the deft hand of this skilful surgeon. A dozen ordinary surgeons could not have bestowed quicker aid and relief to the wounded than this invaluable officer. It was a pleasure to watch the extraordinary devotion of the man, and those who witnessed his work will know what a first-class surgeon can do to alleviate pain and exhaustion during the exciting moments of a battle.

At 2.30 P.M. I left Egginassie, proceeding to Amoaful town, which was captured by the Highlanders at 11.30 A.M.

The road sloped down very gradually for about 500 yards to the lazy, sluggish stream coursing amid marshes and deep mire, on each side of which the heaviest fighting had taken place that morning. The entire distance was strewn with cartridge-cases, and frequently one paused to observe those which the Ashantees had used, which revealed to us the fact that among the weapons employed by the enemy were double-barrelled breech-loading central-fire guns.

Around the Ashantee camp the dead Ashantees were thickly lying, with most frightful wounds, proving that Snider rifles have terrible force and penetration. If other evidences were needed to establish this fact, the wreck of the bush, and the deep rugged rents of great trees, seen all around, did so. Those who undervalue this gun make a great mistake. Its rapidity and ease of manipulation is such that a regiment has to be kept constantly supplied from the reserve-ammunition, while the perfect construction of the Boxer cartridge and its excellent materials, leaves nothing to be desired in the shape of an efficient small arm for any service.

Proceeding up the hill, with our minds dwelling on the great tragedy of the day, we arrived within about 15 minutes at Amoaful.

The town is conveniently situated on the extensive tabular

summit of a low hill, the sides of which slope down into thin threads of ooze, and marsh, and running stream, all clothed in impenetrable underbrush and dense growths of tropical plants and trees.

A broad street or avenue about 60 yards wide running the whole length of the town, a distance of about 400 yards, divided it into equal portions.

The residences consisted of four, or six, or eight small huts, with interiors facing one another, and enclosing small courts ; but none of these came up to the neatness, taste, and architectural ornamentation of Fomannah, the capital of the Adansi kingdom.

As I entered the town the aspect of the principal avenue was a busy one. Under one great tree reclined, on benches or cots taken from the houses, the wounded officers of the 42nd Highlanders. Major Duncan Macpherson, who commanded the 42nd under Sir Archibald Alison, was rubbing his wounded leg, and doubtless thinking what a narrow escape he had from losing his poor limb. Others were detailing the scenes through which they had passed, and the exciting incidents which came under their personal observations.

Among the various remarks I heard as I passed through, were :—

' Ah, Sandy, my lad, it wor a braw fight,' with a rich accent from a Scotch Highlander.

' What a pity we didn't have our kilts and bonnets; how they (the Ashantees) would have speered, and no mistake,' from another Highlander evidently, with which I sympathised.

' By jingo, but they cut a gangway through my beard, anyhow. I say Bill, what will Sally say when she hears how they have spoiled my beauty ? ' said a frank-faced blue-jacket, running his fingers through the evidence of his manhood which garnished his chin, and which truly enough exhibited a ' gangway ' cleared to the throat by an Ashantee missile.

' Oh, hang them ! just look at my figure-head, will you ? ' said another, exhibiting a swollen black eye, and an ugly scar on his temple. ' What d'ye say to that, now ? '

Indeed there were but few men who had not a story of hurts suffered and clothes damaged by slugs.

There were a great many Ashantees killed around and in

the town of Amoaful, and the effect of Saunders's last dead shot was visible at the northern end of the town in a group of six slaves stark and stiff around their dead master, who lay still on his stretcher.   Many too, wounded at the battle, had hurried to the village, and breathed their last quietly within the huts. Yet, though the slaughter must have been fearful, we were disappointed at the results, until recollecting the invariable custom of the enemy to bear his wounded and dead away from the field, and the almost instinctive haste with which the wounded, who are able to crawl away, plunge into the dark bushes for shelter, out of fear of decapitation, with which they generally visit the wounded.[1]

An instance, only one fortunately, of this kind, occurred. A wounded Highlander on the extreme left flank of his regiment, perceiving himself to be wounded too severely to continue in line, sought to retrace his steps to the village of Egginassie to have his wound dressed.   He lost his way, apparently, and was unfortunate enough to be discovered by a party of Ashantees, who overpowered him and cut his head off.   The struggle which the doomed man made for his life was too apparent in the slashed hands and almost severed fingers.

About 5 P.M. Sir Garnet arrived at Amoaful.   The troops bivouacked on the avenue in the open air, as the baggage had been left behind at Insarfu, five miles in the rear, and a series of brilliant bonfire lights marked the positions of the bivouac during the night after the great battle.

The following is a copy of the report of the Major-General commanding, interpolated here just before the publication in book form.

Head Quarters, Amoaful, February 1.

Sir—According to the intention expressed in my despatch of the 30th ultimo, I yesterday attacked the enemy to the south of this town, and drove them with heavy loss from the position they had selected to fight upon, and where I now find that the King of Ashantee had been for some time collecting his army.

---

[1] The losses which the Ashantees suffered have been variously estimated to be between 800 and 1,200 killed, with as many more wounded.   Their entire force with which they engaged the British cannot have been less than 12,000.   The King of Mampon, who commanded the right, was wounded mortally; Amanquatia, who commanded the left, was killed ; Appia, one of the great chiefs engaged in the centre, was also killed.

My whole force, divided into four columns, as per margin, advanced along the main road, preceded by the scouts, until the enemy were met at the village of Egginassie, which was carried by a rush of the scouts at about 8 A.M.   The front column then extended into the thick bush on each side of the road, which was cut and widened by labourers under the Royal Engineers so as to admit of the advance of the guns.

As the leading column advanced northward the left column, according to orders previously issued, cut a path diagonally to the left front, with a view of protecting the left flank of the front column; and as it moved along this path, the right column, closing up, cut a path diagonally to the right to protect the right flank, while the rear column extended, so as to gain touch of the right and left columns, which were designed to follow the flanks of the front column and, should it be outflanked, to face east and west outwards.   My intention was to fight in the form of a square, and so oppose the invariable flanking tactics of the enemy, which their superior numbers would probably allow them to carry out against any line which I could form.

The front column, under Sir A. Alison, found the enemy in great force beyond a swampy stream to the north of Egginassie, and suffered heavily in dislodging them.   They were driven out by the steady advance of the infantry, aided by the fire of Rait's guns.   The large numbers of dead Ashantees at this part of the field, and the numbers of the 42nd Highlanders here wounded, showed the stubborn resistance made by the enemy.   The 42nd Highlanders finally advanced and captured the town of Amoaful about noon, after being more than four hours in action.

Meanwhile, the left column, advancing under a heavy fire, by which Captain Buckle, R.E., was killed while urging on his labourers, occupied the crest of a hill, where a clearing was made, and the enemy driven away from this portion of their camp by an advance of the Naval Brigade and Russell's Regiment.   Colonel M'Leod having cleared his front, and having lost touch of the left of the front column, now cut his way in a north-easterly direction, and came into the main road in rear of the Highlanders about the same hour that the advance occupied Amoaful.   I protected his left rear by a detachment of the Rifle Brigade.   Our left flank was now apparently clear of the enemy.

On the right, Lieut.-Col. Wood was met by a fire which prevented the advance of his column for more than a very short distance into the bush, consequently when the front column took Amoaful it would have become detached from the right column, but that communication was kept up along the main road by two companies of the 42nd, the headquarters and detachment 23rd, and a company of the Rifle Brigade. Long after Amoaful was taken, the Ashantees kept up a heavy fire on the right of the main road, and these troops lay down and replied to it, repelling the enemy, but not without loss.

Up to 1.30 P.M. the enemy kept up a very heavy fire on Lieut.-Col. Wood's column, whose right was extended into the bush, east of the village of Egginassie. But they made no progress, and soon after half-past one an advance of the Kossoes and Bonny men of Wood's regiment drove them away, clearing the flank of the Naval Brigade, and enabling them to complete the discomfiture of the enemy on this flank. By 1.45 P.M. firing had ceased.

At this time heavy firing was heard in rear, and I learnt that another body of the enemy had attacked my entrenched post at Quarman. I sent back part of the Rifle Brigade; but the attack was continued till nightfall, though of course repulsed. Shortly before dark a large convoy of baggage which had been parked at Insarfu during the action, and was now ordered on, was fired upon, though accompanied by a large escort. A number of carriers threw down their loads and ran away, and had it not been for the great exertions of Lieut.-Col. Colley, whom I have placed in charge of my line of communications, and who recovered much of the baggage during the night, more serious consequences might have ensued than the loss of the few loads which occurred. On learning of this affair I took immediate steps for clearing my line of communications, and brought in large convoys this morning in perfect safety to Amoaful. The officers commanding the columns as above-named performed their difficult tasks most excellently, and were efficiently aided by their staff. Lieut.-Col. Evelyn Wood, V.C., was wounded while at the head of his troops.

Nothing could have exceeded the admirable conduct of the 42nd Highlanders, on whom fell the hardest share of the work. As Colonel M'Leod was in command of the left column, this regiment was led by Major Macpherson, who was twice wounded. The two wings of the Naval Brigade, under command of Captain Grubbe, R.N. (severely wounded), and Commander Luxmore, R.N.; the head-quarters and detachment of the 2nd battalion 23rd Royal Welsh Fusiliers, under Lieutenant-Colonel Mostyn; the 2nd battalion Rifle Brigade, under Lieutenant-Colonel Warren; Wood's Regiment, under Captain Furse (42nd Highlanders); Russell's Regiment under Major Russell (13th Hussars), and the scouts under Lieutenant Lord Gifford (24th Regiment), all behaved steadily and coolly under the trying conditions of a fight in dense bush. Sir A. Alison has brought the valuable services of Captain Rait and his Artillery specially to my notice; and Major Home, commanding Royal Engineers, led the advanced working party of the front column, being slightly wounded in so doing.

Commodore Hewett, V.C., R.N., was present with me during the action, and placed his services at my disposal. I received every possible assistance from him and from the following officers of my staff: Colonel Greaves, Chief of the staff; Major Baker, 18th Regiment, Assistant Adjutant-General; Captain H. Brackenbury, R.A., Assistant Military Secretary; Captain Buller, 60th Rifles, D.A.Q.M.G.; Lieutenant Rolfe, R.N., Naval A.D.C.; Lieutenant the Hon. H. Wood, 10th Hussars, A.D.C.; Lieutenant Maurice, R.A., Private Secretary.

The arrangements for the wounded by Surgeon-Major M'Kinnon, C.B., Principal Medical Officer, were excellent, and I regret that the enclosed return of casualties shows that his duties were not light.

It is difficult to estimate the numbers of an enemy in thick bush, but still more difficult to estimate their loss, but the Ashantees opposed to us must have numbered many thousands; and as we have buried about 150 corpses beside the main road only, and as the enemy ran great risks to carry off their dead, their loss in killed and wounded must have been very heavy.

I intend to-day to attack Becquah, which the scouts report occupied by the enemy.

I have, &c.,

(Signed)   G. J. WOLSELEY, *Major-General.*

The Right Honourable the Secretary of State for War.

Army Headquarters, Amoaful, February 1, 1874.

Nominal return of officers, non-commissioned officers, and men killed and wounded in action at Amoaful, on January 31, 1874:

OFFICERS KILLED.

Royal Engineers.—Captain R. N. Buckle, gunshot wound through lung and heart and abdomen (penetrating).

NON-COMMISSIONED OFFICERS AND MEN KILLED.

42nd Highlanders.—Private Thos. Thompson, gunshot wound through head; Private P. Reid, gunshot wound through heart.

Wood's Regiment.—Private Salyah (Kosso), gunshot wound of brain.

OFFICERS WOUNDED.

Royal Engineers.—Major R. Home, gunshot wound of thigh—slight.

23rd Regiment.—Lieutenant G. H. Hutton, gunshot wound of right side—severe.

42nd Highlanders.—Major D. Macpherson, gunshot wound through right leg—severe; ditto right side of neck—slight; Brevet-Major W.

Baird, gunshot wound of left thigh—dangerous; ditto of right leg—severe; ditto of right forearm and back—slight; Captain E. Whitehead, gunshot wound of left thigh (contusion)—slight; Captain A. M. Creagh, gunshot wound of scalp and right side of neck (contusion)—slight; Lieutenant G. B. M. Cumberland, gunshot wound of right arm—severe; Lieutenant A. S. Stevenson, gunshot wound of neck—slight; Lieutenant W. A. Berwick, gunshot wound of right leg and ankle—severe; Lieutenant W. H. C. Mowbray, gunshot wound of left temple and contusion of right arm—slight.

79th Regiment attached to 42nd Highlanders.—Lieutenant R. C. Annesley, gunshot wound of left thigh (contusion)—slight.

2nd Battalion Rifle Brigade.—Major A. Stephens, gunshot wound of wrist—slight; Lieutenant C. D. Sharston, gunshot wound of right arm with fracture—very severe; Lieutenant W. J. Smyth, gunshot wound of thigh—slight.

Wood's Regiment.—Lieut.-Col. Evelyn Wood, V.C., 90th Regiment, gunshot wound of left side of chest—slight.

NON-COMMISSIONED OFFICERS AND MEN WOUNDED.

Royal Engineers	4
2nd Battalion 23rd Royal Welsh Fusiliers	3
42nd Highlanders	106
2nd Battalion Rifle Brigade	6
Rait's Artillery	1
Wood's Native Regiment	16
Russell's ,,	17
Lord Gifford's scouts	6
Royal Naval Brigade, Sailors and Marines	34
	193
Sailors and Marines, killed and wounded at Becquah	5
Total	198

CHAPTER XVI.

BATTLE OF BECQUAH—BURNING OF THE VILLAGE—HUMAN SACRIFICES—
CAPTAIN BUTLER—PRELIMINARIES OF A BATTLE—BATTLE OF ORDAHSU—
DASH UPON, AND CAPTURE OF COOMASSIE.

Coomassie: February 5, 1874.

ON THE 1st instant, the day after the battle of Amoaful, we received news that during the night much baggage was lost, and officers were lamenting the loss of the only change of clothes they possessed. This unlucky accident was caused by the flight of the carriers, who when attacked by the Ashantees between Insarfu and Egginassie dropped their loads and ran into the bush. Later in the day, however, much of the baggage was recovered, though many of the gentlemen were utterly stripped of every article of clothing they possessed beyond what they wore on their persons.

About 11 A.M. the force for the attack on the town of Becquah was mustered in the main street. This place was considered to be a most important one, lying as it did within such a short distance from our line of communication, and Sir Garnet determined to destroy it before moving farther north.

The force to be employed for the work of necessary destruction consisted of the Naval Brigade, a gun and a rocket trough of Rait's Artillery, Russell's Regiment, Lord Gifford's Scouts, and a small detachment of Home's Engineers, and formed an advanced guard under the command of Colonel McLeod of the 42nd Highlanders. The main body consisted of the 23rd Royal Welsh Fusiliers and five companies of the 42nd Highlanders under Sir Archibald Alison, the Brigadier-General commanding the brigade.

The advance moved on to the west of Amoaful by a road which ran in that direction from the north end of the town.

P

It was not a very great distance to Becquah—not more than a mile and a half—so that we were soon creeping quietly towards the entrance of the village.

Presently sharp firing was heard in the advance, and as company after company took up the fire, and as we began to move at a quick pace, we were in time to observe the rapidly retreating forms of the Ashantees as they fled around an angle of the main street, and darted into the bushes. At first they seemed disposed to resist vigorously after they had recovered from their surprise, but the volleys upon volleys shot into them from each company of the advance, soon silenced them, so that the main body was not employed at all. The advance continued its march through the town to its other extremity until the Colonel in command became satisfied that there was not a single armed foe remaining.

After the usual search for 'loot' had been completed, Colonel McLeod ordered the place to be set on fire, and every thatch was in a short time blazing, and the destruction of Becquah became assured.

On the return march, the Naval Brigade, which had the honour of entering Becquah first, followed Russell's Regiment, and the rear was guarded by the Fusiliers and Highlanders.

The following is a copy of the official report which Sir Archibald wrote the same day for the information of the Chief of the Staff:—

Amoaful, February 1.

Sir,—I have the honour to report to you for the information of his Excellency the Major-General commanding, that in compliance with instructions received from him, I this day proceeded in charge of the troops (Naval Brigade, detachment of Rait's Artillery, detachment of Home's Engineers, detachment 23rd Royal Welsh Fusiliers, five companies 42nd Highlanders, Russell's Regiment, Lord Gifford's Guides) to destroy the village of Becquah. The force was divided into an advanced guard and main body. The advanced guard was under the command of Colonel McLeod, C.B., 42nd Royal Highlanders, and consisted of the Naval Brigade, a gun and rocket trough, Rait's Artillery, Russell's Regiment, Lord Gifford's Guides, and a detachment of Major Home's Engineers. The main body, which I retained under my own direction, consisted of the 23rd Fusiliers and five companies 42nd Highlanders. I entrusted the entire arrangements for the attack on the village to Colonel McLeod, with the advanced guard, merely following myself with the main body in reserve. Colonel McLeod's arrangements met with my entire approbation, and the attack on and capture of the village were effected with his usual ability and vigour. He executed it entirely with his own force, and I did not require to support him with any part of the reserve. I enclose his despatch, in which the full particulars are given. When the village was fully in flames, I occupied the entrance and eastern portion of it with the detachment 23rd and three companies 42nd Highlanders. Through these I withdrew the whole of the advanced guard, except Rait's Artillery, which I retained. When they were well clear of the village I filed off 23rd, detachment Rait's Artillery, and three companies 42nd. The Ashantees in small numbers beginning to drop into the village, and threatening to harass the rear guard, I kept a company of the 42nd concealed behind some houses at an angle of the main street. When a considerable number of the enemy had entered and were crossing the street, I wheeled up the company and dispersed them with a few volleys. I then withdrew quietly from the village, and my return to camp was entirely unmolested by the enemy. My best thanks are due to Colonel McLeod, for the very able manner in which the capture and destruction of the village were executed. In addition to the officers mentioned in his despatch, I must mention the perfect satisfaction afforded to me by the way in which the dispositions for the evacuation of the village were carried out by Lieut.-Col. the Hon. S. Mostyn, commanding 23rd detachment, and Major Scott, 42nd Highlanders, who had the particular direction of the withdrawal of the rear guard. My best thanks are still further due to Captain Robinson, my brigade major, Captain Russell, my aide-de-camp, and Lieutenant Fitzgerald, my extra aide-de-camp, for the indefatigable way in which they carried out the staff duties. Colonel McLeod encloses a list of the casualties in the expedition.

I have, &c.,

(Signed)   A. ALISON,

Brigadier-General, Commanding Forces.

The Chief of the Staff, Head-quarters.

On the 2nd instant, the army, leaving the baggage behind, resumed its march towards Coomassie, preceded by Russell's Regiment and Gifford's Scouts.

We were again in the twilight shadows which the tall forest cast over the path, trampling over the relics of property which the fugitives had dropped in their hurried flight, when the screaming shells from Rait's Artillery came tearing after them. Stools, silk-cotton bolsters, native accoutrements, scores and scores of corn rations contained in neatly plaited corn leaves, bestrewed the ground everywhere; they either lay thrown on one side of the road, or we stumbled over them as they littered the centre of the path. Each bundle of Indian corn seemed sufficient to last a man ten days, and we thought that 'fortunate must be the General who can have such simple effectiveness of commissariat,' as we mentally compared the cumbersome ineffectiveness of the Control System to that of compelling each man to carry on his own person ten days' provisions for himself. How easy a thing the capture of Coomassie might have been if Sir Garnet Wolseley commanded 3,000 troops who conveyed their own provisions and ammunition!

However, it mattered not. We might take a little longer time to accomplish the purpose of the Expedition, but it might as well be done efficiently and with comfort to one's self, as to do it speedily to our sore discomfort and inconvenience. Besides, 'the race is not always with the swift,' as was proved at Amoaful.

We were sensible that in the jungle which bounded our road on each side there were a great many dead, not only by the empty litters which we saw frequently along the road as we hurried on, but by the strange peculiar odour of death which filled the air. The road was also tramped down broad and smooth by the hurried throngs who were driven from Amoaful on the fatal 31st of January.

About a couple of hours from Amoaful we came to a stream which had been the scene of a tragedy but a short time before. Just beyond the crossing there were broad patches of mire and marsh, while struggling through which the advance guard was fired on by the enemy. Major Russell instantly deployed his men in line across the road, and commenced pouring a vigorous fire ahead and into the bush on either side. Here Captain Bromhead seconded his superior's efforts in an admirable manner, and the indefatigable Gifford, whom no service daunted, distinguished himself with his Scouts, who penetrated into the bush quite fearlessly. Rait's Artillery coming up, the 'Steel Pen's' 7-pounders began to speak, and the firing with fresh accession of force began to be quite spirited, and for a while the affair promised to be a battle of some importance. But it soon became evident that the terrible defeat at Amoaful was fresh in the memories of the Ashantees, for their musketry soon slackened, and they were flying from before the rapidly advancing men of Russell's Regiment.

As soon as it was discovered that the flight of the enemy was begun, the impetuous Russell urged his men in the pursuit, and onward they strode after the fugitives with long strides, until they reached the village of Agimamu.

The main body did not participate in this action. The road, though clear for them, was marked with the dead fallen in the affair of the morning; and each village had its human sacrifice lying in the middle of the path, for the purpose of affrighting the conquerors. The sacrifice was of either sex, sometimes a young man, sometimes a woman. The head, severed from the body, was turned to meet the advancing army, the body was evenly laid out with the feet towards Coomassie. This laying out in this manner meant no doubt, ' *Regard this face, white men, ye whose feet are hurrying on to our capital, and learn the fate awaiting you.*'

We arrived at Agimamu about noon of the 2nd, and on the morning of the 3rd pressed onward towards the Ordah River.

Between Agimamu and Edunkoo, one mile beyond, the forest was intersected by many paths leading from the main road to the short road. This last had been abandoned by Sir Garnet, who had chosen the longer and more roundabout road we were now pursuing.

Half a mile beyond Edunkoo was another village, called Sakrabah, consisting of two or three huts, and a mile farther we came to Adwabin. Between Adwabin and the Ordah River, distant two and a half miles, the advance was again resisted from behind a series of ambuscades, and after such a hot style that out of the regiment under Russell over thirty were wounded and three were killed; seven out of the detachment of eleven of Rait's Artillery were wounded, and the officers had exceedingly narrow escapes. But the passage through the woods to the Ordah was soon forced, and the river gained by the advance about 10 A.M., whereupon Home's Engineers went gallantly to work to construct a bridge across.

When the main column arrived at the Ordah, we heard the news of the fourth day's fight, and the report of a large force being at Ordahsu, a village situate a mile and a half beyond the northern bank of the river.

Between Sakrabah and Adwabin Sir Garnet had received another letter from the king, imploring him to halt, and so give him time to collect the indemnity, promising him also that the heir apparent and the queen mother should also be given up. To which Sir Garnet replied that he should halt only when the money and the hostages demanded were in his power, as substantial guarantees of his good faith. If they were not in his camp by morning of the next day, Sir Garnet would march upon Coomassie, and burn it to the ground.

We were again then on the eve of a great fight. Events were so crowding themselves into these later days of our march, that we had barely leisure to think and reflect upon the probable issue of the campaign. Every man congratulated himself, as each day shortened the distance to Coomassie, that there were fewer days for him to remain in the country.

Most of my colleagues, with myself, joined Russell's advance, which was bivouacked on the northern bank of the river, whence there was but about a mile and a half of forest intervening between us and 10,000 or 12,000 of the enemy. The main column, under Sir Garnet, remained on the southern bank.

News came this evening from Captain Butler in command of the Akim force, that he was almost ready to join the main column with 1,400 Akims, which was most agreeable news after the discouraging rumour afloat at Fomannah.

But while we were being gratified with the encouraging report, Captain Butler had written a letter the night previous from Tribee, a place to the right, nearly due east of Amoaful, to the following effect :—

Tribee: February 2, 1874.

Sir,—After the despatch of my letter of the 29th and 30th ultimo, a rapid retreat commenced in the whole Akim force. So universal was the panic that during the greater portion of the night bodies of men continued to move south, and early on the morning of the 31st Yancoma was reached and left behind. No attempt was made on the part of the kings to excuse or explain their gross treachery, and I could elicit nothing more tangible than that they were too far advanced into Ashantee territory, and that they must be cut off before assistance could reach them.

My belief is that the statements of a blind Ashantee prisoner taken at Nansuah, coupled with a verbal message sent by King Attah of East Akim, the purport of which I was not made aware of, did much to induce the disgraceful flight.

In the hasty postscript to my letter of the 30th ultimo I had only time to make you acquainted with the fact of the retreat.

No previous warning—no intimation whatever—had been given to me that it was the intention of the kings and chiefs to abandon Akina; indeed, up to the very moment that the retreat began I was moving through the camps, endeavouring to get the Akims to move to Dadiasso *en route* to Amoaful, and I had actually engaged a party of hunters to proceed through the bush to the main line near the latter place. Yet, so complete was the flight that within two hours not a man was left in Akina, and long before midnight every Akim was beyond the Ennoon River, ten miles distant. All this without a reverse having taken place, and after a loss of only two killed and two wounded in the entire force.

I cannot express to you the feelings which which I beheld this total break-up of the movement. Little by little, step by step, I had succeeded in leading these savages to within a very trifling distance of the point aimed at.

It is true that there were many indications of how utterly untrustworthy was the composition of the force under my command; but, nevertheless, at Akina I naturally looked at the task as all but accomplished, and my junction with the main body at Amoaful, within a few miles of Coomassie, as only a question of some hours. Encumbered with baggage, without carriers, save those given me by the kings, and with three of my officers sick from actual fever or its recent presence, I dared not wait at Akina the chance of communication from the main body. I had to abandon the fruits of no little labour at the moment success seemed most assured.

The kings, with about a quarter of the original force, are now encamped close to the Prah at Embronen; they have refused to move to Prahsu. The remainder of the men are scattered or dispersed to their homes.

I will proceed to head-quarters with all speed, and afterwards be ready to undertake such punishment as the Major-General may determine to inflict upon the Akim chiefs.

Captain Paget has had a relapse of fever. He proceeds with carriers to Prahsu.—I have, &c.,

(Signed)   W. F. BUTLER,
Captain. W. F. B.

To the Chief of the Staff.

The above requires no comment, it speaks for itself, and concludes the career of poor Captain Butler on the Gold Coast. All his months of labour and energy were ended in one night, by the total desertion of his entire force, leaving him to return to Amoaful with the escort of policemen that had accompanied him to Akim.

We received also news of the attack made upon Fomannah on February 2, by the same force which, probably, baffled at Quarman and Insarfu, had struck across the road from the right side to that on which Becquah, and Borborassi on the left, stands, and following it south, had made an attempt upon the fortified station on left flank and rear. Some women about to proceed to the stream to the left of the village, discovered an unusual number of men in the bushes, and hastened back at once to give alarm. A few minutes after they made their presence known with a tremendous fire of musketry, and several of their number succeeded in penetrating within a few yards of the hospital (the king's palace). During the beginning of the attack the place was defended by Captain A. W. Duncan, R.A., who contrived to muster some 80 men, several of whom were on the sick-list. Subsequently Colonel Colley (2nd Queen's) took command, who ultimately drove the Ashantees out of the post with great loss.

The loss on the British side was small. Captains Duncan (R.A.) and Dudley North (47th Regiment) were wounded, the latter officer seriously. Three of the soldiers of the 42nd Highlanders, 1 of the Rifles, and 4 men of the 2nd West India Regiment were wounded.

The night before the entry into Coomassie was spent on the Ordah most uncomfortably. It began to rain about dusk, and continued until men, wearied with the day's fatigues, dropped off to sleep despite the rain and discomfort. The discomfort may be imagined when it will be remembered that every man had, according to orders, left his blankets, rugs, changes of flannel, &c., in the rear stations of Amoaful and Agimamu.

The morning of the day, the night of which saw us assembled in the main street of the Ashantee capital, the ever-memorable February 4, was long dawning. It seemed as if the day sympathised with the waning of the once terrible power of the Ashantees, and lagged behind in its course to delay what had now become inevitable, viz., the total rout of the last army of King Coffee, and the capture of Coomassie. If this false and treacherous king had been spending the night imploring the fetiches of the aulic Bantammah to spoil the powder of the white men, to drown them with rain, he must have been specially pleased to observe how faithfully the royal fetiches laboured for him, and as he looked at the clouds on the morning of the 4th it may be he felt still more encouraged by observing the deep gloom of the sky. It is very probable, since there is no end to the superstitious crotchets of an Ashantee king.

Perhaps also the long period of daylight that elapsed—nearly two hours, before a move was made—contributed to impress King Coffee still more with the power of the royal fetiches. The bridge across the Ordah, some fifty yards in length, was having the finishing touches put to it by Home's Engineers, but at 7 A.M. the bridge was completed, and Russell's Regiment moved on, preceded on this day by the Rifle Brigade, who were to have the honourable duty of first trying the mettle of the Ashantees, as the 'Black-Watch' had done at Amoaful. The Rifles were preceded only by the Scouts of Gifford, whose good fortune it seemed to be to have the honour of initiating every battle since the army crossed the Prah.

The advanced guard, consisting of Gifford's Scouts, the Rifle Brigade, Russell's Regiment, and Rait's Artillery, had barely begun to leave the banks of the Ordah when the sharp eyes of Gifford's men discovered the Ashantee scouts about five hundred yards in front of the village of Ordahsu. The first shot on this morning was fired at 7.40 A.M. Henceforward during our march with the main body, we heard nothing but the rising or subsiding volleys of musketry as fired by the combatants. Now and then a more than usually loud volley of the enemy's guns was heard, which was answered immediately by repeated volleys from the Sniders. And while this loud music continued we were marching on with the main body, with only a little gap separating us from the advance.

As each advancing company drew nearer the scene of the combat, it became engaged in it. Throwing themselves down in the road, the men, as they received orders to fire, simply took care to keep their guns level with the ground about a foot above it, and thought more of firing with nervous rapidity than of taking aim at human objects.   I opine that such was the case with the Ashantees, otherwise they may well be pitied by those who know what a deadly scathing storm of bullets swept just over the ground to 100, 200, and 300 yards distance.   This kind of fighting may perhaps be not the most scientific, but it certainly was the only one possible under the circumstances, and the waste of ammunition was amply recompensed by the number of lives saved.

Even Sir Garnet was heard to say, when a remark was made to him about the fearful waste of ammunition, ' It does not matter, let them fire it away; it will prevent the Ashantees from doing any injury while it lasts.'

Such was precisely the fact.   The British soldiers had been severely punished by the enemy at Amoaful, because their training and discipline restrained them from firing until they were fired at.   Hence the casualties which the 42nd High-landers suffered.   The ' Fire low—fire slow,' though reading well in the general orders of Sir Garnet, did not appear to be the best rule that ought to have been made for the guidance of the troops in Ashantee.   With all due deference to Sir Garnet, a better rule would have been, ' A fast fire, and a fast advance.' For the Ashantees in presence of an immovable force seem to be able to stick to the earth they lie on, and fire in that posi-tion as well as most men.   But a rapid advance, and a rapid fire, simply discomposes and destroys them.

The village of Ordahsu was but a mile and a quarter from the Ordah River; it yet took us until 9.30 to get within 600 yards of it.   About this time the main body halted, and we sat down in the road to listen to the awful sound of battle, the ripping, tearing sound of hundreds of fast-firing Sniders, the sonorous volleys of the Ashantee guns, and the harsh barking of Rait's artillery.

While we were seated, listening to these battle sounds, some nervous man ahead of the 42nd fancied he saw an Ashantee stealing southward through the woods, as if he were heading a force for attacking our flanks, and fired, which startling the next to him, made him nervous, and he fired, and in a second almost the alarming firing ran from end to end, before one could make up his mind whether the enemy was attacking the 42nd, or whether the 42nd had cause for firing. For a second this rapid gush of fire, without warning, from a line extending over a mile of road, was one of the most alarming things that happened during that day, Sir Garnet rising, however, in a calm manner, ordered the 42nd to cease firing; and after a second or two, the gallant regiment had become as submissive and as cool as ever, and were the first to laugh at the idea that the 'Black Watch' could have been by any possibility startled from its usual enduring self-control and steadiness.

Sir Garnet at this time received the news that the Rifle Brigade had carried the village of Ordahsu, and his quick mind at once conceived a bold design, viz., that of rushing his baggage immediately after the Rifles.

Colonel Colley, Captain Baker, the Inspector of Police, and Mr. Irvine, chief of the Control, were requested to send the baggage up; the main body of the column parted on each side of the road, and in a few seconds the head of the long baggage train came up at a fast pace, every officer lending his voice to cheer the Fantee porters on.

Within fifteen minutes after the design had been conceived the last porter had passed by Sir Garnet, and the main body of the column closed up its lines again fast and sure. Sir Garnet followed the porters, the main body following him, and thus an impenetrable obstacle to any panic flight of the porters was raised. Ten minutes later we were among the dying and dead Ashantees, some fifty or so slain at the mouth of the village; but porters and main body passed over these into the village of Ordahsu.

Another such scene as that which first presented itself as we entered Egginassie during the early part of the battle of Amoaful, appeared here, viz., the death of Lieutenant Eyre (90th Light Infantry), son of Sir William Eyre. It was a sad event, for young Eyre had made himself universally beloved. The word 'Mother!' was the last sound heard from his lips as he breathed his last, surrounded by groups of sorrowing friends, even while the battle roared its discordant symphony around them.

For two hours after entering the village of Ordahsu, we listened to the battle sounds, which had now become monotonous.

Towards the north of the village there was a dead silence in the woods, but on each flank and rear the musketry continued to bellow without intermission. For the Ashantees, according as Sir Garnet had divined, after being driven from the front or north of the village had surged on the right flank, and then precipitated their numbers on the road, expectant of plunder. Enraged at their ill-luck, they swung round both flanks of their enemy, united, and together they attacked the rear; but the Naval Brigade, nearly two hundred strong, under Captain Grubbe, had added their number to Wood's and Russell's native allies,who were coolly marking their victims as they came to view.

It should be mentioned here that Colonel Wood, wounded by a slug in the chest at Amoaful, and almost invalided on account of it, had implored Sir Garnet to be permitted to join again, which request was granted. Hence it happened that Colonel Wood's regiment was in the action at Ordahsu, and Lieutenant Eyre, who served in the native regiment, was killed.

Sir Garnet now did that which he ought to have done before, but which done even at noon, half an hour before the battle ended, shows better than anything, in my opinion, the audacity of his character, and the quick intelligence of the active and capable General who is prompt to conceive and ready to execute.

He ordered up the 42nd Highlanders, and gave orders to Colonel McLeod, to carry the positions in front, and march straight into Coomassie. Captain Rait's artillery was to cover the attack.

No man is more cool than Colonel McLeod in action. He drew up his men in double file from one end of the village to the other. The famous ' Black Watch ' appeared, though greatly reduced in numbers,[1] to be fit followers of their Colonel. Both Colonel and soldiers mutually understood one another. There was no doubt or hesitancy in either commanding officer or men.

During the brief halt, Colonel McLeod surveyed his men, and then said, ' The 42nd will fire volleys by companies, according to order. Forward! '

Then began the sublime march to Coomassie, the most

[1] 340, on this day.

gallant conduct, and most impressive action of the Ashantee campaign. It was on the 'fire fast' and 'advance fast' principle.

The Highlanders marched out of the village, from the garish sunlight of the open, into the gloomy chasm of the forest, by a road, beset by ambuscades, with a proud military bearing, full of determination and a joyous courage.

Soon after they advanced into that fearful gaping pass in the forest, the enemy opened on them from his coverts. Colonel McLeod shouted out clear and loud,

'Company A, front rank fire to the right, rear rank fire to the left. Forward!'

The companies fired in succession, according to order, front ranks firing to the right, rear ranks firing to the left, and halting not even to deliver their volleys, marched past the ambuscades, the bagpipes playing, and the wild Highland cheers echoing as loud as the musketry; Captain Rait, with his hard-striking artillery hurling his shot and rockets to the right and left of the enemy.

This was a new game of war which the white man inaugurated in Ashantee, and which the Ashantees did not understand. It was out of all precedent. The custom used to be to lie down and adhere to the earth, and fire away for hours until one party or the other expended all his ammunition, or got tired of the tedium of this kind of fighting to try in another part of the field. But this marching past ambuscades with salutes of bullets they did not understand, they became anxious and then panic-stricken, and within half-an-hour after the Highlanders had departed, the impression that something unusual had happened in the front seemed to have been transmitted throughout the ranks of the enemy on all sides.

A loud blowing of horns heard on our right, and in our rear, seemed to announce, 'To your tents, Oh Ashantees! Coomassie is fallen, the battle is lost;' and subsiding notes heard at a distance, sounded like a wail of despair, as the gentle breeze bore them through the forest to our ears.

As the last sounds of the battle of Ordahsu died away, the last clod of earth was thrown over the remains of the young and gallant gentleman who had fallen in the early part of it.

Then the Native Regiments were ordered to advance, the

porters carrying the reserve ammunition and medical stores
were driven after them; the Staff followed, then came the Rifles,
and finally the Naval Brigade.   It was in this manner we
followed the road which the 42nd Highlanders had cleared for
us but half an hour before. A few of the results of their volleys
we saw in dead men lying across or on the side of the road.   We
saw one man who had evidently taken shelter behind a thick
cotton-wood buttress.   A Snider bullet with great penetrative
force had gone clean through the five inches of cotton-wood,
and had slain the man behind it.

A mile from Ordahsu we passed through the village of
Ensiayu, a small cluster of huts nestling among plantain fronds.
Three quarters of a mile beyond Ensiayu we came to Akkan-
wassi, which, during the battle just ended, had been honoured
with the presence of the king, who, seated on a golden stool,
under a temporary shelter made of plantain leaves, was commanding his army at this distance from it.

Early in the morning he had taken his seat here, vowing
that he would behead the first chief who from cowardice deserted
his post.   To the last, apparently, this king of an unconquered
people entertained the belief, that at the eleventh hour
there would be some special intervention of a fetich deity in his
behalf, or his army would drive the insolent white men back,
with woful results to all prisoners he could lay his hands on.

The chiefs seem to have cheered him up to the last.   Their
chairs were all around the shelter under which the king had
been seated.   But the firing had got nearer to His Royal Self,
a thing not on the programme at all; strange sounds of
music, and ringing cheers of advancing white men were heard;
and the strangest thing of all, despite all predictions of Fetich
Ulemas to the contrary, a sacrilegious Snider bullet sang past
his ears; thereupon the poor deluded king was fain to have
himself borne away by fear-inspired slaves, with a volley of
bullets screaming behind him.

The king fled ! the last hope of His Majesty had died away !

But the Highlanders continued their march past Karsi,
meeting with flags of truce on the way—Sir Archibald Alison
on his white mule, and Colonel McLeod on foot, leading them
across the fetid deathly swamp which insulates Coomassie.
While crossing this place, Sir Archibald's mule stumbled, and

the gallant Brigadier fell under the animal into the nauseous liquid, which reeked with human putrefaction. As Sir Archibald has but one arm, it would have been difficult for him to have extricated himself from his dangerous position, had not his Brigade Major, Captain Robinson, immediately relieved him.

In the meantime the main body, consisting of the 23rd Royal Welsh Fusiliers, Wood's and Russell's regiments, Rifles and Naval Brigade, were toiling on hard and fast after the 42nd.

On the road Sir Garnet received a cheery despatch from Sir Archibald Alison to this effect:—

'We have taken all the villages but the last before entering Coomassie. The enemy is flying panic-stricken before us. Support me with half the Rifles, and I enter Coomassie to-night.'

Then came by two different flags of truce, letters from Dawson the missionary, begging in piteous terms for delay for the sake of his life and that of his fellow-captives. A bearer of one of them informed us that the king had left Coomassie the night before for the battle-field, and had not returned to his capital since, but he and his army were known to be in full flight towards Amineeha, a country residence of His Majesty.

Still marching past, we came through and passed by Karsi, and at 6 P.M., the Staff and head of the main body of the column had crossed the swamp and entered a long broad avenue flanked on each side by pretentious-looking edifices of porticoed and alcoved houses.

We were at last in Coomassie!

As we arrived at the market-place, we saw hundreds of wondering Ashantees with weapons in their hands, regarding us most curiously. It was attempted at first to disarm them, but the General, doubtless thinking that at this late hour of the day it was bad policy to begin hostilities, ordered them to be treated kindly and to be left alone.

Turning to the left when we arrived at the market-place, we saw another wide and noble street half a mile long, where the 42nd Highlanders were drawn quietly in line, awaiting the arrival of the General.

As the General arrived in front of them, the Highlanders

uttered their victorious cheers; and soon every straggler and new arrival of the main body caught up the hearty cries, and announced to those far behind not yet arrived in the capital as well as to the wondering citizens regarding us, and the advancing fugitives from the battle-field, the certain FALL OF COOMASSIE, the dread capital of the Ashantee kingdom.

CHAPTER XVII.

February 14, 1874.

THE march of the British Army from Cape Coast to Coomassie culminated in the capture of the enemy's capital on the 4th instant, and in the destruction of it by fire on the 6th. The same day that it was destroyed by fire the army began its homeward march, and that night was encamped at Agimamu, 13¾ miles distant from it.

My colleague, Mr. Winwood Reade, the Special Correspondent of the London 'Times,' and myself arrived at Cape Coast Castle on the 12th from Coomassie. Mr. George A. Henty, of the 'Standard,' Mr. Melton Prior, the Special Artist of the 'Illustrated London News,' and several officers, arrived to-day from the front, whence the reader may infer that the late Ashantee Expedition is virtually at an end.

There remains only to chronicle what occurred within Coomassie after the entry of the army, to describe the city during the time it was occupied by the troops, and its destruction on the 6th, before rendering our verdict upon the Campaign.

After the triumphant cheering which greeted the arrival of Sir Garnet Wolseley within Coomassie, the soldiers were dismissed from line to seek their quarters along the main street. The authorities seemed to content themselves with solely defending the broad avenue along which the troops were quartered.

While we sought our quarters weary with the long march and the fatigue consequent upon the high excitement so long maintained throughout the long day, we met several hundreds

Q

of fugitives coming up by another road from the battle-field around Ordahsu. These men had double-barrelled guns, some excellent Enfields, appearing almost new, while the majority were armed with the long-barrelled Danish guns with flint-locks. One or two individual officers, moved by indignation at sight of these bold knaves appearing in the camp, made an effort to disarm a few of them; but the sturdy resistance they made compelled them to desist, lest at such a late hour blood-shed might be provoked, and a general massacre of them take place.

Had Sir Garnet, however, taken the precaution to set a cordon of guards around the city at the end of each street, with strict orders to allow the entry of every Ashantee who desired to come in, but on no account to permit one man to depart with his weapons, Sir Garnet would have enhanced his success without doubt.

But the young General was ignorant of the nature of the African, and retained his faith in the king's good intentions to the last.

The following letter will serve to show the nature of the communication he made to the king soon after our arrival in Coomassie :—

Coomassie, February 4, 1874.

King,—You have deceived me, but I have kept my promise to you.

I am in Coomassie, and my only wish is to make a lasting peace with you. I have shown you the power of England, and now I will be merciful.

As you do not wish to give up your mother and Prince Mensah, send me some other hostages of rank, and I will make peace with you to-morrow on the terms originally agreed upon.

If either your Majesty, or your royal mother, or Prince Mensah, will come to see me to-morrow morning early, I will treat you with all the honour due to your royal dignity, and allow you to return in safety. You can trust my word.—I am, &c.,

(Signed)　　G. J. WOLSELEY,
Major-General and Administrator of the Gold Coast.

To his Majesty Coffee Calcali, King of Ashantee, Coomassie.

The above letter was sent by the ambassadors who came from the king to hear what the General proposed to do now that he was captor of his capital.

We were preparing to retire to our well-earned couches for the night, which were made under the porticoes or in the alcoves of the rich men's houses which lined the street on either side, when we were alarmed by seeing a great blaze of fire in the neighbourhood of the king's palace. Soon after this another fire started to the right of it, and still another to the left, and before we had barely time to reflect upon its origin, two more fires started on the southern side of the town. Looking towards head-quarters to see whether they were concerned in the matter, we found all the members of the Staff unconcernedly looking on, except Captain Buller of the Intelligence Department, who had volunteered to go and find out what it meant. Captain Buller was presently seen returning from the scene of one of the conflagrations and heard to remark quietly, that he thought he would ' turn in.'

This unconcern of the Staff caused people to think that these conflagrations were of no moment, and resolve to think no more of the matter, but to obtain the rest which we all needed.

But about midnight even the soundest sleepers were awakened by continued cries of some one, who was either suffering extreme torture, or who was undergoing severe punishment. Those who, attracted by the despairing screams, hurried to the scene, were witnesses of a most fearful tragedy. A poor boy, about seventeen or eighteen years old, lately enlisted in the Fantee Police, was being hanged by the neck in the most clumsy and barbarous manner, by the Military Commandant of Coomassie. It seems that at 12 P.M. this lad was relieved from guard,[1] and following other examples of plundering, had gone to seek such plunder as he could find. Soon after entering a house he was discovered with a cloth in his hand by the patrol, and, taken before the Commandant with the evidence of his guilt on his person, was ordered by that stern official to be hanged then and there to a tree about sixty yards from head-quarters on the main street. Those deputed to be his executioners flung a rope about his neck, and began to hoist him up to the branch of the tree above his head. By the omission to tie his hands before hoisting him up, the young

[1] The author's authority for the above is Captain Baker, Inspector-General of Police. The scene itself was witnessed by hundreds.

Q 2

policeman was able to protract his agonies by his hands, and to harrow every man's soul with the cries with which he startled us all from sleep. Half-strangled, he was lowered again, and his executioners, after a fearful scene, were able finally to tie his arms behind his back ; but the bunglers, instead of quickly hoisting him up to end his sufferings, passed the rope from his throat to his mouth to gag him, and by this singular mode protracted the hateful scene fifteen minutes longer.

After this example, which was done to terrify the plunderers, persons far more guilty than this unfortunate were let off with the comparatively light punishment of flogging. So ended that strange night in Coomassie.

For a moment it appeared to the narrator that the frenzy of severe and capital punishments was being communicated to the army, which had hitherto travelled to Coomassie without a blot on its proud escutcheon, as alarming rumours of other capital punishments were being promulgated. These, however, fortunately were not true.

In the morning the evidence of the night's fearful work was still hanging to the fatal tree, and when the native allies and carriers of Russell's and Wood's regiments were about departing on their homeward march, several hundreds of them were marched by this scene, that the memory of the hateful sight might cling to them, and so save them from a like fate.

If it were supposed that this human sacrifice on our side would have the effect of proving to the Ashantees how sincere the British were in their assurances of friendship and esteem for them, Sir Garnet must have been sadly disappointed when he woke that morning, and perceived that out of the crowds of Ashantees, to the number of four or five thousand armed and unarmed men, that were in the city the night previous, there were but thirty or forty men left. They had departed, with their guns on their shoulders, and their treasures on their persons, with their slaves, overnight, and we were left to deplore the faith which had induced the General to trust them and treat them with such consideration.

Had he never relaxed the vigour which characterised him at Amoaful, at Ordahsu, and on that rapid dash to Coomassie, his success would have entitled him to still higher glory as a diplomatic General during most critical periods.

It was tact that was now needed ; a patient calm temper, a perspicuous reason, and a strong determination to act unflinchingly on every occasion that would contribute to exhibit the strength and power of England, and impress on the mind of the cunning king that all art was in vain, and only his sincerity would obtain mercy for him.

Not one weapon should have left Coomassie, nor even one sword, or sheath-knife, nor one ounce of powder, nor anything of the value of a farthing or half a farthing, until the king had submitted himself wholly to the mercy of the conqueror.   Then clemency would have become a virtue, and the king's mind would have become convinced that the English desired only friendship with him.

Through Sir Garnet's omission to guard the capital for the prevention of such a wholesale withdrawal of arms and ammunition, the king was persuaded to believe, as he saw his summer palace of Amineeha surrounded by his still numerous adherents, and his force increased every moment by fugitives fresh from Coomassie, that his state was not hopeless, that he had still sufficient people with him perhaps to create a new kingdom.[1]

---

[1] At the banquet given to Sir Garnet Wolseley and his officers in the Egyptian Hall, Mansion House, London, the General said:—

"The military world has learnt many military lessons in recent years, but the most valuable to us as a nation that has been taught us by the Abyssinian and Ashantee wars is that when you have to appoint an English General to command any military undertaking, it is necessary to trust him ; to supply him with all he asks for; and above all things, to avoid the error committed during the New Zealand war of severing the military command from the diplomacy necessarily connected with the operations. (Hear.) I have no hesitation in saying that had my operations been encumbered by the presence with me of a Civil Governor, or of an Ambassador authorized to give me orders, I do not think I should ever have reached Coomassie." This raises at once the question of what has been contemptuously termed, 'making war by contract,' a fruitful source of recent discussion in the newspapers.

What Sir Garnet states about the bad policy of placing a civilian over a General in actual warfare is true.   The New Zealand war proved it.   Yet I have never seen an occasion which more required the presence of some civilian of the Diplomatic Department to manage the political questions, and to conclude gracefully a campaign, than the occupation of Coomassie by the British.

The military part of the campaign was performed exceedingly well, but the attempt which Sir Garnet and his Military Secretary made to play the Diplomats was a total failure ; and to cover the various gross blunders committed Sir Garnet was compelled to abandon the idea of burning the Bantamma, which was more important than Coomassie, to content himself with destroying the capital, and to

It is certain that much valuable plunder was taken from Coomassie during the night. Assistant-Commissary Ravenscroft, while looking for gin to supply the troops with 'grog,' stumbled upon a house whose courtyard was crowded with objects of value. In the morning, when he visited the house again, he found the whole had been carried away.

We were curious to see the capital, with its most remarkable objects—the king's palace, the place of execution, the great market, squares, &c., &c., as the various reports of its wealth and magnificence had considerably whetted the appetite to put them to the test of personal observation.

In size Coomassie came up to the standard I had formed of it. The streets were numerous, some half-a-dozen were broad and uniform. The main avenue, on which the troops had bivouacked during the night, was about seventy yards wide, and here and there along its length a great patriarchal tree spread wide its branches.

The houses in the principal streets were formed after the style of those at Fomannah. They were wattled structures, with alcoves, and stuccoed façades, embellished with Mauresque patterns. Behind each of the pretentious buildings which fronted the streets were grouped the huts of the domestics, enclosing small courtyards. From one courtyard might be seen a small alley leading to another, where the store-rooms and closets were located. By the general order and neatness of the arrangements in the rear, I am compelled to say that in their domestic life they appear to me to be a very cleanly people.

Coomassie is said to be over three and a half miles in circumference. It is such a 'city of magnificent distances' that one could well believe it.

It stands on a low rocky eminence, consisting of iron-stone, the greatest breadth of which may be half a mile. It extends across this hill, descends into a valley, and occupies a portion

---

hurry back to the Prah. Never was a brilliant success marred so completely by want of tact. and a little patience, as was that of the Ashantee campaign. Yet the presence of a Civil Commissioner, which Sir Garnet thinks unnecessary in war, was the only thing needed to have made the Ashantee campaign an unprecedented success.

Sir Garnet amply supplied the Von Moltke of the war in his own person, but the Bismarck was absent, to our sorrow.

of a neighbouring eminence. The greatest length of the town may be about a mile and a half.

Each house fronting on the great streets had its lower part, as high as the floor of the elevated alcoves, painted an ochrish-red, the upper part was coloured white. In the courts, also, the houses were decorated in the same manner.

Passing down the main street, which was littered with drums large and small—from the great *kinkassi*, which sounded the alarm of battle or the death-minute of a condemned person, to the little drum, the plaything of children during the evening dance—we came to the grove, whence the terrible effluvia issued which caused all men in Coomassie to describe the place as a vast charnel-house.

The grove, which was but a continuation of the tall forest we had travelled through, penetrated as far as the great market-place. A narrow footpath led into this grove, and now the foul smells became so suffocating that we were glad to produce our handkerchiefs to prevent the intolerable and almost palpable odour from mounting into the brain and overpowering us.

After some thirty paces we arrived before the dreadful scene, but it was almost impossible to stop longer than to take a general view of the great Golgotha. We saw some thirty or forty decapitated bodies in the last stages of corruption, and countless skulls which lay piled in heaps and scattered over a wide extent. The stoutest heart and the most stoical mind might have been appalled.

Bowdich, in 1817, speaks of this death-grove, and we heard along the road from Cape Coast hither strange reports of it. M. Bonat says he has seen some two or three hundred slaves slain at one time, as customary after the death of the king's sister, and during his long residence as a prisoner in Coomassie he has frequently seen as many as a dozen slaves executed in the most barbarous manner and dragged to the grove, now filled with the relics of the butcheries which have been going on year after year. If it is true that about a thousand slaves, offenders, rebels, and others are executed annually, we may form an approximate idea of the number of victims which have contributed to swell the terrible death-roll of the Coomassie Golgotha since the time of Sy Tutu, the founder of the present dynasty, established his capital here in the middle of the

eighteenth century.   At the rate of a thousand victims a year,
it would be no exaggeration to say that over 120,000 people
must have been slain for 'custom' since Ashantee became a
kingdom !

From the Golgotha we proceeded to the king's palace.   It
was situated about 300 yards off, and occupied a level area in
the valley or depression dividing the two eminences on which
Coomassie is situated.   The first view of what was designated
as the 'palace' was a number of houses with steep thatched
roofs, clustered together, and fenced around with split bamboo
stakes, occupying an area 400 or 500 feet square, at one corner
of which rose a square two-storied stone building.   In appearance it was like any of those merchants' residences which we
saw at Cape Coast Castle.

The first court we entered was similar, though larger, to
the court of the King of Adansi's palace at Fomannah, already
described.   The lower part of the lofty walls of stucco was
painted red, the upper part white.   The designs, diamond-
shaped, scroll-work done in alto-relievo, were bolder than
anything of the kind yet seen.   The columns were square, with
simple pediments and capitals.   The alcoves were spacious,
probably 14 feet long by 8 feet deep.   Other courts were after
the same style as this, but the columns were circular and
smooth in some, while others were curved.

A tame crested crane was the only inhabitant of the courts :
he trumpeted his assumption of regal authority in the absence
of his master in clear musical sounds, as he stalked about rest-
lessly from place to place.

The alcoves were littered with curious articles.   One contained a large number of war drums, stained with blood, and
decorated with ghastly trophies of wars and triumphs, with
human skulls.   Another contained a number of cutlasses, rusty
sabres without scabbards, accoutrements plated with gold, old
worn-out guns with bands of silver or gold, horse tails, and
wisps of elephant tails, a great number of ivory war-horns, each
with its human jaw-bone ; and in another we beheld any number
of tall umbrellas—of silk, satin, velvet, crimson damask, and
woollen cloth, bespangled with bits of gold and silver, or fringed
with small gold, silver, and brass bells.

On proceeding finally to the stone structure, which is the

king's private residence, we found the interior court and the rooms opening upon it filled with curious but intrinsically valueless articles, while the upper story contained much valuable plunder. The following hastily written inventory taken on the spot will serve better than any description to convey information as to the nature of these articles :—

A breakfast and dinner service of silver, with English cutlery.
Bohemian vases.
Large glass goblets.
Strings of the valuable Aggry beads.
Gold nugget, and bead bracelets and necklaces.
Piles of faded Kidderminster carpets.
Persian rugs.
Fetiches and charms covered with leopard skins.
Leopard skins.
European regimental drums.
Swords, European and native.
Yataghans and scimetars of Arabic make.
Gold and silver-headed canes.
Regalia staffs, gold topped.
Royal stools, beautifully carved and ornamented with gold and silver.
Seven gold masks, each weighing several ounces.
Golden toys.
Damask bed-curtains and counterpanes.
Silken and cotton cloths.
Enormous silken umbrellas.
African teak boxes and trunks.
English engravings.
Glass bottles.
An oil painting of a gentleman.
Silver tankards and cups.
Skull-caps of leather and velvet.
Leathern shakoes.
An old uniform of a West Indian soldier.
Gold decorated muskets.
A sword—a gift from Queen Victoria.
Several knives, with bits of gold on hafts.
Sandals, gold plated.
A pistol or two.
Brass blunderbusses.
Porcelain and china ware.

Common European delf ware.
Prints from illustrated newspapers.
A copy of the London ' Times,' October 17, 1843.
A copy of ' Bristol Courier,' same year.

And so on, *ad infinitum*, of valuable, curious, and worthless things heaped together in every room. An old curiosity-shop could not exhibit a more miscellaneous variety of things than the interior of the king's bed-chambers, his private apartments and store-rooms, contained.

From the flat roof of the palace, which is surrounded by battlements, a very extensive view of the city and the surrounding forest and swamp was obtained.

During the day the prize agents went into the palace and principal chiefs' houses, and proceeded to pack up the most costly things they could discover to sell, for the benefit of the army after arrival at Cape Coast.[1]

Had Sir Garnet Wolseley planted a cordon of guards around Coomassie when he first arrived, and ordered every person desirous of leaving the city to be searched, he might have been able to have secured much wealth of gold dust and valuable plunder. As he neglected to do so, it is unnecessary to say that the troops will suffer in the loss of prize-money.

Beyond the king's palace and the Golgotha nothing else of interest remained to be seen.

Some of the king's messengers, after bringing to Sir Garnet news that the king intended to come to him to sign a treaty of peace, were discovered later in the day giving out arms and ammunition to their fellows to carry away. When asked what they were doing, they replied, in a matter-of-course way, ' We are taking our things away to the king.'

They were arrested on the spot, and all of any consequence found promenading through the town were ordered to be taken up and confined. A confidential messenger of Sir Garnet's informed him that the king evidently did not mean to approach him, as he was even then marching towards the north. Only now, at this late hour, seemed to dawn on the General's mind that the king had never entertained the least intention to make

---

[1] The trophies taken at Coomassie were subsequently sold at Cape Coast, and netted about 5,000*l.*

NAVAL BRIGADE MEN BREAKFASTING IN THE COURTYARD OF AN ASHANTEE HOUSE.

peace with him, only now it seemed clear to him that the destruction of Coomassie was a thing that must happen.

Sir Garnet's letter to the Secretary of State for the Colonies, obtained subsequently, may be introduced here, that the reader may understand his views, according to his own explanation of them :—

Camp Agimamu, February 7.

My Lord,—I have the honour to report that immediately after the despatch from this station of my last letter, on the 3rd instant, I forwarded a flying column, taking with me five days' provisions, and marched upon Coomassie. The enemy's resistance on the 3rd was considerable, very large numbers being in our front. I halted on the river Ordah.

During the course of the day's march I received from the king a letter, of which I enclose a copy, with a letter from Mr. Dawson, also enclosed, and returned at once the enclosed reply.

On the 4th I advanced again at daybreak. The enemy had occupied a position of considerable strength near the village of Ordahsu. This they held against us from about seven till a quarter to two, when they yielded, and, on my advance-guard, under Colonel M'Leod, being pushed on against them, broke and fled in all directions, leaving behind, strewed along the road, the chiefs' umbrellas and chairs, &c., and the food which had been carried with the army.

In the course of this day's march I received from Mr. Dawson two letters, of which I enclose copies. I ordered that the troops should be pushed on at once. No opposition was offered to our entry into the town.

We occupied the town. Mr. Dawson was met, at full liberty, walking in the streets.

I immediately issued stringent orders for the protection of the inhabitants and the safety of the town. But night fell almost immediately after our entry, and in the darkness it was impossible to prevent some of the numerous camp followers from pillaging. The Fantee prisoners had also been released, and, in all probability, were largely engaged in the same pursuit. The result was the outbreak of many fires. Captain Baker, Inspector-General of the Police, and several officers, were engaged nearly all night in the suppression of the pillaging and in putting out the fires. One policeman taken in the act was hung.

I endeavoured immediately on my arrival to communicate with the king through Mr. Dawson, and through every channel that appeared to offer an opportunity. A chief having come into Coomassie, who was said to be sent by the king, I saw him myself, and impressed upon him my wish to spare the town, and my desire to impose on the king no severer conditions than those he had already accepted.

Moreover, I told this man that, now that I had shown the power of England, I was ready, if the king would make peace at once, to accept a small indemnity, and not to exact the half I had previously required to be paid in ready money.[1]

Other messengers were obtained, who undertook to reach the king. To these I entrusted the letter dated Coomassie, February 5, 1874.

All was, however, of no avail.

The men whom I endeavoured to employ as messengers, and who came avowedly as envoys of the king, were found treacherously removing powder and gold dust from the houses.

The whole scheme of Ashantee politics is so based upon treachery that the king does not either understand any other form of negotiation or believe it possible that others can have honest intentions.

Under these circumstances, my Lord, it became clear that a treaty would be as valueless to us as it was difficult to obtain.

Nothing remained but to leave such a mark of our power to punish as should deter from future aggression a nation whom treaties do not bind.

I had done all I could to avoid the necessity, but it was forced upon me. I gave orders for the destruction of the palace and the burning of the city.[2] I had at one time also contemplated the destruction of the Bantamah, where the sacred ashes of former kings are entombed, but this would have involved a delay of some hours. Very heavy rain had fallen.[3] I feared that the streams might have risen in my rear sufficiently to seriously delay my march.[4] I considered it better, therefore, not to risk further the health of the troops, the wet weather having already threatened seriously to affect it.

The demolition of the palace was complete. From all that I can gather I believe that the result will be such a diminution in the *prestige* and military power of the Ashantee monarch as may result in the break-up of the kingdom altogether. This I had been anxious to avoid, because it seems impossible to foresee what power can take this nation's place among the feeble races of this coast. I certainly believe that your lordship may be well convinced that no more utterly atro-

---

[1] This is vacillation, and contributed (in a small way, it is true) to the non-success of Sir Garnet's attempt at diplomacy.

[2] The Bantammah was but 1½ miles from the capital, to which one long street leads.

[3] This rain was but a freak in the weather. Sartorius' experience subsequent to our departure proves this.

[4] The real cause of our panic-like departure from Coomassie, was want of food, which the General's inattention to the Transport and Control Department caused.

cious Government than that which has thus, perhaps, fallen ever existed on the face of the earth. Their capital was a charnel-house; their religion a combination of cruelty and treachery; their policy the natural outcome of their religion. I cannot think that, whatever may be the final fate of the people of this country, the absolute annihilation of such a rule, should it occur, would be a subject for unmixed regret.

In any case, my Lord, I believe that the main object of my expedition has been perfectly secured. The territories of the Gold Coast will not again be troubled by the warlike ambition of this restless power. I may add that the flag of England from this moment will be received throughout Western Africa with respectful awe—a treatment which has been of late years by no means its invariable fate among the savage tribes of this region.

The troops are now on the march homewards, and will embark for England immediately on reaching Cape Coast.

As the mission entrusted to me has now, I may say, been brought to an end, I avail myself of this opportunity for bringing to your lordship's most favourable notice the valuable services rendered to me throughout the war by Lieutenant Maurice, R.A., my private secretary. I would also take the liberty of mentioning the name of Captain Lanyon, 2nd West India Regiment, who for some time performed the duties of Colonial Secretary, and whose knowledge of colonial duties and regulations were of great service to me since my arrival in this country.

The police duties in connection with the recent military operations have been most effectively performed by Captain Baker, Inspector-general of Police. He has rendered the force under my command most valuable service, and his zeal and energy mark him out as peculiarly suited for the post he occupies.

My aide-de-camp, Lieutenant the Honourable H. L. Wood, 10th Hussars, who is the bearer of this despatch, I have the honour to recommend to your lordship's favourable notice.—I have, &c.,

(Signed)   G. J. WOLSELEY,
Major-General and Administrator, Gold Coast.

After the arrest and imprisonment of the chief who pretended to be the king's messenger, Sir Garnet's first duty should have been to have sent the 42nd or the Rifles to occupy the Bantamah—the semi-palace and treasure-house of the King of Ashantee. The distance from the capital was but a mile and a half, and a broad road led direct to it. Until the morning of his departure from Coomassie the troops might have remained in possession, that the least chance of an amicable settlement should not be said to have been thrown away. But the entire day—the only day remaining to him in Coomassie, since he had determined to leave next morning for the sea-coast—was spent in listlessly wandering about the capital, or gazing with a vague curiosity at the miserable remnant of the plunder that had been gathered, and I felt myself sympathising strongly with Sir Garnet as he still hoped against hope for the arrival of the king.

In the afternoon there came a heavy downpour of rain, which converted the streets into broad streams; and this was another piece of ill-luck, both as regarded our comfort and the meditated destruction of the city by fire the following morning.

Night came, bringing more rain with it—a sleety, dropping rain, which lasted the whole night, to our sore discomfort.

In the morning, the Royal Engineers, who had been working hard during the early part of it in mining the stone palace of the King, began their work of destruction by setting fire to the western end of the city, and as they passed from house to house with their fiery torches, the rear-guard, composed of the 42nd Highlanders, moved slowly away before them. At 9.30 A.M. the Highlanders were drawn up close to the swamp along the avenue they had first entered on the evening of the 4th inst., and soon after an explosion was heard, quickly followed by two others, and the king's stone palace, blown up with 125 kegs of powder discovered in the capital, had become an unsightly ruin.

The army experienced the usual discomforts of travelling in Africa after a heavy rain. The streams were flooded : tiny rivulets, that we lightly leapt over as we advanced towards Coomassie, had now become almost unfordable. The road had become a series of puddles, broad expanses of swift running water. The Ordah ford had become unfordable ; only the strong bridge constructed by Home's Engineers yet stood, of which, by careful management, we were able to avail ourselves. It was night when we came to Agimamu, the village at the forks of the Coomassie road where we had halted on the afternoon of the 3rd, but the rear guard did not arrive until the next day.

From the 7th inst. the troops continued their return march to Amoaful, thence to Detchiasu. From the latter place the General's following despatches to the Secretary of State for the Colonies relate what happened.

Sir G. Wolseley also writes to the Earl of Kimberley on the same date :—

Head-quarters, Camp, Fomannah, February 13, 1874.

My Lord,—I have the honour to report that, on the 9th instant, at Detchiasu, I received from the king of Ashantee a fresh messenger, who assured me that the king was most anxious now to accede to all my terms and to make peace. I told him at once that I was quite as desirous of peace myself, and that as the Ashantee kingdom had been already so severely punished, I would waive the question of the hostages, who were no longer necessary, and if the King would send to me 5,000 ozs. of gold-dust as an earnest of his sincerity and as a first instalment of the indemnity, I should be willing to arrange the terms of a treaty of peace with any envoy of sufficient authority whom the king would send to Fommanah, where I would wait till the evening of the 12th, to allow sufficient time for him to communicate with the king and collect the money. Last night I was informed that envoys from the king had arrived at a point about two miles from my outposts, and that they desired permission to treat with me, saying, at the same time, that they would not have come had they not fulfilled all the conditions I had demanded. This morning I received the messengers here, but found that they had only brought with them 1,000 ounces of gold. They declared, however, that the king could not at the moment produce more—that 1,000 ounces was more than had ever been demanded of them before, Governor M'Clean having only required them to deposit 600 ounces as a temporary security, to be subsequently given back to them. I thought it exceedingly probable that additional pressure might have induced them to produce the larger sum ; [1] but I considered that the main point was to obtain the treaty of peace, and that the money being important chiefly as a proof of complete submission, the quantity actually now paid was a matter of comparatively secondary importance. I therefore allowed the envoys to go into the question of the treaty, the nature of which was carefully explained to

---

[1] Historians must bear in mind the above frank admission. It is a pity that Sir Garnet did not think so before commencing his return march ; a greater pity that some kind friend, even though he were a Civil Commissioner, was not near him, to whisper ' patience !'

them. They made objection to two clauses only. First, they professed not to have understood that the sum of money demanded was so large as 50,000 ounces. When, however, I informed them that the king had already expressly agreed to this sum, they withdrew this objection. I fancy that it is very doubtful whether the whole of the money will ever be obtained by Her Majesty's Government; but as the payment of a few thousand pounds cannot be a matter of relatively so great importance as the maintenance of peace, I have caused the wording of this clause to be carefully so framed as to make it clear that the money is only to be paid in such instalments, and at such times, as Her Majesty may direct. The whole question of the money will thus be open for solution in any way Her Majesty's Government may think fit. The second point to which objection was raised is one which has caused me much more anxiety. As I have already explained in my despatch of this same date, I have not felt myself justified in refusing absolutely to allow the King of Adansi and his people, on their earnest petition, to join with the Wassaw people and occupy their lands together. I considered, therefore, that it was indispensable, in order to avoid future trouble, that the King of Ashantee should distinctly acknowledge the independence of the Adansi people. I carefully explained to the envoys, at the same time, that I had taken no steps whatever myself in this matter, that the action on the part of the Adansi people was purely voluntary, and that it was one with which it was impossible for me to interfere. It has been agreed that within a fortnight the king shall send to Cape Coast the treaty signed by himself. I incline to believe that the danger of breaking up, to which his kingdom is exposed, unless we are are on friendly terms with him, will induce him to fulfil, with as much punctuality as he is capable of, the conditions of the treaty. For not only have the Adansi people begged to be allowed to migrate, but I have heard through the King of Adansi that the much more powerful Becquah tribe is quite as anxious to do the same, and only awaits my decision on the case of the King of Adansi, before declaring its intention to join some tribe nearer the coast. These and other indications that only very slight action on our part is required to cause the complete dissolution of his kingdom, will, I believe, be for the fulfilment of the engagements he has now entered into, better securities than any others we could have.

I have, &c.
(Signed)   G. J. WOLSELEY,
Major-General and Administrator, Gold Coast.

Head-quarters, Camp, Fomannah, February 13, 1874.

My Lord,—I have the honour to report that the King of Adansi has applied to me for permission to transfer his whole tribe to the Wassaw country, the Wassaws being willing to receive him and all his people. I told him that there was not land for him in the place he asked first to be allowed to enter, I assured him that I could give him no protection, that he must fight his own battles, and that I could guarantee nothing as to what course Her Majesty might think fit to pursue in the future. But when he declared that he feared the vengeance of the King of Ashantee if he remained here after he had negotiated with me, and when he declared that he and all his people detested the Ashantee rule, and were anxious to escape anywhere from it, I did not think I should be justified in interfering with an arrangement which will, at all events, have the advantage that it will to some further extent render the Ashantees less able to invade the Fantee territory. I feared lest, if I refused, I should be simply handing over the Adansi people to massacre. I have therefore allowed the King to migrate, as he proposes, but on the express understanding that it is an arrangement between himself and the Wassaw people, with which Her Majesty does not interfere. The whole subject has caused me very great embarrassment. I have been influenced in the decision I have arrived at by the consideration that the Adansi people, who formerly formed a part of the same tribe as the Wassaws, are now only imitating the example which was set them some years ago by that portion which now passes under the Wassaw name; and, moreover, before the question had to be decided by me the Adansi and Wassaw people had arranged the matter together, and I should have had to exercise an authority I did not feel myself altogether entitled to employ in forbidding a tribe always so vaguely under our influence as the Wassaws to enter into such an arrangement, calculated as it is to contribute to their national safety. Moreover, at the time the application was first made to me, the King of Ashantee had not given any trustworthy indications of a wish for peace. To refuse, therefore, an important accession of strength would have been, on military grounds, an action scarcely defensible. I very much regret that the point should have arisen at all, because it seems to me surrounded with difficulty. But under the conditions of time and place at which I had to decide, I do not think I could have arrived at any other solution. I hear that the powerful Becquah tribe is anxious to follow the example of the Adansi; but I intend, now that the preliminary negotiations as to the treaty have been completed,

R

to evade, at least for a time, their embarrassing questions by returning to the coast.—I have, &c.,

(Signed)   G. J. Wolseley,

Major-General and Administrator, Gold Coast.

The following is the Treaty of Peace between Sir Garnet Wolseley, acting on behalf of her Majesty, and Saibee Enquie, acting on behalf of his Majesty Coffee Calcali, King of Ashantee :—

Treaty of Peace.

Art. 1. There shall be hereafter perpetual peace between the Queen of England and her allies on the coast on the one part, and the King of Ashantee and all his people on the other part.

Art. 2. The King of Ashantee promises to pay the sum of 50,000 ounces of approved gold as indemnity for the expenses he has occasioned to her Majesty the Queen of England by the late war; and undertakes to pay 1,000 ounces of gold forthwith, and the remainder by such instalments as her Majesty's Government may from time to time demand.

Art. 3. The King of Ashantee, on the part of himself and his successors, renounces all right or title to any tribute or homage from the Kings of Denkera, Assin, Akim, Adansi, and the other allies of her Majesty's formerly subject to the kingdom of Ashantee.

Art. 4. The King, on the part of himself and of his heirs and successors, does hereby further renounce for ever all pretensions of supremacy over Elmina, or over any of the tribes formerly connected with the Dutch Government, and to any tribute or homage from such tribes, as well as to any payment or acknowledgment of any kind by the British Government in respect of Elmina or any other of the British forts and possessions on the coast.

Art. 5. The King will at once withdraw all his troops from Appolonia and its vicinity, and from the neighbourhood of Dixcove, Secondee, and the adjoining coast line.

Art. 6. There shall be freedom of trade between Ashantee and Her Majesty's forts on the coast, all persons being at liberty to carry their merchandise from the coast to Coomassie, or from that place to any of Her Majesty's possessions on the coast.

Art. 7. The King of Ashantee guarantees that the roads from Coomassie to the River Prah shall always be kept open and free from bush to a width of 15 feet.

Art. 8. As Her Majesty's subjects and the people of Ashantee are

BURNING OF COOMASSIE.

From a drawing by Melton Prior, special artist in Ashantee of the ' Illustrated London News.'

henceforth to be friends for ever, the King, in order to prove the sincerity of his friendship for Queen Victoria, promises to use his best endeavours to check the practice of human sacrifice, with a view to hereafter putting an end to it altogether, as the practice is repugnant to the feelings of all Christian nations.

Art. 9. One copy of this treaty shall be signed by the King of Ashantee, and sent to the administrator of Her Majesty's Government at Cape Coast Castle within fourteen days from this date.

Art. 10. This treaty shall be known as the Treaty of Fommanah.

Dated at Fommanah this 13th day of February, 1874.

At Agimamu Sir Garnet met the unfortunate officer Captain Butler, who had been deserted by the Akims, after weeks of ceaseless labour and trouble he had taken to raise a force of them. It is only fair to Captain Butler to permit Sir Garnet to be heard in his defence, as well as in that of Captain Dalrymple, who equally failed to accomplished anything with the Wassaws on the left flank.

Head-quarters, Camp, Agimamu, February 7, 1874.

My Lord,—I have the honour to enclose a copy of a report just received from Captain Butler. Your lordship will perceive that it represents the complete collapse of the expedition under his orders.

From Captain Glover I have received no report since that of the 17th from Opogo, the contents of which I forwarded to your lordship from Fomannah, and which represented him as short of ammunition and awaiting its arrival, after the success which placed him in possession of the town.

From Captain Dalrymple I have received no official report, but I hear that the chiefs under his orders have almost entirely deserted him.

Since the above was written, Captain Butler himself has arrived. I cannot express too strongly my sense of the efforts which he and the officers under his orders have made for the cause in hand. In so far as failure represents an additional trouble, which has to be undergone by a man of energy and resource, and an additional hardship from which he suffers, Captain Butler has had all the bitterness of it. He has failed to induce the miserable chiefs of this district to do all that he, in his fine belief in native character,[1] fancied he would be able to

---

[1] Sir Garnet very frankly declares the cause of Butler's failure. Captain Butler is a gentleman of such large nobility of heart, that he believes that kindness will enable a man to mould as he pleases an African savage. Over-leniency is as bad as harshness, indiscriminate kindness is fatal to discipline. If Captain Butler, with

induce them to do.   But so far as the interests of the expedition under
my orders are concerned, Captain Butler has not failed, but most successfully achieved the very object which I had in view in detaching him
for the work he so cheerfully and skilfully undertook.   He has effected
a most important diversion in favour of the main body, and has detained before him all the forces of one of the most powerful Ashantee
chiefs.

For any purpose for which energy, practical ability, high spirit,
and much knowledge of character are required, I cannot too highly
commend Captain Butler to your lordship's attention.   All these
qualities have been, under the most trying circumstances, displayed by
him upon the present occasion.[1]

In speaking of Captain Dalrymple, I must point out that his position has been in some respects a far more difficult one than even
Captain Butler's.   Captain Butler is not only a man known to the
world at large, but I have on a previous occasion had opportunities
myself of recognising his ability.   He was known personally to all
those officers who were with me in Canada, and of whom many are
here.   Captain Dalrymple came as a perfect stranger to most of us.
The difference is very considerable in the situation of two men, each
in positions of somewhat painful isolation.   Under these most difficult
circumstances Captain Dalrymple has worked with a heartiness and
energy which I cannot too highly commend.

Captain Glover's position before the world is so conspicuous, and
your lordship's knowledge of his ability is so complete, that I scarcely
think it would be suitable for me to report upon him as if he were one
of the officers in any ordinary sense under my orders.   In fact, since
his military operations have been carried out under my orders, I have
received but one report from him, and know nothing of what has been
the history of his force.—I have, &c.,

(Signed)   G. J. WOLSELEY,
Major-General and Administrator, Gold Coast.

By February 22 the troops had nearly all arrived at Cape

whom all sympathized in his misfortune, had been stern but just, he would have
succeeded. 'Always spare the whip, but firmly grasp the reins.'

[1] It must strike the reader, with what exceeding warmth Sir Garnet speaks of
his personal friends. Of his military and private secretaries—the former of whom,
as Sir Garnet said at the Mansion House, is about to write the Official History of
the Ashantee Campaign—the General speaks unqualifiedly.   Of Captain Butler,
who, though he possesses many virtues and many social qualities, Sir Garnet
speaks as though he had made a brilliant success.   But of Captain Glover he has
scarcely a civil word.   Undoubtedly Sir Garnet makes it clear that he is a man of
strong friendships.

Coast, and were embarked as fast as they arrived at that unhealthy seaport, on board the splendid steamers the munificent
British Government, who had watched over them with paternal
care throughout the campaign, had provided for them. The
'Himalaya' troopship conveyed away to Gibraltar and England
the Rifle Brigade ; the 'Tamar' was assigned for the use of
the 23rd Royal Welsh Fusiliers, and the 'Sarmatian' for the redoubtable 'Black Watch,' who had so much increased their fame
in Ashantee land, while Sir Garnet Wolseley and staff left
the Gold Coast by the 'Manitoban.' In due time the entire
force arrived in England, where soldiers and officers were received with honours, and distinctions showered upon them by
Her Majesty the Queen, and a grateful nation.

CHAPTER XVIII.

CONCLUSION.—GLOVER'S EXPEDITION.

READERS desirous of understanding thoroughly what Captain Glover accomplished, before Sir Garnet Wolseley received his despatches announcing his march to, and occupation of Jwabin, the second city in the Ashantee kingdom, might read with advantage Chaps. IV. and VI. of this book, wherein the author describes a visit he made to his camps on the Volta in the month of December. Two weeks after I left him (December 28) he writes :—

December 28, 1873.

Yesterday I was with a force of 12,000 men; to-day I am marching on the Prah, under orders from Sir Garnet Wolseley.

I had just effected crossing the Volta when this order arrived. I wanted a fortnight more to have brought him a force of 16,000 men, but they stipulated to fight the East Bank people first, and this was the programme with which I left England.

This little brief note will be understood also better after reading Chaps. VII. and VIII., wherein Captain Glover is referred to as being ordered to leave the Awoonahs unmolested, and to lead his expedition to the Prah, that he may be ready to cross the river by January 15, simultaneously with Sir Garnet Wolseley's force, Captain Butler's, and Captain Dalrymple's.

In the following briefer note than the preceding, this man of action and energy describes how he fulfilled the peremptory commands of his superior :—

In Eastern Akim, January 6.

Three days from the Prah, which I shall cross on the 15th, I left Mr. Goldsworthy with 17,000 across the Volta.

Captain Glover subsequently writes from Abogoo, on

January 22, by which we are informed that he had crossed the Prah, and was 16 miles north of it, on that date. Sir Garnet at the same date was at Moinsee, 32 miles from the Prah, with the Naval Brigade.

Abogoo, January 22.

We took this place with a rush, and are now waiting for ammunition to come up.

We are 16 miles north of the Prah, and 25 from Coomassie. I shall move the day after to-morrow on Conuam, five hours from this, and shall then hope to open communication direct with the General. I have only 50 spare rounds per man, instead of 250.

On January 21 he had written from Abogoo the following letter to Sir Garnet Wolseley :—

Abogoo, January 21, 1874.

Sir,—I have the honour to acknowledge the receipt of your Excellency's despatches, dated December 31, 1873, January 8, January 13, and your Excellency's communication of the 14th instant.

2. My entire force of Houssas and Yorubas, which are capable of being brought to the front, are now up as per state enclosed, and Mr. Ponsonby, R.N., joined head-quarters on 18th, and Dr. Rowe on the afternoon of 19th instant.

3. Yesterday came up 23 boxes of Snider and 20 boxes of Enfield ammunition, 2 cases of rockets, and 1 trough.

4. This afternoon came in other loads, completing spare ammunition, 458 spare rounds Snider for 350 men, and 55 spare rounds Enfield for 500 men, one 7-pounder gun, and 15 charges for ditto, 4 troughs, and 72 rockets.

5. To-morrow I am expecting 2 boxes containing fuses, friction-tubes, slow-match, &c., 4 cases of shot for 7-pounder guns, 16 cases of shell (common), 5 cases of double shell, 17-pounder gun, and 7 cases of charges for 7-pounders.

6. I regret to have to report that I received yesterday the news of the death of Dr. Paule, R.N., who died at Assoom of fever and jaundice, followed by dysentery, on the 19th instant. The loss of this officer is very much regretted by the officers of this expedition, as well as by myself. His kindness and genial disposition had endeared him to all, and since the 27th ultimo, owing to the illness of Lieutenant Cameron, 19th Regiment, he had been in command of an advanced detachment of force until it reached Assoom on the 10th instant.

7. On the 19th instant came up the chief of Assoom with 120 men. Yesterday the King of Eastern Akim came with 300 more, and to-day some 50 others have arrived

8. On the 17th a broad road leading northwards was reconnoitred for three miles by Lieutenant Barnard, 19th Regiment, with 100 men, but was found to end in farms. On the 18th the same officer and force examined road to Conomo for three or four miles without seeing any-one. Reconnaissances were pushed out five or six miles towards the villages of Bangsu and Jo-Ashi. On left flank and left rear nothing was seen, remainder of force being occupied in unroofing and loopholing houses and cutting patrol paths through forest around village.

9. To-day Captain Larcom, R.N., and Lieutenant Barnard, 19th Regiment, with 100 men, examined country ten miles on left rear, burning two deserted villages, and King of East Akim drove out the enemy from Bangsu, a village eleven miles on left flank, which he burned, having four men wounded. To-morrow Lieutenant Barnard, in command of 160 Houssas and Yorubas, with Mr. Ponsonby, in charge of rocket party and 300 Akims, will occupy Bargens, and day after to-morrow move upon Jaashu for the purpose of dislodging the enemy from that place, and opening communication with Captain Butler, whose force cannot be more than four hours to the westward, in neighbourhood of Atcheyasswah.

10. Jaashi, like this place, is five hours distant from Conomo, which will be occupied by force not later than 25th or 26th, and is three from Coomassie or Daubin. I calculate approximately the position of this place to be in latitude 6 deg. 30 min. N., and 57 deg. 30 min. W.—I have, &c.

(Signed)   JOHN H. GLOVER,
Special Commissioner.

His Excellency Major-General Sir G. Wolseley.

Three days before the battle of Amoaful, he writes to Sir Garnet as follows:—

January 28, 1874.

Sir,—Since date of my last despatch to your Excellency of the 21st instant, Lieutenant Barnard, having with him Mr. Ponsonby, R.N., who was in charge of rocket troughs, and 160 Houssas and Yorubas, with 200 Akims, occupied Bangsu, and the next day drove out the enemy from Jaashu, ten miles south-west of this place; casualties, nil. 25th, 200 Akims more despatched to attack a small village to left rear were repulsed; loss of 2 killed and 4 wounded, 20 Akims only being en-gaged, the rest retreating, Akims reporting that from loudness of the guns they believed the village to be occupied by Cobrah Fuah's army (Western Akim), and requested from Lieutenant Barnard some Houssas; but fearing an attack from this place, which was known to be occupied by the enemy, Lieutenant Barnard sent only a bugler and four men with a red English flag. The men reported that the Akims could not be induced to proceed more than a mile, and on their return were only prevented from bolting to Abogos by a rocket being directed down the road to head them back. The next day Lieutenant Barnard was to have effected a junction with me at this place at 1 P.M., so was compelled to leave Assufoo. He left Jaashie at 10 A.M., but did not reach this place until 3.30 P.M., and carried it after fifteen minutes' firing.

Mr. Ponsonby was hit by a spent ball, which caused him to fall; also two Houssas hit, one seriously. The enemy held the bush or border of the village for about five minutes, until dislodged by two rockets, when they retreated in direction of Coomassie or Juabin.

In the meantime I had left Abogo at 7 A.M., and forming junction with 100 men under Captain Sartorius, who had left the day previous and encamped two and a half hours in advance, I proceeded for Conomo, three miles due east of this place. It was quite dark when we arrived within a mile of this place, and I encamped, intending to attack at daybreak; but a reconnaissance made by the King of Eastern Akim and twenty men found it deserted, and it was immediately occupied by his force—about 400 men.

On the 27th I moved to this place, leaving Dr. Rowe and 200 Houssas and Yorubas with King of Akim to prevent Akim force from going off in sudden panic. The King himself is brave, so is his general, and perhaps 100 out of 700 of the entire Akim force with me.

Captain Butler does not inform me of his line of march after crossing Prah, and the few Ashantee prisoners with me give no information. The slaves taken know nothing except the road from the farm they work on to the village in which they are taken. I cannot see the Akims to reconnoitre, so that unless Captain Sartorius should succeed in getting this despatch to Captain Butler, I do not know how I shall effect communication with your Excellency.

I am awaiting reinforcements of Aquapims with spare ammunition, also King of Crobo with a large force, all of whom are within three or four marches of us, being in lat. 6 deg. 38 min. N., and long. 1 W., fifteen or twenty miles from Coomassie or Juabin.

From such information as I can gather, the enemy will oppose my crossing the river at Assoom, six hours from this.

Excepting 'Abogoo,' every place occupied has been found swept of sheep and fowls, consequently my force has been living on yam and plantains for the last seven days, and the officers have no preserved meat up of any description.

One of my messengers was robbed and flogged by some West India soldiers, consequently the Akims at Assoom refuse to convey letter, and I am too far in advance to influence them.

I shall endeavour to get duplicate of this and previous one of the 21st instant sent to your Excellency by route south of Prah.

Enclosed is a copy of Lieutenant Barnard's (19th Regiment) report of attack and occupation of this place, Odumassie.

It is with much pleasure that I bring to the notice of your Excellency the conduct of Lieutenant Barnard (19th Regiment), a most promising and indefatigable young officer. Captain Sartorius, in command of the advanced force which carried Abogoo, brought to my notice his gallant bearing on that occasion; and he displayed great discretion and judgment while in command of the column which I found in possession of this place yesterday. I therefore hope your Excellency will remember Lieutenant Barnard's gallant conduct on this second occasion when this campaign is finished.

I take the opportunity of bringing to the notice of your Excellency the invaluable assistance which I have received from Captain Sartorius, 6th Bengal Cavalry. It was entirely owing to his indefatigable exertions and great personal influence that the Croboes and Aquapims were induced to move on the Volta when they did, thereby showing the Accras, and enabling me, with the assistance of Dr. Rowe on the seaboard, to cross a large force of 17,000 men over the Volta by December 27; nor could any force have reached the Prah on 15th instant, but for the amount of pressure put upon the king and chiefs of Eastern Akim by Captain Sartorius. I shall, on another occasion, bring this officer's service more fully to the notice of your Excellency, trusting that you will forward them to the Secretary of State for India.

I have, &c.,

(Signed)   JOHN H. GLOVER,
Special Commissioner.

His Excellency Major-General Sir G. Wolseley.

When within 20 miles from Coomassie, at a place called Odumassie, Glover halted until February 8, on which date he received news of the capture of the Ashantee capital by the main column under Sir Garnet Wolseley. He then made a two days' march, and on the 10th despatched Captain Reginald Sartorius, of the 6th Bengal Cavalry, with twenty men to open communication with Sir Garnet.

Sartorius undertook the dangerous mission, and though he was fired upon several times *en route,* he entered Coomassie; and after satisfying himself that the capital was in ruins, and the main column far on its way back to the Coast, he hurried on after it, arriving at Fomannah on the 12th, where he found Sir Garnet, who congratulated both Glover and Sartorius for the spirited way in which they contributed to the success of the campaign. The following is the letter which Captain Sartorius brought from Captain Glover to Sir Garnet:—

Essiemampon, February 10, 1874, 12.30 P.M.

Sir,—I have the honour to report my arrival at this place with 700 Houssas and Yorubas, three 7-pounder guns, three rocket troughs, 204 spare rounds ammunition per man. Last engagement with enemy was on 2nd inst., when they were driven from line of Amoon River. King of Duabin has sent in his submission. I sent his messenger back, ordering king to go to your Excellency at Coomassie. I have with me some 1,600 allies, and 5,000 over River Prah coming up. I suppose myself to be seven miles from Coomassie, and shall halt at Atrirea, the next village, until I receive orders from your Excellency. My men have had only one ounce of salt meat per man since January 18, and that once four days since. Captain Sartorius, 6th Bengal Cavalry, who takes this, has done excellent and hard service.—I have, &c.,

(Signed)   J. H. GLOVER, R.N.

His Excellency Major-General Sir G. Wolseley, C.B., G.C.M.G., &c.

In the following letters we have official despatches which conclude such accounts as we have been able to obtain of the work accomplished by Captain Glover, and Sir Garnet Wolseley's hearty though tardy acknowledgments of his services.

Odumassie, February 4, 1874.

Sir,—I have the honour to acknowledge the receipt of your Excellency's despatches dated Prahsu, 19th, and Moinsee, January 23, 1874. Since my despatch from this place dated 28th ultimo, reporting the occupation of this place and that of Conomo on 26th, I have been endeavouring to open communication with Captain Butler in the neighbourhood of Lake Borsumaque, which I estimate to be not more than nine hours south-west of this. Captain Sartorius went to the south-west on the 29th and 30th, returning on the 31st, finding no road leading westward from Jaahsi, and left again on the 1st inst. See report. The river Anoom, three hours in my front, is occupied in force by the enemy. The Akims from Conomo have felt the enemy three hours in their front on the Jwbin-road and on this side of the river Anoom, while Captain Sartorius, crossing the Anoom six miles to the west, drove them from two camps on both sides of that river, with loss on our side of four killed, eleven wounded, and one missing. Yesterday more ammunition came up, making 145 spare rounds per man for 350 Houssas, and 126 spare rounds for 400 Yorubas armed with Enfields, together with 195 rockets, and 72 rounds of shell and case for one 7-pounder gun.   I am waiting for reinforcements to come up, having only in camp fit for duty 262 rank and file Houssas, and 262 rank and file Yorubas; total 524.   1 am obliged to keep 200 Houssas and Yorubas at Abogoo, in consequence of the unreliableness of the Akims, who withdraw their men without leave or warning, and the road between that place to the Prah had been interrupted because I cannot get the Akims to patrol.   Fortunately it was the Akims who suffered, losing three of their men by a surprise, and being unarmed. The sick-list to-day shows 27 Houssas and 21 Yorubas, besides 9 Houssas and 3 Yorubas wounded ; total 60.   Until reinforcements arrive I cannot move from this without cutting off all communication with my rear, and I have not sufficient ammunition up to run such risk.   Captain Butler, writing on the 26th from Enoomsu, would appear to be on the Anoom or Enoom river, 'Su' meaning in Ashantee upon or on.   Lieutenant Barnard patrolled road leading north-west to-day, three miles towards Anoom river, and observed traces of party who had been foraging.   I am expecting Lieutenant Moore to-morrow with some 1,000 Aquapims and Croboes, who, he reports, are very loth to move forward.   He also reports 2,000 more coming up from south of Prah.   As soon as reinforcements and ammunition come up I shall close in on the main body.—I have, &c.,

(Signed)   John H. Glover.

His Excellency Major-General Sir G. Wolseley.

The next enclosure is the communication of Captain Sartorius referred to in the preceding letter of Captain Glover :—

Odumassie, February 3, 1874.

Sir,—I have the honour to report that, after leaving this place on February 1, on my way to Apinamapu, according to your order, I reached the Anoom river, running east and west, at eleven o'clock noon, after five hours' march, and crossed the river without seeing anything of the enemy.   About a quarter of a mile on the other side I surprised a spy village of about fifteen spies.   They evidently had no idea of any force being so near, for they left their guns, as also some venison they were cooking at the time of the attack.   From this place I continued my march in a westerly direction, and after about two miles came upon the enemy, who were in force ; in a short time they fled, having wounded Serjeant Major Alli and one Houssa (neither of them severely).   I still, continued my march west for about a mile, and came upon an empty village of some ten houses.   No one could have occupied this place for some time ; I therefore did not expect any more fighting, and sent back a serjeant, with twenty Houssas and twenty Yorubas, to you with my report so far. After sending off this party I continued my march still west, and again came upon a large camp of the enemy, having had a morning fire for the last two miles from village. I forced my way on about three miles; the road all this time was along bank of River Anoom. My guide then reported that he had lost his way, so we had to return nearly three miles, and struck a path leading north north-west. This we followed for some five miles, till we turned. I therefore found the guide was again at fault, and, as it was getting late, and the troops had had a very hard day's work, I returned to bank of river, three-quarters of a mile from the last Ashantee camp; this being the most open, and with river on right flank, and therefore easiest held against any attack, and encamped for the night. Next morning I sent out a patrol of 40 men, Houssas and Yorubas, with order to patrol the road leading north from enemy's camp for one hour, meaning to follow myself a little later and find out where the path led to. I had no idea at the time that they would find the camp re-occupied by the enemy so close to me; the serjeant in charge had the usual orders; in case he met the enemy in any force he was to return. About fifteen or twenty minutes after they left I heard heavy firing in their direction, and immediately fell in with my detachment and followed them. Just as I was starting a corporal came from the patrol to report the enemy were in great force. I had not got more than half a mile when I met some Yorubas, who said the enemy were too strong and that the patrol were retreating; my detachment got into confusion and I halted them. Directly the rest of the patrol came in. I had intended retreating to last night's camp, as it was a much stronger position, but Captain Larcom, R.N., who was with me, thought that this would cause a panic, so I took his advice and threw out a line of sentries front and rear. Captain Larcom placed his rocket trough on pathway leading to enemies' camp. While this was going on the Ashantees came down to attack, singing their war song; but when within sixty yards one of the Houssas shot a man, and I sounded the advance at the same time. The enemy bolted at once, and fearing an attack on my rear, as all my men were in disorder, I returned to my position It was then reported to me that fifteen men were missing, and that they had crossed the river and were coming down the other bank to join me; one Houssa reported killed. I then made up my mind to remain where I was till 12 P.M. (four hours), that being the time when the return party, sent to you the day before, was expected by me, and also to give the missing men time to rejoin. At twelve my intention was to force my way through the enemy back to head-quarters, or, if I found them too strong, to cross the river in the night and return by the south bank. This would have been difficult on account of the depth and breadth of the river, and therefore only to have been done as an extreme measure. Whilst waiting till 12 A.M. to advance, two Houssas and Yorubas came across from the other side. At nine or half-past heavy firing was heard from the other side of the enemy's camp. We at once advanced and came up in time to meet Lieutenant Barnard's force, who had had a running fight for some distance before reaching the Ashantee camp near us. I again halted till a quarter to 1 P.M., so as to give the missing men time to return, which they did, with the exception, I am sorry to say, of four Houssas hilled, and seven Yorubas, who had crossed the river and made their way back to head-quarters. The Ashantees had cut off the heads of the three Houssas, but had left them behind on the advance of Lieutenant Barnard's force. These heads were pluckily recovered by Mr. Phillips, a missionary teacher, acting as my interpreter, and were sunk in the river by the serjeant-major of my detachment before we left. The enemy were in large force—I should say about 800 or 1,000—and attacked us in a very determined manner. Their loss must have been considerable, for they left traces of blood, as also some stretchers.—I have, &c.,

(Signed)   REGINALD SARTORIUS,
Captain 6th Bengal Cavalry.

Captain Glover, R.N., Special Commissioner.

The next enclosure forwarded by Captain Glover to Sir Garnet is the copy of a letter addressed by the former to King Quahbira Fuah, which is as follows :—

Head-quarters, Adoomassie, February 7, 1874.

King Quahbira Fuah,—News reached me this morning that you recrossed the Prah eight days ago, and that Captain Butler had left you to join the General. I shall endeavour to open communication with you from Joahsi and Prahsu, both of which places I destroyed ten days ago. I have with me 660 soldiers, guns and rockets, and 3,000 Akims, Croboes, and Aquapims, and 2,000 more expected in a day or two. You should immediately recross the Prah, and endeavour to meet me either at Odumassie or nearer on road to Prahsu. It is quite impossible that King Calcali can leave the General and come down to surround you. Do not spoil the good name you have gained by your previous bravery, but join your force to mine and go on with me to the General, upon whose line of advance I am ordered to close. Captain Butler has not communicated with me : but in the event of your receiving any orders from the General counter to my request you

will at once comply with the General's order.   You may expect to hear
of my force being in your front before you receive this.

Faithfully, &c.,

(Signed)   JOHN H. GLOVER,

Special Commissioner.

To King Quahbira Fuah.

The next letter is from Captain Glover to the Chief of the
Staff:—

Koraman (Quarman), February 14, 1874.

Sir,—I have the honour to acknowledge the receipt of your communication of the 13th instant, acquainting me that the King of
Ashantee has paid the first part of the indemnity demanded, and that
the Major-General is going to sign the treaty of peace.   In compliance
with the orders of his Excellency the Major-General Commanding, I
shall leave this place to-morrow *en route* for Prahsu, on my way to
Accra.   I regret that your letter of the 9th instant, to which you refer,
has not reached me up to the present time.   I shall require to-morrow,
for the provisioning of my force, 1,000lb. salt beef, 2,000lb. bread,
500lb. tobacco, and 25 gallons of rum.   I have the honour further to
request that you will be good enough to cause seven days' provisions
for 1,000 men, and a ration of rum for the same number for the same
time, to be detained for me at the depôt on the River Prah.

I have, &c.,

(Signed)   JOHN H. GLOVER,

Special Commissioner.

Colonel E. R. Greaves, Chief of the Staff.

A second despatch from Sir Garnet Wolseley to the
Colonial Office, also bearing date February 22, is also
published:—

Government House, Cape Coast, February 22, 1874.

My Lord,—I have the honour to forward a copy of a despatch
dated 20th instant, from Captain Glover, R.N., describing the concluding phase of the operations carried out by the force under his
immediate command, in accordance with the instructions originally
conveyed to him by me on December 31, as reported to your lordship
in my despatch No. 1, of 1st ultimo.   Captain Glover has conducted
these operations with great skill and ability, having overcome all the
numerous difficulties that he had to encounter with the zeal and energy
for which he is well known.   Captain Glover reports to me that the
Eastern Akims, who accompanied him, proved to be utterly worthless,
having evinced the greatest cowardice throughout the campaign.   The

fact that he was not able to bring the large native force into the field
north of the River Prah that he had at one time reckoned on, is, therefore, in my opinion, a fortunate circumstance, as useless and cowardly
allies are always a serious clog upon all military operations. The
diversion effected by Captain Glover's force in favour of the main army
operating direct upon Coomassie under my own immediate orders
contributed materially to the success that has been achieved in this
war, and I have great pleasure in recommending Captain Glover and
the officers serving under him to your lordship's most favourable consideration.

I have, &c.,

(Signed)　　　　G. J. WOLSELEY,

Major-General and Administrator, Gold Coast.

Captain Glover's letter enclosed in the preceding runs as
follows :—

Mansu, February 20, 1874.

Sir,—Ammunition having come up during the two previous days,
completing spare ammunition to 200 rounds per man, I left Odumassie
on the morning of 8th instant and bivouacked for the night in Ashantee
camp, on north side of River Anoom, from which the enemy had been
driven by the two detachments commanded by Captain Sartorius, 6th
Bengal Cavalry, and Lieutenant Barnard, 19th Regiment, on 2nd instant,
as reported in previous despatch ; here we found corn and corn flour,
which had been left by the enemy.   The next day we halted to breakfast in a large camp capable of containing 4,000 to 5,000 men, where I
was joined by Lieutenant Moore, R.N., and some 4,000 Akims,
Aquapims, and Croboes. Here it was evident the enemy had left in hot
haste, and for miles along the road we found stools, calabashes, corn and
corn flour, which the enemy had thrown away in their flight. A
wounded straggler from Captain Sartorius' force, who only came in the
day before I left Odumassie, had approached the enemy's camp during
the night, and had heard them discussing the hopelessness of fighting
Europeans and Houssas, as it was evident that it was not a force of
Akims with whom they had been engaged. After a march of ten
miles, principally northwards, along a hunter's path with much bog and
fallen timber, I bivouacked for the night on left bank of a river (name
unknown) ; the next morning I marched north-west about five miles,
to a village on the south of Essianimpon, from which at noon I despatched Captain Sartorius with escort of twenty men to open communication with your Excellency, supposed to be at Coomassie, and at 2.0
P.M. finding myself one mile distant from River Dah, I halted at Akinna,
until I should either receive orders or hear from Captain Sartorius. At

3.0 P.M. next day I crossed River Dah, and halted for the night at Apraymassie, and at 2.0 P.M. on 12th instant entered Coomassie. I then learned that messengers had passed down from the king accepting your Excellency's terms and carrying a first instalment of indemnity. I have since come up by easy marches to this camp, where I arrived at 9.0 A.M. this morning, *en route* for Anamaboe and Accra.

At Moisee, at the foot of the Adansi-hill, on the 15th instant, a messenger from the king of Coomassie arrived bringing with him fourteen ounces of gold, and a gold dish, as a present to me, and requesting that I would order the king of Eastern Akim (who they said, had gone to Juabin to attempt to detach the king of that place from his alliance with the king of Ashantee) to return at once to his own country.

I returned the money and dish to the king of Coomassie and sent the king of Akim a copy of Colonel Greaves' letter to me of the 14th instant, and directed him at once to recross the Prah and return to his country with all his people. I took the opportunity of sending by this messenger also a letter to Mr. Bannerman directing him also at once to break up the camp at Abagoo and march to Accra with the detachment of Houssas and Yorubas left in Abagoo, taking with him the sick which had been left at the various stations on the march up.—I have, &c.,

(Signed)   J. H. GLOVER,

Special Commissioner.

His Excellency Major-General Sir G. Wolseley, C.B., K.C.M.G., &c.

The following is a portion of a note sent by Captain Glover to a friend in England, which was written by him at Yan-Coomassie (Fantee) February 20, when only 30 miles from Cape Coast Castle; and a more fitting conclusion to this chapter and the book than the following generous acknowledgment by Sir Garnet Wolseley of Captain Glover's valuable services and brilliant success, cannot be written:—

En route for Accra; but I intend, if possible, to go straight to England, and am going to Cape Coast Castle to see the General.

You may perhaps know that the troops had to leave Coomassie, having no provisions and without the General having heard from the king, and that it was much to the relief of his mind that I had occupied Coomassie after he had left.

I give you a copy of his telegram:—

'Congratulate you most sincerely on all you have done.'

Captain Sartorius, whom I had sent on, writes:—

'The General just drank your health.'

I was fighting up to the 2nd instant, when the enemy was driven from the line of the Anoom River and Coomassie was taken on the 4th.

The General found the Ashantees so tough that my force was supposed to be destroyed, and our marching considered wonderful.

I had 6,000 men over the Prah, and marched 4,600 into Coomassie.

The day after the king sent me 140 ounces of gold on a gold salver, begging me to order the force I had left on the Anoom River to recross the Prah.

I returned the gold and complied with his request.

I also received the submission of the King Iwabin, and have lost ten killed, and have one officer and twenty-four men wounded. So ends my share in the Ashantee Campaign !

Cicero said very wisely that ' war should be so engaged in that nothing but peace should appear to be aimed at.' The Ashantee war was undertaken with that aim, as all the world knows, but it was marred on the threshold of perfect success by a series of accidents. First: By the fire which raged in Coomassie the night following the entry of the troops into the capital. Secondly: By the omission of the British commander to guard what he had captured. Thirdly : By the impatience of Sir Garnet, and his intense desire to hurry from the capital, because he feared that the rainy season had begun, which, of course, would have been calamitous to the white troops.

The first contributed to mar the success which was almost assured, by the evident impression on the king's mind that the destruction of his capital was already begun. Whether the fire was the handiwork of some trucculent Ashantees, or an accident arising from permitting the Fantees to plunder, matters not.

The omission or neglect of the British commander to secure what he captured, served to furnish the king with a force which he believed to have been utterly destroyed, and a confidence in his fortunes which, but a few hours before, were at the very lowest ebb.

The king's confidence in his fortunes, so quickly restored, was soon lost again when the news of the arrival of Glover between Jwabin and Coomassie came to him; but the impatience of Sir Garnet and his anxiety to return to the coast, prevented him from reaping the results of his brilliant success as General commanding the British Expedition ; and this is proved by the immediate despatch of messengers after him by the hard-pressed king, and the voluntary gift of 1,000 ounces of gold, with a prayer for peace.

The return march was begun on February 6. Five days later Captain Sartorius rode through Coomassie accompanied by only twenty men. The next day Special Commissioner Glover marched 4,600 native allies into Coomassie.

If Sir Garnet could only have stayed five days longer, peace might have been made, which, as all will agree, was a desirable ending of the Expedition. Russell's and Wood's native regiments, with Rait's artillery, were well able to have kept Coomassie against all odds, and the white troops might have departed at once for the Coast.

Such historians of the Anglo-Ashantee Campaign as Sir Garnet's own personal Staff may furnish, will naturally take a eulogistic view of things, and no doubt will lavish eulogistic paragraphs on the genius of their Chief. This will be but natural. Sir Garnet has spoken well of his Staff, which no doubt they all deserved, and the historians of his Staff will, as a matter of course, speak well of him. But a good advice to an historian, which I have striven to remember in this book is —'Ne quid falsi dicere audeat, ne quid veri non audeat.' (Let him not dare to say anything that is false, nor let him fear to state the truth.)

The last act of Sir Garnet's was a culpable one in more than one respect, and it is a fortunate thing that it had no evil results, which it might very easily have had.

Glover, about to cross the Volta to crush the Ahwoonas with 16,000 men, received peremptory orders from Sir Garnet to be ready to cross the Prah on the 15th. If the Commissioner was expected to be there by this date, he had no time to lose, and he obediently set about fulfilling his commands. But it was impossible on the short notice he received to march the entire force of native allies he had been at so much pains to collect, and he accordingly set off for the Prah with his faithful Houssas and Yorubas, numbering about 1,100 men. By the time however he had arrived within such dangerous proximity to Coomassie as he was on the day he received the information that the capital was destroyed, Glover's force numbered about 1,400 men.

The error which Sir Garnet committed in not disarming every Ashantee whom he found in Coomassie the night he entered, might have had most fatal results for Glover. Sir

Garnet permitted quite a little army of fugitives to gather about the king; and if he had been made of different material, no doubt that would have inspired him to retaliate on the comparatively weak force of Glover, and by this means have retrieved the disasters of Amoaful and Ordahsu.

To the first error he adds another one, by leaving Coomassie without communicating with the man, whom we might say, he had despatched from the Volta to an almost certain doom. Fortunately, however, the Fates willed it otherwise, and Sir Garnet has escaped blame. The king turned out to be such a coward that he had no heart left in him after the defeat of Ordahsu; thus he never attempted to snatch the prize which lay so close to his grasp. But such an act as Sir Garnet's abandonment of his faithful and brave ally could not be done in Europe without being quickly taken advantage of to the ruin of the party concerned.

However, 'all is well that ends well.' The campaign has been a success, and the brave and deserving have been rewarded by their Queen and the nation. A baronetcy and a pension have been conferred on Sir Garnet Joseph Wolseley, and another decoration, the Grand Cross of St. Michael and St. George, has been added to the many he already possesses. Nearly all the officers have been promoted, the gallant Lord Gifford has received the Victoria Cross, and Sergeant M'Graw of the 42nd for his conduct at Amoaful has received a like distinction.

But there is one man who deserves a Victoria Cross, which the newspapers have not recorded as yet, and he is Captain Reginald Sartorius, of the 6th Bengal Cavalry, for that daring ride of fifty-five miles with a force of twenty men, through an enemy's country, while numbers of the foe lurked behind bushes, whence he was fired upon twice. It was a remarkable and unique act of daring bravery, which had good results. It calmed the mind of the general who must have been naturally anxious about Commissioner Glover's fate, it assured him also of the extent of his success, and was the means of impelling the king to despatch messengers with the first instalment of the idemnity and to abjectly sue for peace.

Some reward should also be given to the gallant Glover, who lent such aid to Sir Garnet Wolseley, by distracting the attentions of a larger force of the enemy, for compelling the king of Jwabin by his presence between him, and his ally King Coffee to send in his submission to Sir Garnet, and for extricating himself with such credit to himself and force from a most dangerous position, thus insuring the complete success of the British arms in the campaign in West Africa. Glover has already received the thanks of Parliament, but if he were permitted to subscribe himself Sir John Hawley Glover, Her Majesty's most faithful and dutiful servant, very few Englishmen who admire courage and success in arms would begrudge him the honour, or say that it was given to an undeserving man. *Detur digniori.*

Seeing that Sir Garnet Wolseley was the Commander of the British Expedition to Coomassie, it would be but right before the reader and the author parts, that he should be permitted to sum up this most veracious history with the following words, which he wrote to the War Office :—

Our success does not lie merely in our having defeated the enemy's army, and occupied and destroyed his capital, but in our having thoroughly established in this kingdom a wholesome fear of the British power and a knowledge of the advantages of an alliance with Her Majesty.

Three days after we have destroyed his capital and his palace, and commenced our homeward march, the king sends to make peace, and gives a substantial guarantee of his good faith.

Five days after the same event an English officer, accompanied by only twenty men, is able to ride unmolested fifty-five miles through the heart of the invaded kingdom, while one of the principal tributary chiefs beseeches permission to dissever himself from Ashantee, and cast in his lot with the tribes allied to the British Crown.

I venture, therefore, to hope that the object for which her Majesty's Government intrusted to me a force of British soldiers has been attained, and that the exertions of the officers and men under my command have not been made in vain.

END OF PART I.

COOMASSIE AND MAGDALA

PART II.

LORD NAPIER OF MAGDALA, G.C.B., ETC.

From a Photograph by Mr. John Watkins, Parliament Street (by permission).

CHAPTER I.

A BRITISH CAMP—THE 'BEST FELLOW IN THE WORLD'—THE REQUIREMENTS OF A GENTLEMAN—INTRODUCTORY.

A MOST extraordinary and novel sight to me, presented itself as I landed upon the bunder at Zoulla. Thousands of half-naked coolies were shouting and chanting a barbaric song while they worked under as hot a sun as ever blazed in the tropics, and hundreds of uniformed superintendents, armed with long *courbaches*, were coercing the labourers under their charge to work. The braying of hundreds of donkeys, the neighing of horses, the whinnying of mules, the lowing of thirsty kine, the shrill shriek of two anomalous locomotives, the noisy roll of ricketty cars as they thundered to and fro, caused the scene to appear at the first impression as if a whole nation had immigrated here, and were about to plant a great city on the fervid beach of Annesley Bay. The mountainous piles of stores covered with tarpaulins, the long warehouses with their roofs of brushwood, filled to the utmost with the *matériel* of war, and the noble bay crowded with majestic transports, steamers, men-of-war, great sailing packets, tiny tug-boats, elegant little yachts, and innumerable Turkish kanjeahs, from Mocha, Jeddah, Souakim and Massowah, flitting about with their swallow-winged sails, only served to heighten the illusion.

As I knew nobody as yet of the English army, it was necessary I should scrape acquaintance with some person who could initiate me to British camp-life. Recollecting I had a letter of introduction to a commissariat officer, given to me by Major S——, I bade my servant Ali remain near a certain large warehouse until I should return to him. I hastened on to where

I observed a group of lofty bell-tents, pitched close to a wooden building where numbers of white-robed and turbaned Indians were running about in seeming great hurry. Asking a soldier who stood guard near the largest and most luxuriously furnished tent, ' Which was Captain Z——'s quarters ? ' he directed me to a lofty marquee which stood not very far off, surrounded by a hedge of baubool bushes, the very picture of exclusiveness.

On arriving at the indicated tent I made the occupant aware of a visitor being present by scratching on the canvas. I was immediately invited to enter by a person who was lying full length on a sofa, dressed in the cool morning costume of India, which consists of a striped cotton shirt and a voluminous pair of drawers, which they call pyjamas.

' I came to see Captain Z—— of the Commissariat, sir ; ' said I, surprised at his nonchalance in the presence of a stranger. ' Are you the gentleman ? ' I asked.

' Yes, I am the gentleman,' he replied, slightly lifting his eyebrows. ' Who are you, and what do you want ? '

' I am the bearer of a letter of introduction to you from Major S——,' said I, at the same time bending forward to hand him the letter.

' Hum ; ah ! to be sure ; Major S—— ; aw ! let me see. Won't you sit down ? Excuse my indolence ; this country is so hot that it melts the marrow in a fellow's bones ! '

He had half-risen when he commenced to deliver this apology, but directly relapsed into his former attitude with a deep sigh of relief, turning an almost helpless look upon me before he read his letter, which told as well as volumes of the anguish he had suffered in rising.

Soda water and brandy having been called for by my host and served by a dark-featured native to the great relief of both of us, I asked if I could procure a tent and rations.

' Oh, yes,' replied the Captain ; ' easy enough. Make out your indent. No, let me see. First, you will have to go to Major X—— and get an order for your rations and a tent, after which you will be pretty comfortable.'

Major X——, he further told me, was Acting —— —— of the Force at Zoulla, ' a very nice gentleman ; splendid fellow ; first-rate chap ; do anything in the world ' for me ! Wishing to see this paragon of an officer and settle my business,

I bade Captain Z—— a 'good morning,' telling him I should see him again before long.

I was about to depart when the Captain bawled out, 'I say, you; can you dine at our mess, and would you please consider yourself as an honorary member of ours while you are in camp? We have a fine set of chaps, all perfect gentlemen. There's A—— of the Commissariat, B——, an old sailor, now Bunder Master; then there is C—— of the 3rd Light Cavalry, D—— of the elephant lines, and lastly, we have E—— of the Bombay ——. Do come, will you? Be sure, now! dinner sharp 9 P.M. Ta-ta, ole fellah.'

'Certainly, my dear Captain, with the greatest pleasure. *Au revoir.* Ta-ta, ole fellah!' and out I departed to find the quarters of that 'splendid chap' Major X——.

The Major's tent was fully a mile from Captain Z——'s tent, and during the painful walk there, with the hot sand burning my shoes, and the wind coming in hot fervid gusts in my face, I began to experience the discomforts of a tropical life, and I fully sympathised with the gallant Captain in his endeavours to alleviate the miseries that tortured him by imbibing cooling drinks.

Upon entering the Major's tent I made known my presence by a cough, which attracted the attention of a gentleman whose face was garnished with an abundance of whiskers and who, while he advanced towards me, popped a huge eye-glass to his eye, looking at me in a very inquisitive manner. The mute but critical interrogative I answered by instantly informing him who I was and stating my business, presenting my credentials at the same time, which he examined with the greatest care. After a satisfactory perusal of the letters, he enquired, with much anxiety faithfully depicted on his features, what I should like to have. 'Some brandy and soda, cherry brandy, port wine, lemonade?' The very names of these grateful refreshments uttered in the unctuous accents of the hospitable major, were enough to revive a fainting pilgrim.

His servants—quiet, well-trained fellows—glided in at the sound of his voice, bearing trays with most delectable drinks on them. The Major taking it for granted that, being an American, a little brandy would be of infinite service to me, poured half a glass-full into my tumbler, filling up with sparkling soda-water, cool and fresh. We toasted each other, and then the Major carefully screwing that eyeglass—apparently indispensable for right judgment—into his right eye, commenced to question me concerning American affairs: was very positive I was a 'Finnegan,' and only came to enter Theodore's service—to all of which I answered in the best possible humour. Major X——, abrupt as was his manner, was, I must confess, a gentleman entirely devoid of any affectation, save one—that of using an eyeglass. He was a famous man in India, and his sketches of 'Life in the Jungle' were most artistically executed.

In the afternoon my tent was pitched close to Captain Z——'s, and Ali, my servant, was sent to draw five days' rations, as I intended to march to the front next morning. A baggage mule, and an Otago saddle and gear were also given to me, having received which, I imagined myself pretty comfortable, and in first rate trim for marching.

Had I been in America, or following some army in Europe, I should have certainly been very well off; but in Africa it was a different thing, and a very rude awakening out of my illusion was the consequence of my want of experience on a British campaign and the knowledge of proper outfit.

'I say, old boy!' shouted Captain Z—— to me; 'are you off to-morrow? How are you off for horse-flesh and servants?'

'Well, my dear Captain, I am but poorly provided. I was not aware of my necessities before I came here, but I now find I am rather short in everything.'

'Gad; that won't do, you know. A gentleman should never be short of indispensables you know. How many servants have you?' he asked.

'I have only one;' I answered with humility.

'Good Heavens! only one; and how many horses!'

'One horse and one mule.'

'Pooh, pooh, that will never do. You must have a cook, a hay-cutter, a cutcha wallah (groom), panee wallah (water-carrier) to wait, and then you must have one horse for your personal baggage, one horse for your cooking utensils, one horse to carry your rations, one horse to carry your tent, and two riding horses; for without these things, you know, egad, it is

MAP
OF THE PORTION OF
ABYSSINIA
Traversed by
THE BRITISH EXPEDITION in 1868
from Annesley Bay to Magdala
to accompany
H. M. STANLEY'S
'COOMASSIE & MAGDALA'.

SCALE OF ENGLISH MILES.

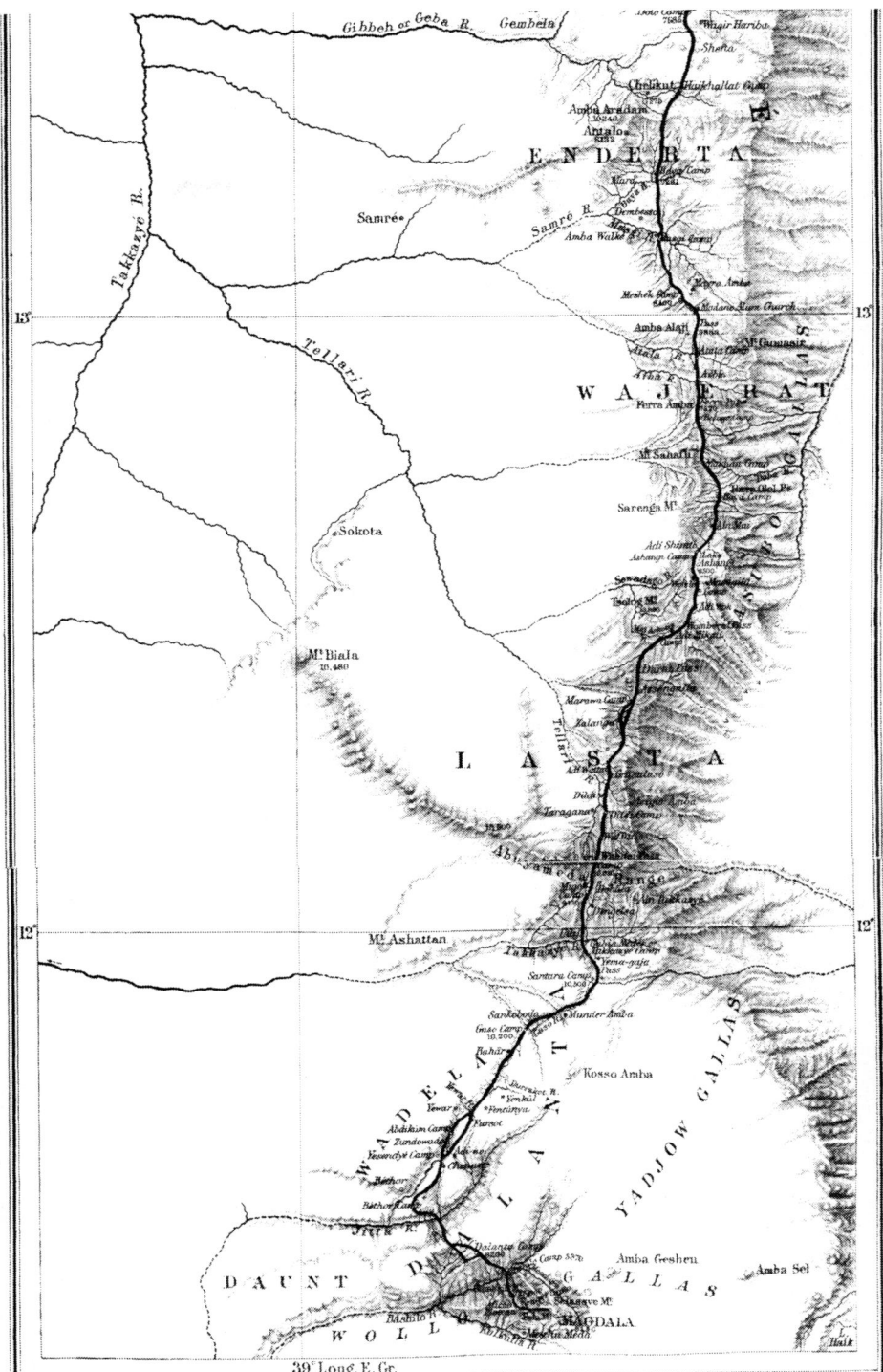

Gibbeh or Geba R. Gembeia

Doto Camp 7686

Wajir Hariba
Shella

ENDERTA

Chelikut Hayshallat Guns

Amba Aradam 1024d
Antalo 8585

Debra Tabor Camp

Mard..
Dongolo
Dembeas

Samré R.

Amba Walka Mahago Guns

Mayu Amba

Meshek Camp

Madane Musa Church

13° ENDERTA 13°

Amba Alaji

Haila Adigerat Camp Ali Gumsait

Atba R. Haila

WAJERAT

Ferra Amba

M¹ Saharti

Chiguin Camp

Amba Bet Ba ..Camp

Sarenga M¹

Ba-Han

Arli Shindi

Ashango Camp Ashangi Aalage Aalanga

Sewadago

Tsalog M¹

LASTA

Dunk Pass Dildiyalla

Marawa Camp

Zalamba

Tellari R.

M¹ Biala 10,480

Sokota

Atbi Gumbata

Dildi

Taraganat Dildi Camp

Abi Camp

M¹ Range

Wandach

Ain Bahana

12° M¹ Ashattan 12°

Takasaze

Santara Camp

Zebit

Dokowa Camp Yema-gaja Pass 10,508

Sankebota Guso Camp 10,300

Munder Amba

Bahir

Rosso Amba

WADELA

DALANTA

YADJOW GALLAS

Dansqet R.

Tewar

Yenkus
Fentinya

Abkikim Camp

Pamot

Zundomada

Yesandys Camp

Adeges

Charaka

Bethor

Bethor Camp

Jitta R.

Dalanta Camp

Camp 5800

GALLAS

Amba Geshen

Amba Sel

DAUNT

DALANTA

Selassye M¹

Basheio R.

MAGDALA

Tenta Medi

WOLLO GALLAS

Haik

39° Long. E. Gr.

Stanfords Geog: Estab! 6.8.7. Charing Cross

London: Sampson Low, Marston, Low & Searle.

simply impossible to live in this country, and people will only look down on you, you know,' said he compassionately.

'That's so,' chimed in the rest, ' to be somebody, you must have these things.'

'Really,' said I, feeling as if bound to say something in my own defence, 'I have been on many an expedition, and accompanied many an army, and from the experience gained from previous travels, I imagined I knew what I should want, and what would be necessary. When I first started for Abyssinia I thought Sir Robert Napier would not halt long anywhere, but would march on as rapidly as possible, and in such a case, gentlemen, unnecessary baggage would be an encumbrance.'

'Ah, but you do not mean to say that a dozen servants are not necessary or half a dozen horses too many, do you ? If you do, you know nothing about the country, or an Anglo-Indian army and its requirements.'

The above was a portion of the conversation held at dinner. I had seen enough during the day to make me aware that servants were very plentiful in camp, and from the life English officers lead in India, one can well imagine how extremely necessary they are. Another reason for their multiplicity is, that an Indian servant when employed, has a certain task or business allotted to him, and that one only will he perform ; for instance, the syce will only look after your riding horse ; the cutcha wallah after the baggage animal ; the panee wallah will but carry water ; the hay-cutter will but cut hay, and so on. Each of these servants would consider it derogatory to his dignity to perform any other duty than that for which he was engaged.

It will not be out of place to introduce here an introductory explanation of the events and causes which led to this assemblage of British soldiers on the Zoulla sands.

An Abyssinian prophecy, uttered ages ago, and handed down from generation to generation, had gathered strength, until the people had come to believe that the fullness of time had arrived, and grew accustomed to expect a Messiah. It became known to an ambitious young fellow, named Kussai, the only son of a widow, a kousso seller dwelling in Kuara, west of Dembea. The prophecy rang in his ears day after day, and month after month. It haunted him in his dreams and in his waking hours, until he finally became imbued with the belief that he was the chosen instrument who should deliver Canaan from the Moslems.

The prophecy was as follows:—

'And it shall come to pass that a king shall arise in Ethiopia, of Solomon's lineage, who shall be acknowledged the greatest on earth, and his powers shall extend over all Ethiopia and Egypt. He shall scourge the infidels out of Palestine, and shall purge Jerusalem clean from the defilers, he shall destroy all the inhabitants thereof, and his name shall be THEODORUS.'

Kussai, being a kinsman of Dembea's governor, Dedjatch Comfu, was accorded instruction which otherwise, in his plebeian rank, he would never have received. Upon finishing his education, which was limited to learning the art of reading and writing, young Kussai was persuaded to enlist under Dedjatch Comfu's banner, with a promise of swift promotion. Great qualities became immediately apparent in him. His conduct as a soldier was such as to call forth his kinsman's warmest admiration. Comfu promoted him to a captaincy, and he began to regard him as one of his most faithful adherents. Distinguished valour, and a quickness of comprehending the best points of a battle field, which he exhibited in a fierce engagement in the neighbouring province of Begemder during an expedition there, accelerated his further promotion, and endeared him greatly to the governor, insomuch that he gave Kussai his favourite daughter in marriage, and appointed him governor of a district.

After a short period of faithful service, during which he felt his way carefully, Kussai suddenly threw off the mask, and declared war against his kinsman and benefactor.

Dedjatch Comfu mustered his troops and marched against the rebel. In a pitched battle his army was defeated with great slaughter, and he himself killed by the hand of Kussai; who thereupon marched upon Gondar, the capital of Dembea, numbering some sixty thousand inhabitants, as some travellers will have it, which city capitulated without bloodshed to the conqueror. Upon this event he proclaimed himself governor of the province, marched an army he had trained to the other districts, and subjugated them in succession, replacing the district governors with faithful captains of his army.

He was now fairly established upon his gubernatorial chair, and by his heroic conduct in the field, and the extraordinary celerity of his movements, ascribed by some of the most superstitious to supernatural ubiquitousness, and after a year of peace, following up the principle of training soldiers to war, and inuring them to a martial life, he found cause to quarrel with the governor of Begemder. Forthwith, he marched his army into the country, and after repeated victories in that mountainous country, the Begemder troops, commanded by the governor, were conquered, and all put to the sword except the governor, whom Kussai consigned to one of the state prisons of that province. Upon the news of his victory spreading abroad the chiefs of the neighbouring provinces made common cause against him.

Before proceeding to meet his enemies, who were indeed numerous, he augmented his own army considerably, by proclaiming his name to be Theodorus (pronounced 'Todoros' by the natives), of the line of Solomon, and declaring that he was the Messiah who the prophecy foretold should come and destroy the Mahomedan nations.

This proclamation had a wider effect than even he could have imagined. Deserters from the other provincial armies flocked to his standard, and considerably elated by his prospects, he marched to give battle. This tremendous assumption of 'Todoros' or Theodore as we must now call him, the governors who were inimical to him at first affected to laugh at and regard in the light of a good joke; but when they found their legions reduced to mere skeletons by wholesale desertion, and heard through their spies of the growing power of their ambitious enemy, they altered their tone. Some laid down their arms and resigned submissively to him; others advanced to meet him, but were all defeated, and those who were not slain in battle were sent to keep company with the governor of Begemder, in the living tomb in the Begemder mountains. Thus one province after another was subjugated until the whole of Abyssinia, with the exception of Tigre, claimed Theodore as master. It was after these victories that he assumed the name and title of EMPEROR THEODORUS BY THE POWER OF GOD. This took place in the year 1851.

When he had made himself master over all Abyssinia, he was about thirty-five years old, in the vigour of lusty manhood, active as a panther, brave as a lion, beloved, nay, almost adoréd by all war-loving soldiers.

Having acquired from historical and traditional accounts of past governments, and from his own experience, a knowledge of the intrigues and dangers which beset an Abyssinian crown, he conceived a distaste for life in the city of Gondar. He said to himself, ' I will have no capital ; my head shall be the empire, and my tent my capital.'

His talents and military ability enabled him to successfully repress all revolutions. The Gallas were many times defeated, but they were such roving nations that he was unable to make his victories good. He could only make his name dreaded, and command peace with them by his prowess.

Merchants were invited to Abyssinia, and manufactures were encouraged. European workmen hearing of his liberality, flocked in numbers to Gondar, where they received employment, favour and riches. Military men were also regarded with special favour, and many were the travellers who were invited to reside in the country.

Consul Plowden was a great favourite with him and resided with Theodore five years, during which time the Emperor strengthened his power and government in every possible way. But though he was undoubtedly a great soldier, far in advance of his people, he knew not the way to make his victories subserve for the interests of the country. No sooner was he conqueror of a rebel province than he was compelled to overrun another, and in the war he was continually harassed, until gradually his whole nature changed. He became embittered at the ingratitude of the people, whose welfare he so ardently desired and for whom he laboured so assiduously.

Some Russian and German engineers came to the country and they requested service with him. They told him of the wonderful cannon and monstrous mortars they could make, until his imagination rioted in the belief that the time was drawing nigh when he could undertake his contemplated conquest of Egypt, Palestine, and Arabia. Differently constituted from the fabulous Brobdingnag Emperor, he only saw the means to effect his great and darling project by these deadly weapons, and he at once closed with the offer.

Large powder manufactories were erected, Greek merchants were commissioned to buy muskets for him, and foundries were constructed for the casting of cannon, and all augured well for his prosperity.

Theodore regarded the Viceroy of Egypt as his most deadly enemy. Aggressions were frequently committed by the Egyptians, and captured Abyssinians were taken down the Nile and sold as slaves to the Pachas of Cairo and Alexandria, while many were shipped to their worthy brothers in Constantinople. Nor were these the least of the crimes committed by the Egyptians, of which Consul Plowden made the Secretary for Foreign Affairs aware.

When Mr. Plowden was appointed Consul, it must be distinctly remembered that it was to Abyssinia he was accredited, and his position therefore was with the reigning monarch. The monarch happened to be the Emperor Theodore. Through his kindness of heart and desire to cultivate friendship and amity between the English Government and Theodore, he won the latter's affection and good will. His duties in the interior of Abyssinia, were to ' watch and counteract foreign intrigue; to suppress the slave-trade, and to promote commerce as much as lay in his power.' Towards these objects, he urged upon Theodore to send an embassy to England, that the Legation might be able to see for itself how advantageous commercial relations between the two countries would be. But the Emperor was so perpetually engaged in war that he found it impossible, so he said, to pay the attention to it that the subject demanded.

Consul Plowden was killed by some of the rebel chiefs' forces while on a journey to Massowah in March 1860. Theodore mourned for him greatly, and punished the rebels who were the cause of his death. For his kindness to the Consul, Queen Victoria sent several presents to him, among which was a revolver, whereon the following was inscribed on a silver plate upon the stock : ' Presented to Theodore, Emperor of Abyssinia, by Victoria, Queen of Great Britain and Ireland, for his kindness to her servant Plowden, 1861.' These presents were carried to him by Plowden's successor—Captain Cameron—a gentleman

T

who had formerly served in India as a commissioned captain in Her British Majesty's Service.

Theodore still followed up the idea of manufacturing cannon, and one after another was turned out of the foundry, tested, and proved to his unbounded delight to be terrible weapons of warfare, by their tremendous noise and the huge shot that they hurled out of their wide mouths. Certainly the day was coming when he would be able to make war upon the cursed Egyptians.

Egypt, under Abbas Pacha, was a very poor disorganised country, and it was not until Said Pacha ascended the viceregal throne that it visibly improved and began to be the compact power that it now is.

On February 9, 1862, Captain Cameron, the new Consul gazetted for Abyssinia, arrived there, presenting himself before Theodore with his credentials, and, better than all, with rich presents. The Emperor was dressed in royal robes and surrounded by gorgeously apparelled functionaries of state.

The same kindness and courtesy extended to Plowden was shown to Cameron, and all things for a while went well. Soon after, the Rev. Mr. Stern and Rev. Mr. Rosenthal, missionaries, arrived and were equally well received. Other missionaries also found their way there by-and-bye from the Scottish Mission and the Society for Promoting Christianity among the Jews. They were all welcomed by the Emperor.

It was at this time that his cruelties began to be notorious. The disaffections of the provinces embittered his life greatly, his whole nature underwent a violent change. He also began to lead a very intemperate life, and in his drunken fits his atrocities were absolutely fiendish. Brands were impressed upon the foreheads of deserters, and traitors were laid on the ground and stakes driven through their hearts.

Severe punishments for treason are common enough in the history of civilised nations ; but Theodore practised his cruelties upon innocent people. The least suspected had his back flayed with the *courbach*, or had his stomach ripped open. People were crucified and shot without mercy. As a consequence his name began to be execrated by all his subjects.

A short while previous to the arrival of the British Consul in Abyssinia, Earl Russell withdrew British protection from the

Abyssinian Convent at Jerusalem, at the same time that a commissioned agent was en route to the Emperor of Abyssinia, whose favour was certainly worth cultivating, while the Government incurred the expense of sending an agent there; but Lord Russell thought otherwise.

Consul Cameron, according to instructions, followed in the footsteps of his predecessor, and urged the Emperor to agree to an amicable treaty between the two nations, and entreated him to send an embassy to the Queen of England.

Theodore finally did consider upon the subject. He wrote a letter to Queen Victoria, wherein, after the usual grandiloquent phrases of Abyssinian Emperors were passed, he declared his intentions of attacking the Turks (or the Egyptians, which are precisely the same in the Abyssinian language,) for their unprovoked attacks upon his people at Bogos, and for their repeated depredations, which were as unwarrantable as they were unjust, and for many other things all clearly told and succinctly set forth. In the same letter he expressed a hope 'that lasting good will may exist between their two countries, which must redound to the glory and advantage of both,' and at the close of the missive he requested Her Majesty to prepare the means whereby his ambassadors may reach England, 'for so soon as he shall be made acquainted that all is ready, the embassy shall proceed to the sea coast.'

Consul Cameron, while on his way to Massowah, was stopped by some rebel chief, but he contrived to send the letters of the king by a native courier.

The letter arrived in England safely in February, 1863; was received by Earl Russell, opened, read, thrown upon the table, docketed, and in the pigeon-hole it rested.

The Consul was released after a short confinement among the rebels; and he afterwards interested himself in the suppression of slavery, until he became a very thorn in the side of the Pacha. Representations and complaints from the Pacha to Earl Russell were immediately attended to. Said Pacha was condoled with for his inconvenience, and Cameron was told to deport himself to Massowah, an Egyptian trading port near Annesley Bay; and there 'to follow the occupation of his consular duties.'

Not long afterwards, as the war in America caused great distress in England, and a new field for growing cotton was eagerly sought after, the Foreign Office sent instructions to Cameron to visit the Soudan and report on the prospects and capabilities of that country for the purpose.

The visit of Cameron to an Egyptian province was interpreted wrongly by Theodore. When they met again the Emperor questioned him in a cold imperious tone as to where he had been, to which the Consul replied, that he had been sent to the Soudan to find out about the prospects of cotton-growing and trade. Theodore asked him if he had brought him an answer from the Queen to his letter, which he had sent nearly a year before. Cameron replied in the negative, and Theodore flew into a passion, and said that since 'the Queen could send him to visit his enemy the Turks, perhaps to conspire against him, and could not write a civil answer to a civil letter, then he (Cameron) should not leave him until the answer came. Thus in July, 1863, Cameron became a prisoner. Not long after the answer from the French Government arrived. Very unsatisfactory, it seems, were the contents, for the French Consul Munsinger was hustled out of Abyssinia immediately.

Three months after his imprisonment Cameron received an answer from Lord Russell.

But the excellent old statesman had forgotten one important point—there was no allusion to the request of Theodore, nor even a word in reference to his letter. Maddened at the slight, the Emperor ordered Cameron's servant to be beaten. At the same time, the missionary Stern's two servants were beaten so cruelly, that they both died the following night. The poor missionary, horrified at the spectacle, put his hand on his mouth to repress the rising cry of horror. This simple movement was understood by the suspicious Emperor as a revengeful threat, and he at once cried out to his men, ' Beat that man, beat him as you would a dog; beat him I say.' The soldiers at once fell upon him, threw him upon his face on the ground, and they beat him with their sticks until he fainted. From this month of October Missionary Stern dates his long imprisonment.

In the employ of the Emperor was a Frenchman named Bardel. He was particularly conspicuous for his hatred to all Englishmen, and he suggested to Theodore that he should open the missionaries' trunks and make a search among their papers to find out their real business in the country.

All Europeans in Abyssinia, particularly Englishmen, were regarded with suspicion by Theodore. He ascribed the silence of Queen Victoria to the representations of Stern, and the indifferent letter from the Emperor Napoleon to European intrigues. Indeed he had cause to think so.

The Vice-Consul, or assistant to Cameron at Massowah, was a Captain Charles Speedy, a giant of a man, fully six feet six inches in height. After a year's trial of consular duties at Massowah, he gave them up and started for the interior of Abyssinia to shoot elephants. He was accompanied by a young cousin of his named Kerens, an Irishman by birth. It seems they were both highly successful in the pursuit of 'big game,' and the Telamonian Ajax, from his gigantic size and strength, acquired quite a reputation. When Cameron commenced the term of his imprisonment Speedy arrived at the Emperor's court. His herculean proportions excited the admiration of Theodore. He was invited to serve him in the capacity of a centurion. Speedy won his favour, but not to the same degree that Plowden did formerly.

The Emperor had lost much of the impulsive kind-heartedness that was once his prominent characteristic. In a battle the English centurion distinguished himself so much by his desperate courage and great strength that Theodore made him a commander of a thousand men and called him by the prenomen of Basha—' Basha Felecca'—or the *Speedy* commander. Speedy by and by, became ambitious, or at least Theodore's Ministers thought so, and many intrigues were put in operation to oust the white favourite. Basha Felecca had a Turkish scimitar, the gift of an Indian Rajah, with him. With this weapon, a pure Damascus blade, he used to amuse Theodore with splitting sheep in two from head to tail, which feat no Abyssinian could imitate. By such strength of hand-work he kept Theodore's attachment to himself for a long time.

After residing with the Emperor some eighteen months, Speedy was despatched on an expedition to Gondar. While on the way thither, he was met by a dozen men, under the leadership of Ras Feet Ourarie Guvrie, who for some reason—probably jealousy—hated him thoroughly. He was commanded by the Ras to give up his arms and return with him to Theodore, who was at Debra Tabor. Speedy demurred, and invited the Ras to come and take his arms, warning him at the same time that if he came within reach of his arm he would split him in two with his scimitar in the same way as he had seen him split sheep. The Ras said that he did not want to have recourse to violence. ' Well then,' replied Speedy, with an ominous flourish of his keen weapon, ' return to your master, and tell him that so soon as I finish my duties at Gondar I shall return to him.'

Basha Felecca went on his way, and Ourarie Guvrie returned to the Emperor's camp where he enlarged upon Speedy's obstinacy, and gave in detail a long story of the meeting and conversation he had had with him, which, according to Guvrie, was not complimentary to His Majesty.

In the course of time, after he closed the Gondar business with full satisfaction to the Emperor and himself, Speedy presented himself before Theodore at his camp on Debra Tabor (Debra means mountain), and the following dialogue is said to have taken place between them :—

Speedy. Well I have come back.

Theodore. I see you have ; why did you not come when I sent for you ?

Speedy. Because I did not believe you could reward faithful services by punishment.

Theodore. What, did Ourarie Guvrie tell you that I was going to punish you ?

Speedy. He did.

Theodore. Then he told you more than I knew myself.

Speedy. That may be very true ; but I have finished your business at Gondar well and faithfully ; I now ask for my pay and a discharge from your service.

Theodore. Oh, no, that cannot be. I cannot part with the best soldier in my army.

Speedy. King, you may be a great man, you may have a heart and reason ; but though I am not a king, I have a heart and reason also. When I came here first you requested my service, promising me good pay and promotion. I agreed to serve you ; I have served you faithfully. I have been on the battle-field, and with your own eyes you saw my conduct, and promoted me there and then, but no money have I received from you since I have been with you. The bread and salt I have eaten I purchased with my own money ; the servants I had I paid out of my own money. While on my road to do your bidding, you, listening to lies told in your ears by such fellows as Ourarie Guvrie, sent men to fetch me back, for what purpose you know best. I demand my money and discharge.

Theodore. And where will you go to?

Speedy. I will go home.

Theodore. Suppose I choose that you shall not leave the country?

Speedy. Then somebody will get killed, for I will fight, and you know I am not an infant.

Theodore. Will nothing persuade you to remain with me?

Speedy. Nothing. I am tired of the treachery and intrigue that I have seen around me.

The Emperor, finding that nothing would move Speedy's determination to leave, paid all the money that was due to him, gave him a horse, a shield and spear, and took an affectionate leave of him.

Captain Speedy, ex-Basha Felecca, took Kerens along with him to Massowah. Young Kerens went to London with a letter from Consul Cameron. Speedy, roving in search of fresh adventures, like another Quixote, went to New Zealand, where he was made a captain of militia and distinguished himself in the Maori war. We shall hear soon of him again.

Kerens was despatched back to Abyssinia with a letter from the Foreign Office to Cameron. He arrived in Abyssinia in November 1863. The letter made no mention of having received any communication from Theodore. This long silence, regarded by the emperor as a purposed insult to his dignity, rendered him fiercely indignant; and the consequence of Lord Russell's diplomacy was, that Consul Cameron, the missionaries, and attendants were put in chains.

Kerens was also very unfortunate, for he brought a carpet as a present to the emperor, and on the carpet was a picture of Jules Gerard, in Arab costume, about to kill a lion. Theodore fell into a rage at what he regarded as an insulting absurdity. To his oblique vision, the lion about to be shot by Jules Gerard represented the 'Lion of the tribe of Judah' (the Abyssinian crest) about to be conquered by the Egyptians, and Kerens was therefore unmercifully flogged and put in chains with the rest.

Some efforts were made to release the prisoners by philanthropic Englishmen. Mrs. Stern, on being informed of the captivity of her husband, petitioned the Queen to send a letter under the sign-manual to Theodore. Lord Shaftesbury kindly interested himself in their behalf, and presented the letter to the Premier, but it was returned with a ' I don't think it advisable that the Queen should write to the Abyssinian Emperor.'

Continued persistence in this harsh policy, if policy we must call it, brought a concatenation of punishments, indignities and hardships, upon the heads of the persons who were languishing in captivity. Theodore became more morose, more cruel, and writhed under the insult to his high imperial dignity, offered to him, too, by a woman. He avenged himself upon the captives, and most lamentable are the tales of their life in bondage.

With money a man may do anything almost in Abyssinia, and ' Samuel,' a chief who was a great favourite with Theodore, and who had followed Cameron to and fro through Abyssinia before his incarceration, was found amenable to bribes. For money, Aytoo Samuel would undertake to take letters to Massowah, or secure trustful couriers to convey them for him. In one of these letters thus sent, Cameron explained at length the causes of his captivity and his sufferings, and said, ' There is no hope of my release unless a letter is sent as an answer to his Majesty's.'

This letter of Cameron's found its way to the ' Times ' newspaper, and then went through all the ramifications of the British Press. The ' Times ' thundered and uttered indignant censures upon the imbecility of the Government, which were echoed with more or less noise by innumerable papers. The Minister succumbed before the storm of indignation. He immediately advised Her Majesty to write. The letter was written and sent off.

Mr. Layard, M.P., while pursuing his search for antiquities on the shores of the Tigris, had a secretary who acted as paymaster to his workmen.

This secretary was the Armenian Hormuzd J. Rassam.

When the author of ' Nineveh ' arrived in England ; he did not forget his paymaster. Rassam in the meantime found employment with Colonel Merewether, Political Resident at Aden, after the denunciation by the Press, of the policy of the English Ministry. Rassam confidentially wrote to Layard and stated his readiness to convey the Queen's answer to Theodore.

The grateful antiquarian bestired himself to get Rassam appointed as the messenger. He was successful, and Rassam was the man who was sent off with the long-delayed letter. He was accompanied by Lieutenant Prideaux, and Surgeon Blanc, both British officers, and his title was 'Envoy Extraordinary,' rather a high office for an 'antiquarian scribe' and paymaster.

When Rassam arrived at Cairo, in February 1865, there was discovered that an important feature was omitted in the construction of the state despatch, viz. the Queen's signet. The messenger was recalled. With the document complete, the envoy started anew for Abyssinia, having received instructions to render an explanation of the delay favourable to the English Government.

After due time the envoy, whilom paymaster, arrived at Massowah. At this place Rassam remained nearly one round year, for what reason is not very obvious.

During these long weary months the captives languished in horrible durance. Things went on with them from bad to worse. They were fed on bread and water, beaten, chained to a wall. Cameron was tortured. He says, 'Twenty Abyssinians tugging lustily on ropes tied to each limb until I faint. My shoulder blades were made to meet each other. I was doubled up until my head appeared under my thighs, and while in this painful posture, I was beaten with a whip of hippopotamus hide on my bare back, until I was covered with weals, and while the blood dripped from my reeking back, I was rolled in sand.' Yet Rassam lingered at Massowah. In consequence of debates in Parliament, wherein many things leaked out not very complimentary to Layard's protégé, Lord Russell found himself obliged to commission Mr. Palgrave, the noted traveller, to proceed to Abyssinia to endeavour to effect the release of the captives, and Rassam was recalled.

The recall found the envoy in Massowah. He at once set out for Suez, and from there he telegraphed to the Consul General at Alexandria that Cameron was released. The glad news was flashed along the cable to England, where it was received by all with joyful hearts. Palgrave's mission was ended of course, though he had arrived at Cairo. The presents which he had with him he gave to Rassam, who immediately started for Massowah to deliver them, probably to the chiefs who should accompany the released Consul to that port.

From Massowah the envoy proceeded to the interior, accompanied by his companions Blanc and Prideaux, and never halted until he arrived at Debra Tabor, January 28, 1866. ' Illustrious Potentate, may you live for ever ! ' was the salutation of the man chosen to represent England before Theodore, Emperor of Abyssinia. The letter of the Queen was not what it ought to be ; it was not respectful enough, and Cameron was not released, but Rassam, Prideaux and Blanc were added to the batch of captives Theodore already held. Rassam was not beaten however, he was too cunning a diplomat for that, and what was lacking of respect in the Queen's despatch, Mr. Rassam voluntarily supplied. Theodore's savage bosom was a welcome receptacle to flattery ; and so much kindness as it was possible to exhibit to a prisoner was shown to the Envoy. Cameron's condition was bettered also through Rassam's influence.

Three months after the detention of the mission, Mr. Flad arrived in Alexandria, and gave the first information of its fate. Subsequently he returned to Abyssinia, and he also became numbered with the captives.

Theodore perceived his power gliding away surely from his hands. Neither his strength, his ability, the prestige of his former prowess, nor his cruelties, could keep his provinces intact. One by one the governors rebelled. The Gallas mustered their forces, and attacked him in the rear. For a time Theodore bore up stoutly, but the many reverses which he suffered elated his enemies and discouraged himself. But, wherever Theodore pitched his camp there he was impregnable, because of his cannon. His European workmen had manufactured for him twenty cannon, some of them of monstrous size. Monster guns became an infatuation with him. Twelve and fifteen inch mortars were what his soul delighted in. The roar of their discharge, he calculated, would affright his enemies. The reputation of these huge guns spread abroad through the land, and no army was found bold enough to attack him in his camp, but they rebelled all the same, and his cruel punishments, as he was a rigorous martinet, drove his soldiers away from him by the hundreds.

We are now come to the circumstances through which war was declared. The causes have been plainly set forth. The English prisoners were still in captivity. The mission for their release had failed. Where would it all end? Let us see.

In 1866 the Conservative party came into power; the Earl of Derby became Premier; Lord Stanley, his son, filled the Foreign Office chair. Before proceeding into extremes the Ministers made one more effort to procure the release of the prisoners by proposing a ransom. This also failed. Theodore, seeing how anxious the British Government was to get the prisoners, thought, as a natural sequence, that the English people would do anything rather than have them kept in captivity. One part of Theodore's supposition was correct—the English people would indeed do anything rather than the captives should long remain in durance; go to war even, but never help him to keep the throne he had disgraced. And so the Government declared war.

It is remarkable that there was not the least enthusiasm shown by the people of England upon the announcement of war. Its prospects were discussed calmly, nay philosophically. The least approach to anything like a kindling interest in it brought down upon the unfortunate editors a host of letters croaking about omens, signs, and dangers. One correspondent signing himself 'True John Bull' declared it his opinion that not a man of the army would ever return from the country; the 'British Taxpayer' saw visions in the heavens; the farseeing 'Hertfordshire Incumbent' prophecied that the rain would sweep every trooper into the sea; the 'Jolly Briton' affirmed most positively that the British troops could never capture Theodore; while the 'Cambridge Student,' having in his mind's eye Trojan's works, asked whether the general that was to be was to build bridges across the ravines; remarked that if so Abyssinia was not Greece, nor, as a self-evident axiom, was Greece Abyssinia, and he could not imagine how ravines four thousand feet in depth could be crossed without bridges. This intelligent person also observed that English generals were certainly superior to Trojan, but even they, adopting Trojan's tactics, could never hope to bridge such wide, deep gaps in the earth as Abyssinian ravines.

In short, there was no known horror, danger, or disease, which was not prophesied would befall the Expedition. The tsetse, the cerastes, the deadly adder, were to be amongst the blankets and clothes of the troops. Physicians warned the soldiers to beware of a certain small pink-headed fly; not to touch honey on the roadside ; not to taste any water found in old wells ; to look out for ophthalmia ; to keep a flannel belt round the waist; to apply instantly for medicine when afflicted with symptoms of diarrhœa, and to make the best they could of the society of hyenas and hippopotami, the common pests of the country. Dysentery, disease, and discomfort were to dog the footsteps of the British rank and file.

Men who had travelled in other countries proposed multitudinous absurdities, and people who had never travelled anywhere except to the London theatres, drew wise corollaries from their own experiences, which they offered *gratis* to the Foreign Office. Again, loyal Englishmen offered their services to release the captives by their own unaided efforts. They were all confident they could perform the task without the least danger or inconvenience to themselves. They would undertake to do so simply for the fun of the thing. Others ventured to hope that the Foreign Office would accept a plan which could not fail of success—that of taking a company of cavalry and dashing into the camp of the Abyssinian Emperor, take them off by force ; even the 'Times' special correspondent avowed his belief that were Colonel Merewether sent with a squadron of cavalry the whole thing could be accomplished in less than a month.

Captain Smelfungus wrote to the 'Times' and Lord Stanley stating that an English army could not exist in Abyssinia without an auxiliary force of Houssas to assist as an irregular detachment, to scour the country for forage, and to be sent as scouts ahead of the regular troops. He offered his services as Colonel of the force. As Houssas are only found near Lagos, the Foreign Secretary could not see how Houssas were procurable in Abyssinia. But why follow the improbabilities and impossibilities further ; are they not all chronicled in the Abyssinian Blue Book ? Where was the army to come from ? It was decided that they should all be Indian troops, European

and Sepoy regiments serving in Bombay, and England should
only supply the stores and baggage animals. Accordingly, Sir
Seymour Fitzgerald the Governor of Bombay, and Sir Robert
Napier, Commander-in-Chief of the Bombay army, after being
notified of the decision of the Foreign Office, made their preparations. The army was to consist of about 12,000 soldiers;
2,000 men for the sea depôt at Zoulla, and Senafe on the
Abyssinian Highlands, 2,000 men for the post of Antalo, half-
way between Zoulla and Magdala, 2,000 men to keep open
communication, and 6,000 men for the marching column.

Officers were despatched to Spain, Egypt, and Syria, to buy
animals. The great arsenals of England were to furnish the
cannon, saddles, chains, and halters, tents, and stores of all
kinds. Hospital-ships were to be sent to Annesley Bay for the
reception of the sick, and men-of-war were to guard the port
of Zoulla.

The plan of the campaign was laid down by Sir Robert
Napier with great distinctness. There was no possibility of
failure if once Theodore could be found. Sir Robert, himself,
chose the regiments, and though the other presidencies—
Madras and Bengal—were anxious to furnish some small portion
of the necessary quota, the General thought that, to ensure
harmony, it would be best the troops should all belong to the
same Presidency. The point of debarkation was to be chosen
by Lieut.-Colonel Merewether, the Political Resident at Aden,
a most gallant, brave, and skilful officer. He surveyed the
whole Abyssinian coast. Annesley Bay, though the furthest
from Magdala, was chosen for its fine commodious harbour, its
proximity to the mountains, and for its water supplies.

The distance from Tajurrah Gulf to Magdala was 150 miles;
from Amphilla Bay, 250 miles; Howakil, 325 miles; and from
Annesley Bay the distance was a little under 400 miles.

The next step was to procure interpreters who could speak
both the English and the Abyssinian languages. The Revd.
Louis C. Krapf, an Abyssinian traveller, was selected for one,
but he was rather elderly for a campaign. Mr. Henry Dufton,
another traveller, professed to know something of the language,
and was appointed to the Intelligence Department to help the
Commissariat and the Quarter-Master General with his few
phrases. His duties were somewhat anomalous. Lord Stanley

was finally made acquainted with the name of one who was thought would prove the right person—Captain Charles Speedy, formerly the ' Basha Felecca ' of Theodore's army, just then serving as a militia captain in New Zealand. A telegram was sent to him, *viâ* Bombay, to report himself to Sir Robert Napier. The invaluable services which he rendered the English army will be found clearly detailed in the following pages.

In the month of November 1867, the Advance force of the Abyssinian expedition landed upon the beach of Zoulla.

Now, reader, having been so explicit upon the series of accidents that led to this modern Crusade, and having traced the consequences of mis-diplomacy, and prepared you by a brief outline of the preparations for the campaign, and having launched the Advance fairly upon the shores of Annesley Bay, I propose that we go together to this fabled land, and follow the adventurous soldiers and gallant old Chief to the planting of their banners upon the heights of Magdala.

CHAPTER II.

ANNESLEY BAY—COLONEL MEREWETHER.

ANNESLEY BAY is formed by an island named Dissee running parallel to the coast of Abyssinia for a length of twenty miles, separated from it by the Bay, which is between one and five miles broad.

No better choice could have been made for a pioneer than Lieut.-Colonel W. L. Merewether, a cavalry officer serving in India, and the Political Resident at Aden. This gentleman, probably from previous intimacy with Hormuzd J. Rassam, as soon as he was made aware of his captivity, bestirred himself to release him. He sent money to the prisoners, he bribed Abyssinians to carry despatches to Magdala and Debra Tabor, he wrote several articles in the most influential London journals, he published a treatise or two, calling on Great Britain to save herself from disrepute in India, and threatening that the knowledge of the captivity of the English envoy at Magdala, once known among the Mahomedan tribes of Hindostan, would be the means of exciting a universal rebellion among the subjects of the Queen in the East.

Merewether, a most valuable officer, who had proved himself brave and skilful, dashing and intrepid, in various battles in the East, was selected to command the Pioneer Force, which was to land on the coast of Abyssinia, to institute enquiries among the inhabitants, and to prosecute an examination into the difficulties to be met, and means of access to the country. Young and energetic, he lost no time in scouring the coast to seek a landing-place, whereon was planted the Cross of St. George for the first time by his own hands.

The Abyssinian coast bordering the bay, was a stretch of sandy beach, fourteen to fifteen miles wide, extending from

Massowah twenty miles above, to far below the southern extremity of Annesley Bay.   About three-fourths of the distance from the embouchure of the bay to its extreme southern end, on the low sandy beach, just four miles from the village of Zoulla, Lieut.-Col. Merewether pitched his Head-quarters.   A more inhospitable looking-spot could hardly have been found. But this spot was chosen for two reasons : first, because it bordered a fine deep bay, where the *matériel* of war could easily be landed ; secondly, because a great gap, or a lengthy chasm, splitting the Black Highlands in two, allowing free access through its cavernous recesses to the summit of the table-land had been found—the entrance to which lay exactly fifteen miles direct west of Merewether's Head-quarters.

Colonel Merewether upon landing had despatched his letters, containing the fruits of the rigorous researches and the results of his investigations, to the Governor of Bombay and Sir Stafford Northcote of the India Office in London ; but unfortunately he had painted the prospects far too glowingly.   According to this statement water and wood were abundant, grass was plentiful, the verdure was most refreshing, the Pass to the table-land was comparatively easy, and the country was rich with game of all kinds.

After the receipt of the Colonel's despatches the gentlemen at the India Office in London and the authorities at Bombay vied with each other in getting ready the expedition for this bright and heavenly country.   Sir Robert Napier, Lieutenant-General, was chosen as chief of the force.   By his advice and counsel the number of men necessary to accomplish the undertaking was fixed upon.   The various regiments were collected together at Bombay before starting, and reviewed by the commanding powers, whereat glowing speeches were declaimed.

So much were the expectations of the powers at Bombay raised, that they actually wished to send the soldiers at once to Zoulla, with only a few days' rations ; they limited the supplies in each and every particular to a starving amount, and had it not been for the dilatory policy and firm resistance to the nonsense of civilians on the part of Sir Robert Napier, grave and disastrous evils must have ensued.

The Advance Brigade—consisting of Head-quarter wing 3rd Light Cavalry, 10th Bombay Native Infantry, 3rd and 4th

COLONEL SIR W. L. MEREWETHER, C.B., K.S.I.

Companies Native Sappers and Miners—was finally landed at Zoulla, as a garrison and a nucleus for the expected force. With this brigade there were two correspondents of influential English journals—the London 'Times' and the 'Daily News'—and the sufferings which the reporter of the 'Times' underwent, are they not graphically recorded in the columns of that journal ?

There was but one Brigade landed at Zoulla when several transports conveying some thousands of mules and camels and horned cattle, without muleteers or drivers to attend them, arrived in the bay and began to discharge their living cargoes. When landed on the beach the animals, unhaltered, roamed around in search of pasturage after their sea voyage, and not finding either grass or water, wandered about in the wilderness of baubool shrubs and juniper bushes, until they fell down and died by the hundreds. Many of them were stolen by the Shohoes, and had a smart, cunning horse-dealer been present with a vessel at Massowah, he might have been able to convey the mules and camels away back to India nearly as fast as they were landed in Abyssinia.

It was not until two or three thousand animals had died or got lost, that the commanding officer began to perceive that a remedy for all this mortality was necessary and easily procurable.

The steam transports then in the bay were detained, and set to work at once to condense water. A despatch boat was sent to Bombay for a dozen American condensers and two or three score of Norton's American pumps.

Three or four weeks after the order for condensers had been given, they arrived at Zoulla ; at the same time several thousand coolies arrived in ships. These were set to work at once to prepare the piers and bunders for the reception of other heavy stores about to be landed.

The piers and bunders were completed in an incredibly short space of time. Then the Stone Island was constructed for one of the largest condensers, and a trough was made to convey the fresh water to the water-butts inshore, where it was collected and stored for use.

The arrangements now determined upon, were carried on with vigour ; more camels, and mules, and horses, were continually arriving, and though the mortality among the animals was still great, it was evident that it was greatly lessened by the water supply. With these new shipments of animals, which came from Suez, and which had been sent there from Spain, Italy, Syria, and Egypt, came Turkish and Arab muleteers. This was another unfortunate and most unwise act of those in command.

Any traveller who has had any dealings with these men, knows well the terrible annoyance and disagreeableness which Turks and Arabs provoke.

Shortly the troops under Colonel Merewether, who were pushing ahead to occupy the Highlands were made aware that their baggage and commissariat animals had disappeared with their drivers.

The wholesale desertion of men and animals took place daily, until the commanding officers of detachments, bound for the front, were made aware by frequent losses of such kind, that armed and disciplined soldiers were necessary for other purposes than to march in solemn column in the front, and that it would be an extremely wise thing to detail a few to guard the rear and baggage. Experience in all things is dearly bought, and the English army had rusted so long in their barracks in Bombay and Chatham, that very many things appertaining most essentially to campaigning had been forgotten.

Troops were being concentrated rapidly at Zoulla, when, after a reconnaisance of the Pass ahead, the general, Sir Charles Staveley, decided that the head of the Pass—Senafe—a village situated on the highlands, should be occupied, and be the second depôt on the route to Magdala. A small force was already in possession of the place under Merewether, but as the situation was reported healthier than Zoulla, and fresh water was in a plentiful quantity, and grass, and straw, and barley, and oats, were procurable from the natives, the major part of the force was also sent on to Senafe.

A passage to the Highlands had been made comparatively easy after a great deal of labour. In order that readers may understand it thoroughly, and to simplify its description, I will state that the Pass of Koomaylee was a deep ravine, cut by the force of torrents across several ranges of mountains from the Highlands of Senafe, which rose 7,000 feet above the level of the sea. This pass or ravine descended for a distance of 63 miles to the beach of Zoulla, along the bed of a river which, during the rainy season, was swollen to a powerful body of water.

In the summer season, however, it was dry, and at the time the pioneers passed through, its whole course was choked with great boulders and *débris* from the hills. For the passage of the troops the most unobstructed portions of the ravine had been selected to form a wagon road, but in some places—as the Sooroo, the narrowest part of the pass—a road had to be cut through solid granite. To the Engineers, who had charge of this work, great praise is due. Taking into consideration the time which they had at command for this work, no better road could have been made. The soldiers marched, and baggage, animals, artillery, and wagons, passed over with the greatest facility to the heights at Senafe.

In about two months everything was ready for an onward movement, and Sir Robert Napier arrived. After reviewing all that had been done—the erection of warehouses, bunders, pier-heads, condensers, and the railway—he passed on to Senafe to enjoy the cool mountain breezes.

A short railway had been laid from Zoulla to the extreme end of the camp, a distance of five miles. Two locomotives had been landed, and these proved very serviceable in removing the vast and rapid accumulation of stores to more convenient sites and to different commissariat sheds along the line.

The next morning after my arrival in Abyssinia, before anyone was stirring in camp, my servant Ali and myself left Zoulla in excellent condition for rapid marching, but scantily furnished for a long campaign.

My ideas concerning Abyssinia and the English manner of conducting a campaign were yet crude, as will no doubt be perceived as I proceed.

From Zoulla to Koomaylee was a distance of 14 miles. The route, after passing the camp at Zoulla, lay over a desert of sand, dotted at intervals with clumps of baubool shrubs, furze and juniper bushes.

After passing the native village of Zoulla, we began to as-

cend wild and uncultivated eminences, overgrown with low
shrubs, and broken here and there into deep furrows by the torrents which plough their resistless way during the rainy seasons.
Now and then we crossed the dry bed of one of these spring-
time springs, choked up with round stones, and then were compelled to ascend a steep hill, and shortly cross another deep
gully channelled in the soil.   Not a drop of water could be
found anywhere, and we learned at Koomaylee, that no water
fit to drink was found on any part of the beach ; that Colonel
Merewether had been misled when he reported that water was
plentiful, from the fact that the rainy season had but just
closed before he arrived there.

He had found standing pools in certain places, but those in
a few days had disappeared entirely, owing to the sandy nature
of the soil, which had soon absorbed moisture.

After about five hours' toilsome march we arrived at the
base of the mountains, and mounting the summit of a small
plateau, the camp at Koomaylee became visible.   Here was the
entrance to the Great Pass which cleft the mountain in two,
and admitted the invaders into Abyssinia.

A great, wide, yawning gap was visible flanked on each side by
rising ranges whose precipitous sides were clothed in tangled
brushwood and kolquall trees.   Grey, weather-beaten rocks jutted
here and there out of the mass of dark verdure, with lichens and
mosses clustering about their sides.   Above the nearest range of
mountains rose others loftier, grander, till the summits of the
furthest seemed to kiss the very clouds, and these appeared to
enwrap and fold those within into a vast amphitheatre, through
the centre of which the Pass ran its black and gloomy way.
Had this deep gorge not been thus temporarily inhabited, a
more mysterious looking spot could scarcely be found.

The site of the camp was three-fourths of a circle.   Two
great hills at either end stood like Titan warders to guard the
entrance.   These two great hills were connected by narrow
ridges with the lofty heights beyond.

Before going within the amphitheatre thus formed we
turned around to bid our adieus to the sea and view the prospect.
No brighter one could have been imagined.   At our feet expanded a plain covered over with baubool bushes and other
shrubs, which hid the sand from view.   Beyond the dark

green an expanse of white, glistening, burning sand; beyond this again were the tents of the camp, looking like snow-heaps scattered far around; further off a noble fleet of ships riding at anchor in the bay, each shroud and rope seen distinctly, so clear was the air, so cloudless the sky; bounding their domain and sleeping sluggishly under the noonday sun, was the island of Dissee. From our elevated position, the eye skimmed across the island and traced the blue and placid surface of the Red Sea, burying itself in the distant horizon. Twenty miles or so to our left, the flat-roofed buildings of the Arabic town of Massowah rose to view, nestling in groves of green trees.

A few steps forward, and we were hidden within the confines of the dark Pass, severed from what we fondly imagined home, friends, civilisation, and by a ridge of hills which interposed themselves as obstacles to the view.

Norton's American pumps had worked wonders in the camp at Koomaylee. Water spurted up in fine style. The sickly camels and mules enjoyed the precious beverage, and under the influence of the cool shadows flung over the camp by the stupendous heights above, and the soft and pure airs which blew down the Pass from the Highlands, they improved in a very short time. The sick soldiers in the hospitals recovered after a short residence here from the languor and general debility a few weeks stay at Zoulla had generated in their frames. In fact the whole camp appeared different from Zoulla.

The natives, who were of the Shoho tribe, flocked around the camp with their goats, their sheep and cattle giving a truly pastoral appearance to the scene. The goats climbed the grey rocks on the heights above, overhung with tufted trees, nibbled the creepers and clusters of shrubs growing under their umbrage; the sheep bleated in melancholy tones below; the cattle lowed, and the swart and naked Shoho children shouted lustily or played upon their reed pipes.

Bazaars had already assumed vast proportions on the ground that but a few short weeks since was the home of howling beasts of prey. The booths—made of straw mats—were laid out in two parallel lines to form an open street, and, but for the flimsy material of the dwellings, it would well have borne comparison with a street in Pekin. Looking up the bazaar, we commanded a view of the multitude of commodities with which it was supplied. Nets of onions, tobacco in leaf, and brown heaps of the fragrant weed grown in Latakiah and imported from far Stamboul, were temptingly arrayed on wooden trays, in vari-coloured paper; figs strung together on rushes, and figs in round cases; pipes with amber mouth-pieces, pipe-bowls with quaintest figures engraved on them; common English clay pipes, narghilehs from Grand Cairo, and hubble-bubbles from Persia and Delhi; jars full of black olives, pocket glasses, in red pasteboard cases; pickles, gherkins, sweets, jellies, cognac, claret, soda-water and innumerable other things from Paris and London.

Those things not to be found here were of no use to soldiers on a campaign.

There were no books, no lamps, no newspapers, no post-office— people never perceived the want of them—while dates, figs, nuts, raisins, and all sorts of liquors proved to be very saleable articles, and there was an abundance of these things. The bazaar was often frequented by the Shohoes and camp followers. Here the dark-coloured maidens of that country came unadorned and unclothed; here the wild children romped childishly gleeful in complete nudity; here the camp followers, men from Berbera, Donakil, Aden, from up the Persian Gulf and the jungles of India, dispensed with superfluous clothes, and sought the shelter of their open booths to loll in undisturbed freedom.

As I intended to make forced marches to overhaul the advance force, then at a station called Attigratt, I made out an indent on the commissariat department, and sent my boy Ali to draw the rations it called for.

It must not be thought for one moment that we received what was demanded by the indent; the thieving Parsees who had charge of the distribution of rations managed to retain very often one-half of what was our due. Very often, also, as we advanced into the interior, we were deprived of rice, ghee, sugar, tea, coffee, rum, potatoes, and vegetables, and at one time we were two months without either of the above-named rations.

When Ali returned from the commissariat it was nearly sunset, and instead of the light-hearted boy that he usually was, he appeared much depressed and very sorrowful. On asking the cause of his gloomy looks he answered in his broken English :—

'Arab man tell me in bazaar Ingliliz man kill me in Habesh land, me want paper, me go back to Suez.' And here he broke down and cried.

It seems that an Arab, restless and uneasy under the strict discipline of the English camp, had been telling him how English officers were so cruel that they generally flogged their unfortunate servants to death, and Ali's fears were so excited at the idea of leaving his bones in Habesh land—an Arabic name for Abyssinia—that he had determined to apply to me for permission to return to Suez, where I had picked him up. I managed to soothe and calm his fears for the time by bribing him to go with me as far as Senafe.

On leaving Zoulla, I had not burthened myself with a tent, satisfied that I could purchase one whenever I wished it, and deemed one really necessary. I was partly right and partly wrong. Had I procured my horses, and mules, and tents at Zoulla, I should have been more independent, but the rumour of the dash to be made upon Magdala, had completely upset my better judgment, and caused me to wish to hurry on at railway speed. But as the weather was mild, and I possessed a treasure of a buffalo robe to protect me from the dews, so long as I was marching I felt no inconvenience—I rather liked it than otherwise.

After a light supper, cooked by my nimble adjunct Ali, I lighted a cigar, and, folding my buffalo robe around me, lay down to think—not to sleep—I was too excited for that.

At early dawn we were in the saddle, and plunged into the yawning Pass. We left the camp, with its confused murmur breaking the stillness of the morning, far behind. We were led around abutments of rock, through jungles of mimosa and laurel bushes, past patriarchal sycamores, whose wide-spreading branches seemed alive with birds of gayest plumage. Round acute angles formed by forward hills, up along the base of precipices of sheet rock, and then through paths cut through dense jungle, we wended our way.

Anon the summits and slopes of the hills flanking our defile had undergone a magical change. From a dull grey they had assumed an orange hue, which rapidly discolouring, became silver. Gradually the silvery tints overspread the western slopes, and stole up the eastern side. Above, in the very centre of the heavens, the sun—a ball of liquid fire—was set, pouring his fiery rays remorselessly on our heads, denoting mid-day. At the same time we knew by the noise of voices that the next station, Sooroo, was near, and as we turned sharply to the right, round an abutment of earth and rock, the tents of the garrison appeared in view.

Dismounting from my horse, I sought out a shadowy spot, under a giant mulberry tree, and spreading a carpet there, lay down to rest. Ali first attended to the animals and provided a most bounteous feed for them, then a bright fire was made, the coffee kettle was next brought out and filled with water, and as it was mounted upon the glowing embers by its very extreme homeliness it evoked a smile of contentment from me, and while it rung its merry bubbling noise, the rich aroma of the coffee stole upon my wandering senses.

The two hours siesta over, we pushed on again to the next station, Undel-Wells.

For three miles each side of the Sooroo, the most difficult and dangerous positions of the Pass were seen. For six miles the traveller passes through a very narrow defile, flanked by walls of sheet granite soaring up in all the glorious majesty of a height of 800 feet on each side. Between these seeming infinities of stone we crawled on, not certain but that a rock might become displaced, or a branch of the pine which crowned their lofty extremities might be swept down into the yawning chasm. The least murmur of the human voice sounded in this awful depth like thunder, and the tread of the horses feet like artillery rumbling over a bridge. For a height of ten feet above us, were to be seen traces of the water which surged down the pass during the rainy season. Its extreme width was not more than twenty-five feet, while its narrowest was barely fifteen. In some places it looked as if Titans had been employed for centuries in chiselling and channelling the solid granite to pre-vent another deluge of the country. With all the thousands of ravines, and gullies, and fissures emptying into the Sooroo defile, for a distance of fifty miles, some idea of the vast body of water thus collected may be had, and the terrible force of the whirling torrent may be imagined.

Out of the darkness, out of the gloomy depths of the Sooroo, we emerged at last, into a more extended and much wider defile. Instead of the perpendicular walls of stone, lately passed, a hundred fantastic-shaped hills presented themselves, with their gently sloping sides covered by woody kolquall, groves of firs and pine, with low brushwood, juniper and furze. Then we travelled between ridges of stupendous mountains, with their crowns cut into shattered pinnacles, of dun-coloured rock, until we came to masses of ribbed rock and earth lying diagonally along the bottom of the defile, looking as if they were cast down purposely by some mighty power. Now and then the eye was attracted by the crystals, sparkling like diamonds when the slanting sunbeam lit up the cyclopean masses of granite and quartz ; and then again, the vision wandered to the splintered peaks and sharp and ragged outlines which they presented against the pure cerulean tints of the sky.

As the glorious sunset flushed the loftiest crags and tinted the boldest hills with all the prismatic lines of the rainbow, away at the furthest extremity of an avenue of firs, appeared the welcome fires of the camp of Undel-Wells, the termination of our double march on that day.

Another long day's march through scenes like those already described brought us to the base of a stupendous mountain up which the military road wound itself until it reached a gradually sloping ridge. Along this ridge we continued for half a mile, when, forcing our way up another steep hill, we arrived at last on the table-lands of Abyssinia, the altitude of which is fully 8,000 feet above the level of the sea.

A wide scope was here given to the vision. A great plateau slightly rolling, dotted here and there with isolated cliffs of bizarre aspect, honey-combed with numerous caves and dismal looking holes, split into wide yawning fissures, seemingly by volcanic action, torn into masses of the most unsightly crags, were things seen at the first glance.

At the base of the very highest, a mass of grey rock, upright and frowning, like another Acro-Corinthus, was Senafe—just a mile from the edge of the Highlands—the first village of any importance we had seen in Abyssinia.

Directly in front of the village, a few hundred yards off, and under the shadow of the tall grey rocks, in the fast approaching twilight, and covering a gentler eminence, was the camp of the advance force. Behind the eminence, and protected by a battery of field-pieces, mounted so as to command all approach, were the commissariat and the mule lines. Here and there on grassy knolls, some gentleman idler or privileged officer had planted his tent, so as to have the full benefit of the bracing breezes, the enjoyment of the charming prospect, or that exclusiveness Englishmen in general so delight in.

At respectable distances from the white tents were flocks of goats and broad-tailed sheep, attended by dusky urchins clad in dressed goat-skin, thrown very lightly over their backs, who from afar surveyed with silent wonder and admiration the red coats, bright steel muskets, and the prancing horses of the British soldiery.

As I had heard it said that it was extremely cold at night at Senafe, I determined to apply for a tent at once, and full of that prudent resolution, I directed my steps to the Commanding Officer's sumptuous marquee, which stood in the centre of the camp.   The commanding officer was General M——, with whom I had an interview, of which it will be sufficient to say that I left him with 'good evening,' and an assurance that if ever I rested in his tent again it would be because he had sent for me.

From General M——'s tent I galloped away with feelings of no great love for Englishmen, when I was accosted by a smartly-dressed officer, who asked me if I was in search of anyone.

'Oh, no ; no one in particular,' said I, with a laugh ; 'I am simply looking up a hostelry, where I can stay during the night.'

'Hostelries, my dear sir, are not very plentiful around Senafe, but there are several good-natured fellows who would be glad to take you in.'

'Oh, I believe you, sir, most certainly I do ; but, could you name anyone in particular ?'

'Why, yes; there is Captain Smelfungus.' (Readers of Sterne's 'Sentimental Journey' will not have forgotten that name). 'He has a big tent and is—'

What !' said I, 'is the famous Captain Smelfungus here ?

I have heard of that gentleman everywhere. First at Alexandria then at Suez, Zoulla, Koomaylee, and now he turns up at Senafe. Can you direct me to his tent? He is such a remarkable character that I must make his acquaintance while there is a chance.'

' Oh, yes; you see that tent on that little knoll all alone? you will find the captain there. As I see you are in a hurry, I will wish you good evening.'

' Good evening, and many thanks to you,' I responded heartily; and then putting spurs to my horse, I soon found myself before the captain's tent.

Coming down the slope from camp-ward, on horse-back, was a short, bushy-whiskered man, dressed in a brown suit, with a gun stuck in a saddle-bag, swinging very ostentatiously a couple of ibis he had shot, from which it was evident he had been out on a sporting excursion. Dismounting, and delivering his horse and the ibis into the charge of a couple of servants, who stepped officiously forward to aid him, and turning round, he for the first time saw a stranger in me, and stepping up with the blandest of smiles and gracefulest of apologies, this great man—this remarkable man—demanded of me, if he could in any way be of service to me.

' I came here to seek Captain Smelfungus, sir,' said I; ' I presume you are the gentleman.'

Drawing his podgy fingers through his bushy whiskers, and allowing his face to expand into another exceedingly bland smile, he said: ' Yes, sir; I am the gentleman.'

Smoothing with open palm all traces of the discomfort that had lately put me out of temper, pulling my beardless cheeks in imitation of his whisker-pulling, forcing a tremendous smile, and adopting his style of *empressement,* I informed him that I had heard so much of his talents, his charming eccentricities and his glorious achievements, that, I could not retire to rest without first paying my respects to him.

' Ah, sir,' said he in feeling accents; ' you do me too much honour; but may I ask to whom I have the pleasure of speaking?'

I gave him freely the pleasure he sought. At the declaration of my being an American he pricked his ears, rolled his eyes, and almost ruined his whiskers; but when I modestly hinted that I was a newspaper correspondent, indeed, I never imagined that such beatified feelings as seemed to possess him could be so eloquently expressed by signs, for his emotions apparently were such that words seemed to choke him.

He declared that I was a fellow-countryman—that I was his brother, not only by nationality, but by profession. His tent was mine, he affirmed, his servants were mine also, his cooking utensils were mine, and his rations were mine. What more could a gentleman of the Press wish? Here was a home, here were servants, plenty of cooking utensils, an abundance of rations, and an estimable countryman and brother.

'But come in, my friend; enter my canvas domicile. Why stand we out in the twilight, when a supper meet for the gods is spread ready for the guest.' After delivering his magniloquent, but hospitable address, my new found friend and myself entered the tent, where, in truth, a most savoury meal was laid out. At the same time my precious Ali arrived with his two animals, and these were immediately relieved of their burdens, and also fed out of the captain's abundance.

Feeling necessitated to say something, I said blandly—

'Pardon me, Captain; but you hinted something to me about your being connected with the Press, if I mistake not.'

In what way the great Captain Smelfungus made out his claim to be regarded as a very important and Protean Member of the Press; or what high commission from the British authorities he was charged with, I am not at liberty to divulge. I became from that time the messmate of Smelfungus, the recipient of all his old stories of life in India, and of service in the Confederate army; and much could I tell of his eccentricities and little failings, but the sacred claims of hospitality condemn me to be silent. It is enough to say that I never was more instructed, amused, or astonished than I was that first evening at Senafe. It was at a late hour of the night when we both rolled ourselves in our rugs and finally dropped asleep.

The sun was high in the heavens when I awoke next morning. The gallant Captain was in his brown suit, on bended knees, above a lengthy despatch box marked 'Sir X. Y.,' bristling with the many knick-nacks that so distinguished a gentleman should be supposed to possess. Now he tenderly handled 'stars,' 'orders,' in silver and gold; then a curious dirk which he exhibited to me as a present from the Maharajah of Putiala; then a scimitar given to him, so he said, by the Nizam of Hydrabad; a fancy Malacca cane from the Guicowar of Baroda; an amber necklace from the Begum of Bhopaul; a sword presented to him by the Duke of Cambridge; then a pair of wide-mouthed duelling pistols, presented to him by General Benedek, two days after the battle of Konigratz; a silver-mounted Colt's revolver, a parting gift from General Robert E. Lee, and hosts of other things too numerous to mention.

'Is Sir X. Y. a relative of yours?' I asked him, after again glancing at the box.

'Yes, he was a first cousin of mine. He lately died, and left me a handsome estate worth 7,000l. a year, situated in Dorsetshire.'

'Oh, indeed! a very handsome property that, and no mistake; but gentlemen well-connected like you, Captain, always drop into something handsome.'

'Yes, that's so,' said he, and then in a careless tone added, 'but I have half-a-dozen very pretty little estates like that, in different parts of the United Kingdom.' . . .

I asked him when he intended to leave for the front.

'To-night, by all means,' said he.

'Then that arrangement will suit me perfectly, for I am in first-rate trim for rapid marching.'

'I see you are,' said he with a smile, 'but I have quite a host of things, and I doubt very much whether my five animals will carry what I have; but it is an axiom with travellers, that what they cannot take with them they can always leave behind.'

Captain Smelfungus was indeed well provided with every requisite of camp equipage. He possessed a large American wall-tent, and a small tent d'abri, any number of camp kettles, and dishes, besides being well supplied with provisions; curry-powder, pickles, cracknel biscuits, brandies, a keg of patent punch, and to cap all, five servants to wait on him.

Throughout the day the Captain seemed busy with his baggage, ever restless, ever shifting from one place to another, now attending to his horses, now packing one thing carefully away, destroying another as superfluous, and so long as he was in this state of incertitude so long would the falcon-eyed Abyssinian urchins stay in the vicinity to eye his preparations for departure, ready to swoop upon the trash which is invariably left when a person is on the march.

It wanted half an hour of sunset before our little caravan was fairly on the way; and, having heard that a magnificent view could be had from the top of a peak standing close to the gap which led down to the pass of Sooroo, I started in that direction to see it.

When I had actually surmounted the summit of the peak, and prepared myself to behold the scene, I looked towards the east, and beheld such a one as painter never sketched.

Those who saw it will be for ever haunted by that one moment when the flashing glare of departing day lighted up the Adowa peaks, and bathed the endless chain of rising hills in a sea of light, before the sun dipped beyond them to shine on other worlds.

CHAPTER III.

AN ADVENTUROUS JOURNEY BY NIGHT—AN ABYSSINIAN AREOPAGUS—NATIVE
ORATORY.

THERE is nothing to interest a man on a night travel. He is
continually striving to penetrate the gloom around him, looking on into dim vacancy.

On this our first journey in Abyssinia Proper, and by night
too, our thoughts and fancies were anything but agreeable. It
was scarcely dark, and we had not been on the road an hour,
before we perceived ourselves followed by troops of animals,
bounding upright, giving vent to angry snarls, and flitting
past with supernatural swiftness. Shortly, a chorus of unearthly yells broke the solemn stillness of the night, which
was caught up and echoed eagerly all around us. Another
deep silence prevailed, to be in a short time, however, broken
by a savage roar, uttered evidently in our immediate neighbourhood. It gave us all an involuntary start. I had heard
frequently the lion, the tiger, and the panther ; but I was positive that this one we heard was neither, and, thinking so great
a traveller as the Captain should know, I propounded the question to him. The Captain, I remarked, had been riding
forward previous to the sharp cries that preceded the roar,
like a man bound to clear the way for the troop behind, couch-
ing a large boar spear in true knightly style ; but he had
gradually slackened his pace afterwards, until he was riding
alongside with me.

'What animal is that, Captain ?' As I asked the question
my own fugitive presence of mind returned instantly, as if I
felt emboldened by my own voice.

'It's a hyæna,' answered that gentleman.

'Is it a dangerous animal?'

'Very,' was his reply.

'Will it attack us?'

'No; unless we attack it.'

'But suppose you try his mettle with your spear?'

'No, thank you,' replied he, loftily; 'he is not worthy of my steel. I mean to reserve this spear for gorillas and wild boars, or lions and hippopotami.'

I urged him no further; and feeling, I suppose, that his presence gave me courage, he never quitted my side until we reached Goom-Gooma four hours after.

The hyænas' and jackals' cries sounded close to us the whole night, and it was evident that they followed us, or that they were in plentiful numbers around us.

Around the camp at Goom-Gooma resounded their hideous cries; and, as we were new and strange to the country as yet, the Captain exhibited a tender thoughtfulness for our safety by having our tent pitched in the centre of the camp, where we felt perfectly safe from any attack of wild beasts.

Before we retired, however, we crowded around a blazing camp fire, for the nights were very cold on the table-land of Abyssinia, and, attracted by the friendly blaze, several of the garrison came also to see and greet the new arrivals. The Captain attracted the most attention by his brown garb his cool off-handedness and his military title, with which everybody was made acquainted before we had been five minutes in camp.

The busy stir of the camp and the neighing of our animals, picketed close to the tent, awoke us to the duties and realities of another day, far advanced already; and quickly we stepped from under our canvas roof out into the delicious summer air to enjoy another scene, which called forth our warmest admiration.

The camp at Goom-Gooma was situated in one of the most lovable, delightful little vales that could possibly be found. So enthusiastic was one young Scotchman upon its beauties, that he said it eclipsed Scotland's fairest dell in its pastoral and quiet loveliness.

Above, on a rocky platform, half way up the cliffs that

nearly surrounded the valley, was the Shoho village, from whence the name Goom-Gooma was derived.

Let us climb up and enter for the first time an Abyssinian village. What a queer place! From a distance how picturesque! but close by, how very disenchanting! A huddle of low, flat-roofed, mud houses, populous with queer inhabitants, dressed in queer style. Every available patch of shadow exhibits a cluster of greasy swart heads of male and female children. The youngest people were all naked; the oldest people scantily clad. They looked at me with all their might as soon as I stood in their village square. In the centre of the square was a *Bema* where the patriarch sat, surrounded by the other dignitaries—the sub-patriarch, or the next oldest; the lawyer or orator sat on his right; and the man that was talking —for there was a man talking—was the lawyer, near whom sat a young man, whom I judged to be the plaintiff. It was a court, then—that square—or an Abyssinian Areopagus, and a trial was being held. Though I was such a remarkable-looking stranger, the elders, and lawyers, and court were too much engrossed in their own business to look at me long; but the little ones formed themselves into a circle around me, and kept their innocent jet-black eyes inexorably fixed upon me. My cigar attracted particular attention; the slightest roll of my eyes, or a wink, had a certain charm for these aborigines. A cough set them all to pondering earnestly, and the sound of my voice was an unaccountable phenomenon. Then the colour of my face was a thing they endeavoured to analyse. The formation of my hands set them to meditate upon the principles of comparative anatomy. But I am forgetting the Areopagus.

I have often wondered what were the words the speaker used to cause him to spread his arms, and stamp his feet, and spring forward so, uttering some sentences with such wild emphasis.

From the Abyssinian Areopagus—long may it stand!—I proceeded, followed by the whole village, to examine more closely the material of their huts. They were built of reddish mud, detached from each other; the walls were from six to ten feet high, roofed over with poles on which bamboos were laid; then a layer of straw, over which earth was laid to the depth of half a foot or more.

x

Occupying the scant shade afforded by the projections of the houses were the wizened old grandmothers, toothless and bewrinkled, wearily crooning in the low doorways some ditty that had just crossed their eccentric memories; while in the low doorways might be seen the very oldest great-grandfathers, centenarians, profoundly meditative.

From other parts of the village, where the cry 'Hail! the conqueror cometh' had already preceded us, came scores of women and children, and, as I turned my eyes around, I perceived that the tattooed and swarthy matrons had also squatted themselves around us. Encouraged by the presence of their mothers, fifties of round-eyed little urchins rollicked and tumbled in the dust, and piled it into tiny castles and pyramids, like our own more elegant children do at our watering-places with the sand on the sea-shore. After viewing all that was worth seeing—that was not much, to be sure, but then it was a matter of necessity—I descended the hill, and found the Captain ready for another day's march.

Hastily swallowing a cup of tea which my prudent Ali had left for me, and lighting a cigar, I mounted my horse, and, bidding adieu to the vale and refreshing scenery of Goom-Gooma, I followed my party.

CHAPTER IV.

A SANDSTONE PLATEAU—STRANGE FREAKS OF NATURE—EXPLOITS OF SMEL-
FUNGUS AT FOCADA—AN ABYSSINIAN CHURCH AND A FUNERAL.

WHEN I overtook the Captain and our little caravan, I found him in an extremely bad temper with a servant named Hassan, who seemed indisposed to proceed further.

The black-looking country wore, in truth, a very portentous aspect. It was destitute of all vegetation, save a few dried shrubs. We rode over a solid sheet of sandstone miles in extent. To our left was a labyrinthine ravine, a gully channeled in the rock, which ran along for many a mile, forming a thousand acute angles; and on our right, forty miles off, rose the Adowa peaks, dark and undefined.

Four hours' travelling brought us to the base of a hill which rose high and grim, like another Gibraltar, above the sterile plateau we had just crossed. On all sides it was rendered easily defensible, on account of its scarped cliffs, the lowest point of which was about three hundred feet above the level ground.

Excepting a conical peak a thousand yards' distance to the south of it, there was no spot that commanded it, and from its vast size I judged it was capable of accommodating a garrison of 10,000 men. Why such a place as this, by nature impregnable, was not occupied by Theodore, Emperor of Abyssinia, was a source of much speculation to me at the time; but it was not until after I arrived at Antalo that the politics of the country became intelligible to me.

On the south-eastern extremity, a ridge of rock a mile and a half in length connected this fortress-like hill with a range of rolling mounds covered with short firs and juniper. On either side of the neck, which at its widest part was not a mile wide, were ravines of the perpendicular depth of 600 feet, gradually expanding into broad valleys, rich in meadow lands and groves, and watered by running streams. Game of all kinds was found in these valleys. On this neck of ground was the village and church of Focada, and close to it, on a rising ground, were situated the tents of a company of cavalry, the garrison of the place. Water had been obtained here by sinking several wells, and from these was drawn a sufficient quantity to fill an extensive reservoir, whereat the animals were watered.

We arrived at Focada camp two hours before sunset, where we speedily forgot our toils in copious draughts of the finest Bohea, and savoury platters of curry and fricandeaus cooked by my epicurean friend the Captain.

Focada was so very pleasant and interesting, that the Captain and I, after protracted consultations, agreed to stay there two or three days, as no movement of the advance was expected for a week longer.

The first day was occupied by my distinguished friend and companion in overhauling his miscellaneous baggage, which was in a sad pickle after the march from Goom-Gooma. Several bottles of chutney and curry powder were found to have been broken by the jostling they had received, and the volleys of oaths which fulminated about the ears of his unlucky servants would have gone far, if scientifically directed, towards blowing Magdala to pieces.

Fortunately my remarkable and extraordinary friend who was my messmate was an ardent admirer of the chase, and a great part of the time we halted at Focada he spent in hunting. At night he would return, and the feats of hunting he performed according to his relations would fill a very respectable column. The first day he shot two panthers, eleven species of the gigantic toucan, seven dozen pigeons, one hyæna, one jackal, and one hare, out of which he brought home—that I, his comrade and best friend, might behold them—one scraggy-looking hare, three shattered pigeons, and one immense bird which he called the gigantic toucan. The last-mentioned bird measured five feet from the tip of the posterior feathers to the beak's point. On questioning Ali, who had accompanied him, the boy said he didn't see the Captain shoot any more than he brought ; which probably was owing to Ali's defective vision.

On the second morning of our stay at Focada we heard a great noise which seemed to issue from the village from whence the name of the camp was derived. The village of Focada was situated on a hill a little to our left, commanding the neck of ground on which our camp was situated. At its base was a church—the first Abyssinian church to be seen on the road from Senafe to Antalo. This church, surrounded on all sides by trees, stood almost on the brink of a deep gulf or chasm, which, starting from a point a few yards beyond the church, gradually widened into a broad and extensive valley. Between the village and the church lay the little graveyard of the natives who dwelt there.

Our tent was pitched half way between the village and the camp of the garrison, so that we could have a good view of the camp and village. The noise and shouting drew near, and we observed the head of a column of people of both sexes, coming down the hill at a slow pace; every now and then halting to rest, giving utterance to some most ear-piercing shrieks and doleful howls. After enquiring as to its cause, and ascertaining that it was a funeral procession, I went up to get a clear view of the ceremony.

There were about a hundred and fifty people, old and young, in the procession, preceded by a few priests. These ministers of the Abyssinian Church, clad in the usual apparel of a well-to-do Abyssinian—viz., a cotton robe, with a broad scarlet band running along its entire length—were conducting themselves in a frantic manner. They tore off their turbans, pulled at their hair, and then, folding their hands across their bosoms, looking inexpressibly miserable. Well-thumbed parchment books, written in the Arabic tongue, and illuminated with all sorts of grotesque figures and devices, were in their hands, out of which they now and then selected a long prayer to some of their favourite saints, and chanted it with strong lungs, with a nasal twang, in the most lugubrious manner. This ceremony went on for the space of a few minutes, when they again started on their march, the priests bending their bodies. At every few paces they performed various genuflections, and lifted up their voices in long-drawn howls, expressive of extreme agony.

When they drew near the grave, the grey-headed elders who were carrying the corpse, carefully folded in its cerements made of the cotton cloth which it wore during life, advanced to the edge of the excavation, and lowered the body slowly into it; then the priests chanted a prayer which, interpreted, read as follows :—

'Werkena, son of Yasous, who was the son of Tekee, is dead Rejoice, oh ye people ! he has gone to his rest, with Abraham, with Isaac, and with Jacob. Let us pray for those who still live, and pray for the soul just gone to doom. From vengeance and stern judgment pray that his soul be delivered. How can the souls be delivered from tribulation ? By long prayers. Pray, then, that he be sheltered by Father Abraham, that he may walk in safety by the side of Moses, and the prophets. Amen, and Amen !'

The grave was then closed, an oblong mound of earth raised over it faced round with rocks, and a long slab of slate stone was placed at each end. The company then scattered, the major part following in the wake of a man who discoursed merry tunes on the oboe or hautbois. From the newly-made grave I strolled to the church just mentioned. One of the priests, who had been chanting lustily and laboriously at the late funeral, promptly offered to be my escort and cicerone through the building for a slight consideration.

The consideration had to be paid first, in the shape of a new Maria Theresa dollar, which is worth eight cents more than the American dollar.

A tower-shaped construction, overarched by the branches of two fine sycamore trees, admitted us into the building, set apart for the solemnisation of Christian rites and ceremonies.

Of a certainty nothing in the shape of a church, from a cathedral down to a backwoods meeting-house, found in any other quarter of the globe, save Africa, could have presented such a scene of filth and slovenliness as the building dedicated to Holy St. Michael, at Focada. At only a single step from the door, within the structure, one realised the unutterable forlorn-ness and abjectness of the place.

On the third morning of our arrival at Focada, we packed up, and started for Attigratt, the next station, which was said to be ten miles off.

The labours of the pioneers on this road were to some purpose. A more difficult piece of ground for the march of an army could hardly have been found. Every mile of it was across an almost solid bed of rock, whose jagged points shot upward in every direction. The levers and blasting tools were very necessary here, and much skill and tact were requisite to convert the most easily convertible portion of the rocky plateau into a road. The road lay also through forests of scrub oak, tamarisk, and kolquall. When it was completed it presented a very fair specimen of military engineering. It was fifteen feet in width, and made as smooth as the circumstances of its surroundings would permit. Marching columns with their baggage could move easily along, and even the carts of the lowland train could travel from the sea-coast to Attigratt.

A mile or so beyond this village we came to a plain, or a valley rather, to which we descended by a gentle declivity. In this valley we observed the first green corn visible on our march. It was an extensive patch, well watered by a system which answered its purpose very well, the furrows being fed by a small stream.

In the midst of this broad and smiling scene a cluster of low-roofed houses was visible, whither I directed my steps. Larks sang merrily above ; the corn was all in a ripple and flow before the western wind, and the tremulous heat danced upon it, as upon a sea, or like the shadows of moving waters.

At my approach, nut-brown girls with cunning chignons at the back of their heads crowded up to meet the pale-faced strangers, with as much noise and vivacity as a bevy of school-girls returning home from school in any American city. Naked urchins came wriggling and wheeling, delighted, no doubt, at the opportunity of exhibiting their gymnastic powers before discerning persons. I requested some ' haleeb,' or milk, for which, I said, I would be willing to pay the usual tribute— a bright new ' gourshi,' or dollar.

A young girl, graceful as a Hebe, bore the ' haleeb ' to me, and out of it I drank as deep a draught with as much relish as that which the Sun-god drank at the hands of the herdsmen of old Admetus.

We came in sight of Attigratt, the head-quarters of Sir Robert Napier, about noon.

The distance that we had marched was so short that we made up our minds to travel another march before halting and endeavour to catch up with the pioneer party under Colonel Phayre. The Captain had shot two dozen pigeons on the road, to present to Sir Robert Napier. As he departed to the camp, I struck off to the right to view the village of Attigratt.

Attigratt is situated on a gentle eminence in the middle of a plain. The feature of the place is a castle built of adobes. At one end of it is a square tower, surmounted by four turret-like projections built on projecting rafters. To civilised eyes it seemed more like a caricature than a house for defence and built for strength.

At the time we passed by this castle, it was inhabited by the wife of a chief who was in the power of Theodore. This lady considered herself inconsolable because her princely husband was absent, and had vowed the foolish vow that she would never allow the sun to shine or the rain to beat upon her head again until her lord should return to cheer her heart.[1]

Near the stronghold was a church—undoubtedly the finest I have seen in Abyssinia, save one at Chelicut. It was surrounded by magnificent kolquall trees, whose shade was truly grateful, and cast a kind of soothing calm about the place. The priests—very respectable old fellows, with turbans, larger than ordinary, rising high and steep above the face—came and greeted me with many a salutation and salaam.

My desire to visit their church was intimated to them with the usual backsheesh, and a pantomimic explanation of the reason of my gift. This pantomime one learns very easily, and is a great aid to acquaintance, making, when thorough in it, the luxuries of the country—such as milk, eggs, ghee, bread, and oats—flow in upon you so long as your treasury box permits.

The priests understood the use of the gift, and I followed them as they proceeded through the grove in the direction of the portals, then wide open. Through a clean courtyard, and then to a square entrance leading to the interior, the priests shuffled themselves along, as if walking were an exercise not much in their province.

After passing through the entrance or porch of the building,

---

[1] For the satisfaction of readers, I will state that the chief was delivered from bondage at the capture of Magdala. The author arrived at Attigratt the same day that the chief came in sight of his castle walls.

we descended one step into a lobby or hall running around the entire circumference of the church. The walls of this circular building were constructed of stone, plastered over with an inch of mud laid as evenly as possible.

On this altar was painted with queerest devices the Abyssinian cartoons. First, there was the picture of St. George for ever threatening the dragon writhing under his horse's feet, with an admiring group of noble ladies and gentlemen viewing, with intensest satisfaction on their faces, the death-throes of the man-devouring enemy. The interesting crowd of ladies and gentlemen with their red, blue, and green garnitures, and enormous feathers, was supposed to represent Pharaoh and his court. The second cartoon illustrated in a remarkable way the 'Last Supper,' whereat thirteen apostles were supposed to be partaking of the feast. The third represented the crucifixion of poor St. Peter, whose agonies must have been intense, so faithfully had the native artist endeavoured to paint the contortions of the man undergoing such a cruel martyrdom. Others there were, sketched on the wall, who suffered torments in their martyrisation for their faith. Before I could view with my sacrilegious eyes the *Sancta Sanctora* of the church, another dollar was demanded from me, the tribute was paid, and the priest led me into the Holy of Holies—the interior of the charmed circle, round which I had been promenading preceded by the cicerone.

There was nothing very edifying in the sight—a rough desk, with a large Maltese cross of brass fronting it, before which the priests offered their orisons when officiating on the Sabbath, kept every Saturday. There was also a big book very like a Bible, bound strongly in leather, and clasped, lying on a stand. The floor was not very clean; in some places it was so littered with straw, that I imagine some unholy vagrants must have laid down to sleep there.

From the church I stepped out into the open air, and from the eminence whereon the church and castle was built the entire plain of Attigratt was visible. In the centre of it, like snow-heaps, stood the cleanly white tents of the English army, who had marched thus far under their General to release their countrymen.

Now and then a stray laugh could be heard, very English-like, so hearty and sonorous it was—a side-splitting laugh, sounding at that distance as if a strong frame were in a laughing paroxysm.   Guards and sentries in scarlet uniform, with gleaming steel in their hands, paced backwards near some official head-quarters, or with watchful eye guarded the commissariat containing the real sinew of the campaign.   Officers in splendid uniforms, with rattling sabres and jingling spurs, hurried to and fro on horseback, conveying orders to different regimental head-quarters; horses and mules, tethered in rows by the hundreds, neighed and brayed hungrily for their scanty rations—for it was now noon-day, the period when they are all fed and watered.   Scores of curious natives stood in groups round some sight extraordinary for them, and I daresay were commenting wisely on the queer things they saw.

A few miles beyond this exciting and strange scene on an Abyssinian plain, the bold mountains, curved into bays and headlands, loomed up like a coast of cloudland veiled in amethyst light, serene in the noonday sun.   An indefinable sense of melancholy crept over you while you looked.   Beyond this, again, in tier upon tier rose the dim mountains, sublime and mysterious in their lofty height, serving as an effectual barrier between the Abyssinians and the cruel tribes of the Donakil plains.

After gazing on this scene I bethought myself of my servants wending their way uncaring and uncared for ; and, though I wished to make myself known to Sir Robert Napier, I thought it better, as the Advance Brigade would march upon the morrow, to proceed on my way and endeavour to keep ahead of them and close up with the pioneer force, who were then far in advance making the roads.   So I galloped on, my horse's heels kicking up a cloud of dust, to the intense delight of the rabid urchins who rushed forward before him as he approached them, with much of the same defiance and daring on their features as our more civilised children at home exhibit when with mad audacity they await the approach of the thundering locomotive, and then bound off, barely escaping being crushed to death.

My reveries and visits to the grotesque castle and church of Attigratt had consumed some time, and had enabled my servants to proceed several miles before I caught up with them.

From Attigratt the road led through a grove of fir and juniper, and clumps of the kolquall, very picturesque to look at, but withal very lonely, and suggestive of suitable places for ambuscade by the robber bands of which we had heard often from the timid natives, who mysteriously hinted of ambulant Gallas waiting in secret places for the unwary.

No such bands, however, did we encounter on our march that day, and we arrived at a leafy grove situated chiefly under the shadow of amorphous and massive rocks, over which our route lay the next day. In the centre of this grove we pitched our little tent, and against a fallen sycamore log built our camp fire. The cooks prepared their pots and pans and dishes. Ali took the leathern bottles to a spring of clear fresh water. The syce, though tired and sulky after his long, weary march over the rough rocks, vented his ill-humour by volleys of objurgating adjectives freely bestowed upon the horses and himself, but finally led the animals to the water; and I, part lord and master of the encampment, revelled on the warm blankets and buffalo robes, and waited patiently for the arrival of the Captain, who had already arrogated to himself, undisputed by me, the chief control of our caravan.

In the midst of an intensely interesting soliloquy, the Captain's squat form suddenly broke the perspective of my vision radiant with unusual good humour.

'By the bye,' said he, abruptly, 'what have you for dinner? Let's have a roaring dinner fit for the Immortals; I will be cook, and you be bottle-washer. What shall we have, eh?' And, raising his fingers, he proceeded to enumerate delectable dishes, which were concoctable from the box of stores in his possession.

'First of all we'll have mulligatawny soup: and for *entrées*, we'll have—let me see: a buffalo hump chop; roast hind-quarter of a kid, supplied by the commissariat to-day out of pure regard for me; roast duck—one I saved from a lot I shot to-day. Oh, yes; and then there are the ibis. By Jove, we'll have hashed ibis, boiled commissariat beef, rice, onion sauce, and curry, all of them to be finished up with real Mocha coffee; and before tumbling in for the night we'll have a drink together from my visitor's flask, eh! What say you, old boy?'

'Glorious!' said I; 'and, to use the words of the Bard of

Avon, may good digestion wait on appetite, and health on both ! '

'Bravo! ah-haw! ah-haw! oh my!' shouted the old boy, rubbing his bald head, and pulling his side whiskers, and kicking his stumpy legs quite frantically, through an enthusiastic impulse.

Mohammed, Hassan, Habibullah, and Ali were at once set to work. The fire roared and crackled, and the servants flitted backwards and forwards; and as day settled gradually and kindly into night, and the towering cliffs and lonely plain expanding from view, teeming with green groves, became bathed in the glow of the declining sun, delighting the eye, supper, or rather dinner, was announced.

Every dish was delicious. We both had good appetites and healthful digestion, and we lingered over the feast till far into the night—the Captain recounting marvellous adventures, hair-breadth escapes, love stories truly romantic, gallops over glorious plains by moonlight in India; stories of his Nilotic experiences, wherein Fatima and Zuleika figured conspicuously; Don Juan scrapes with Miss Something; tales of forlorn hopes at Sebastopol; of a white face resembling the Captain's lying for two weeks in the trenches at Inkermann; of being resuscitated from that two weeks' death-like sleep, and a four months' lying in bed at Scutari Hospital; of his gallant Arab ' Emir,' of the matchless Nedjed breed; of his exploits as *sabreur* on the Texan prairies against Federals and Comanches; of bush-fights in miasmatic swamps, and many other wild, interesting, marvellous, astounding novelties. Are they not all written on the books of the recording angel? who, no doubt, is well acquainted with phonography; otherwise, hard indeed must have been his task to write down all the interesting stories of my dear friend and brave messmate, Captain the Honourable Mr. Smelfungus.

From Attigratt to Mai-Wahiz is a distance of fourteen miles. On this stage we passed through an exceedingly wild country. For three hours our road lay over a vast platform of sandstone rock, and through its extreme ruggedness it proved very fatiguing to both men and animals. But half way to Mai-Wahiz the scene changed. Another plain similar to that near Attigratt, of the same arid features, expanded before us, with its dead level only relieved by a few green clumps.

To this plain we descended, picking our way daintily to avoid unsettling the toppling boulders on the hill flank.

As we descended, the sandstone platform over which we had been travelling receded sharply to the right into lofty and scarped sierras, rifted and split into many a fissure. These rifts and abradations form excellent coverts, from whence issue numerous carnivorous animals every evening on their nocturnal ranges in quest of food.

The poorest class of peasants, like the ancient Troglodytes, also find these cavernous recesses in the cliffs admirable habitations, and a protection against cruel chiefs and unjust masters. But, growing bold from impunity, and waxing stronger and insolent with numbers, the peasants generally select a chief from amongst themselves, and become the oppressors instead of the oppressed. After amassing wealth through the success of their depredations, they not unfrequently take to building a town upon the top of the most inaccessible hill, and fortify it with palisades and moats, whence they issue out periodically, and pounce upon a confiding farmer and take from him his wives, children, cattle, and substance.

This is the history of most of the towns in this neighbourhood, and, in fact, of those of the whole of Abyssinia. What the Abyssinians are to-day, not many centuries ago the warlike British barons were; or the Highlanders who used to make many a wild foray upon the more peaceful burghers of the Lowlands; or the robber counts and barons of the Rhine.

I was led to make these reflections by seeing the numerous fortified towns, built like eyries on the numerous conical hills rising far above the level of the plain, in every direction. The warning voice of the watchful native warder bawled out in unmistakeable accents of anger if we advanced near the base of the fortified cone with the intention of purchasing any necessaries, such as eggs, butter, milk, whey, or bread. Near each hill was a well belonging to the village on its summit, and these wells were the only places where water was procurable on the march. We generally camped near some pools of very dirty water—compelled to do so, because there were no others; and it was considered unsafe, until the army had passed through, to place much confidence in the Abyssinians.

Instead of stopping at Mai-Wahiz, as we had designed when we started in the morning, we pushed on to Ad-Abaga, where a plentiful supply of water was to be found.

Ad-Abaga is situated in a basin-like valley two miles in extent, surrounded by a range of heights, crowned by villages inhabited by very predatory and fighting-loving people; who strange to say, though all more or less related to one another, manage to live on wonderfully good terms with themselves, though not with the rest of the outside world.

We camped one night at this place, and pushed on to Dongola next morning, a station situated just twelve miles from Ad-Abaga. This march lay through a very interesting country. Much labour had been employed upon some portions of it to make it available for the coming army. For the first three miles we travelled through a very tame and uninteresting country, mere up-hill and down-hill work, with a village of conical huts, standing on the summit of every eminence, proudly exclusive from the people of the plain.

Then we journeyed over a road cleared through a perfect jungle of low brushwood. Towering above the shrubs were tall columnar juniper and fir trees, their branches darkly waving and crooning under the strong gusts of wind which at certain hours of the day blew from the Eastern seas across the Donakil desert; then through a wilderness of boulders, some of gigantic size, some small, all lying around in the utmost confusion imaginable; anon through a dark gorge where, on the summits of the flanking rocky walls, a small band might put to flight an army; and again through a forest of shrubs and through long avenues of noble sycamore trees, in whose branches disported the monkeys, whence issued the warbling notes of the butcher-bird, and whence the gay clock-bird flung his glad piping abroad, as if to welcome the white strangers who were invading their country.

Sadly at variance with these scenes were the buff-coloured masses of granite, gneiss, and the dark patches of slate-stone looming up grim and silent, whereon no bird hopped, whence no lively song of the feathered choristers issued.

Passing rapidly by these scenes and turning round an angle, the towering baobab was seen, whose branches were heavy with scores of Abyssinian monkeys, who, upon our approach, snarled their disapprobation, and irreverently retired up the slope of the nearest mountain, barking and growling their discontent.

We passed the baobab trees also, and our path wound its way around the slopes of a mountain so high that its summit seemed to strike the clouds, and of a depth so profound that the head became giddy on beholding it. We hugged the slope and resolutely turned our gaze away, to admire rather the varieties of rock of which the mountain is composed ; to look at diagonal layers of porphyry, half-buried masses of scoria and soapstone ; to examine the clay, slate, and quartz running in serpentine veins far up out of sight ; to pick up cornelians and pieces of jasper stone ; and thus we travelled until the green valley of Dongola was visible, in the centre of which two or three tents were to be seen like lime-heaps pitched in the midst of luxurious grass.

From Dongola, after one night's rest, we journeyed on to Agulla river, a distance of ten miles. Before the sun was fairly up we were on the road, having contented ourselves with a simple cup of tea before starting again up steep hills, rough with craggy obstructions. Over rolling plateaus, past smiling pastoral valleys, by robbers' roosts and peasants' caves, we marched ever southward. The sun rose; its rays pierced the seeming belt of mist ahead of us, mantling it in tints of exquisite loveliness, and flushing it with delicate hues.

When the mist disappeared a line of dark hills clothed with kolquall distinctly made themselves known to us by their height. We crossed the horizon that we saw an hour previous only to see another cloud-fenced range, grimmer in aspect, bleaker and more terrible, and inconceivable as the sea to him who has not beheld Ethiopian mountains.

We were aware that stretching and expanding far away and wide as space lay other lands beyond that, and unknown deserts, where yet no man had trod. Still the sun beat remorselessly on our heads; still the same scenes repeated themselves over and over again.

Patient and uncomplaining, the animals keep their regular walk, seemingly unmindful of the heat or the blows heavily bestowed upon them by our cutcha wallahs.

Smelfungus was ahead of the little column ; his wide-spreading solar topee shadowed his head and face, and nothing of either was visible; nor could I guess whether he was sleeping or dreaming, or with eyes wide awake diligently observant for a chance to do or say something to break the dead, unjoyous silence, while I brought up the rear.

My horse looked as dejected as I felt sure I must have looked. I had examined the blue sky, and even winked fiercely once or twice at the fierce sun, and scanned the horizon dozens of times ; I had looked in vain for something wrong in the harness of the mules ; I had watched carefully the conduct of the servants, to find cause for reprimand, and show who was master; I had mentally sketched scores of caricatures on the squat broad back of the Captain, and imaged himself sitting in his huge solar topee rowing on Lake Ashangi, for he had promised seriously to do it.

What could I do to relieve the distressing tedium of marching ? The question resolved itself.

With my eyes wide open in the broad hot sunlight, I fell to day-dreaming—most beatified of luxuries, fancying something novel and strange, something exciting and interesting. A queer thought struck me, and I laughed ; at which Ali, my bold surrujee, turned on me a pair of large black eyes, surveying me in mute astonishment, as if asking ' What can you find to laugh at, master, in Habesh Land ? '

The Captain turned his bullet head around, and the green topee swung around like a turn-table. He looked upon me with the greatest astonishment ; I blushed, looking, however, far away in advance of him. I shouted ' Camp! Camp!' Wonderful resuscitation !

The mules brayed, our horses whinnied, the Captain trolled out a song, ' La-la-la-ety,' positively the first tune I ever heard him sing ! Our Arab surrujees and Hindostanee cutcha wallahs crooned out most doleful notes out of four pairs of lips, expressive of great relief and heartfelt thankfulness, to be increased when we arrived at camp, which was on the banks of a river, the first river we had seen in Abyssinia, under the shade of a sycamore, near the ruins of a temple.

The ruins near to which we were encamped were those of a temple erected by King Lala-ba in the eleventh century to St. Dorcas, a Greek saint. This was one of several built by that pious monarch in different parts of the empire. It had been

CASTLE BETWEEN MAI WAHIZ AND AD-ABAGA.

falling gradually to decay on account of the incursions made by the Mahomedan tribes into the province of Enderta, through a portion of which the river Agulla runs ; and, upon King Sabagadis' accession to the throne of Abyssinia, that monarch heard it rumoured that much treasure was buried in the place by his predecessor, and at once ordered search to be made. This search, of course, necessitated the demolition of the building. The treasure was found, however, and the ancient fane remains in precisely the same condition as the workmen of the greedy king left it. Two pillars only are left standing now, but in some portions the marble pavement is still intact. The river, called solely by that name from courtesy, as rivers were so scarce in the country we were travelling, was a rapid little stream, only ankle deep in its deepest parts, and but a few yards in width, into which our animals and ourselves plunged, and where we revelled for half an hour or so.

At the camp of Agulla we found Colonel W. L. Merewether and Mr. Clements Markham, the distinguished geographer of the Expedition.

Colonel Merewether, it will be remembered, was the pioneer officer who first landed in Abyssinia to discover a feasible route to Magdala. A gentleman more suitable for the work in hand— after seeing him face to face—would, I judged, have been difficult to find. He seemed incarnate activity, and the very embodiment of promptitude ; he had such extraordinary perceptive faculties, so quick and decided a spirit.

Colonel Merewether was a Colonel of the Scinde Horse, but acted as Political Resident at Aden before coming to Abyssinia. Rassam was a former secretary of his while at Aden, and it was through the influence of Layard of Nineveh that that Armenian gentleman was appointed as envoy to Theodore. That selection, in my opinion, was most unfortunate ; but Merewether adhered to the belief that no better could have been found.

The Colonel was the very soul of politeness and affability ; and it was delightful to hear a man who could talk without any affectation, and act in a natural way.

During the three days we rested at Agulla the Captain went shooting. At nightfall he generally returned with dozens of guinea-fowl, quail, ducks, ibis, wild geese, and turkeys. He

Y

said he had shot three wild boars, but I never saw one bristle of them though I should much like to have seen them.

On the fourth morning we struck our tent, packed our baggage, loaded our animals, and started on our journey to Antalo.

Very uninteresting was the march from Agulla river to Antalo. We arrived at Dorello, 19 miles distant from Agulla, on the evening of the first day; Haikhullut, 9 miles, the second day; thence to Antalo, 14 miles, on the third day.

CHAPTER V.

NAPIER'S PROCLAMATION—DIPLOMACY—MAJOR GRANT—MEETING BETWEEN
NAPIER AND PRINCE KUSSAI—AFFECTING SCENES.

WHILE the redoubtable Captain and myself were travelling to
Antalo with what cheer and good humour we might, an interesting scene was being enacted a few marches back, in which
an English general and an Abyssinian prince were the principal actors ; and no better place can be found in this book to
describe it than in this chapter.

His Excellency, Lieut.-Gen. Sir Robert Napier, Commander-in-Chief of the Abyssinian Expedition, on landing in
Abyssinia, issued the following proclamation to the governors,
the chiefs, the religious orders, and the people of Abyssinia :—

'It is known to you that Theodorus, King of Abyssinia,
detains in captivity the British Consul, Cameron ; the British
Envoy, Rassam ; and many others, in violation of the laws of
all civilised nations. All friendly persuasion having failed to
obtain their release, my Sovereign has commanded me to lead
an army to liberate them. All who befriend the prisoners or
assist in their liberation shall be well rewarded ; but those who
may injure them shall be severely punished when the time
shall arrive for the march of a British Army through your
country. Bear in mind, people of Abyssinia, that the Queen
of England has no unfriendly feelings towards you, and no
design against your country or your liberty. Your religious
establishments, your persons, and property shall be carefully
protected. All supplies for my soldiers shall be paid for. No
peaceable inhabitants shall be molested. The sole object for
which the British force has been sent to Abyssinia is the liber-

ation of Her Majesty's servants and others unjustly detained as captives; and as soon as that object shall have been effected, it will be withdrawn.

'There is no intention to occupy permanently any portion of the Abyssinian territory, or to interfere with the government of the country.

(Signed)   'R. NAPIER.'

A copy of this well-worded and modest proclamation was received by Dajatchmatch Kussai, Ras of Tigre, who in plain Anglo-Saxon is entitled General Kussai, Prince of Tigre. On the receipt of the letter the Prince manifested his admiration of it, and desired to see the writer, or an envoy from him, who should have the power to arrange for a meeting between the writer and the Prince; as he, the writer, would be compelled to travel through his country to Magdala, and it would be highly politic for both sides to be friends. The following is the letter sent by Prince Kussai to Sir Robert Napier, which to my taste is as well worded as the English chief's :—

'IN THE NAME OF THE FATHER, THE SON, AND THE HOLY GHOST.

'Letter sent by Dajatchmatch Kussai, head of the Chiefs of Ethiopia, to reach the Chief of the English soldiers.

'How are you?

'Very well?

'By Christ's grace I have recovered the throne of my ancestors, of Mikael, of Welda Selasse, of Sabagadis. Of old we are house friends, from Consul Salt till down to Plowden. I expected to receive a letter from you; but, as it has been retarded, I have sent myself. I know not what you have come for; if I knew, it would please me. We are home friends. I am sending by Muroja, the son of Atu Waiku, who knows my language and yours. I am sending (what I have in) my heart, and you too send to me (what you have in) your heart. In the year 1860 from Christ in the time of John the Evangelist, in the month of Hadar the 18th written Wednesday.

(Signed)   DAJATCHMATCH KUSSAI.'

Head seal of the Chief of (Levi of India) Ethiopia.

RUINS OF AN ANCIENT GREEK CHURCH AT AGOOLA.—MR. STANLEY'S TENT.

In the above, Prince Kussai, we are led to believe, has laid bare his heart—in short, sent (what he has in) his heart; and Sir Robert Napier is politely requested, in the name of the Father, the Son, and the Holy Ghost, for (what he has in) his heart.

For purposes of polity—and the English General will show himself, as we proceed, to be not only a warrior chief, but a diplomat of the first rank—a gentleman, Major James Grant, C.B., of Nile Exploration notoriety, was sent with much backsheesh to the head of all the chiefs of Ethiopia.

Major Grant was well received as the English chief's envoy—treated to the best, and shown every civility, and a meeting was finally arranged to be held between the English and the Tigrean chiefs.

Sej Muroja was sent as the Tigrean ambassador, and he was received with all due solemnity by the solemn-looking General Napier and his solemn-looking staff, all looking specially solemn and decorous for the occasion. Complicated military evolutions, sham fights, long rolls and much trumpet-blowing and drum-beating were shown to him ; and the decorous Tigrean ambassador departed on his way back to his chief, thoroughly perplexed and mystified by what he had seen, but certain of one certainty, that the 'Feringhees' were the most remarkable people he had ever beheld—which was very true. But it seemed to those suspicious of his friendly intentions that Prince Kussai was a kind of a Fabius Cunctator.

He sent messengers again, and from afar viewed the result. The hearty and kind reception which they one and all received finally decided the cunctative Kussai, and a day was named and a place appointed, whereat the two 'home friends' should meet.

On the evening previous to the day of the meeting, Sir Robert Napier moved on with the 2nd Brigade to occupy the station called Ad-Abaga. The Tigrean chief was to have encamped at a place called Haussein ; but he did not, and gallant Sir Robert was compelled to cool his heels at Ad-Abaga until the savage potentate, who arrogated to himself the title of 'Chief of the Chiefs of Ethiopia' should arrive. He waited eight days ; and, when his patience was well-nigh exhausted, another messenger,

arrived at a most happy moment, and announced the arrival of
the king at Haussein.

On the next morning Sir Robert Napier sallied out of his
camp, in the comfortable basin-like valley of Ad-Abaga, with a
column consisting of 500 infantry, 300 cavalry, a small party of
the Royal Engineers, and four 12-pounders—a most compact,
soldierly little force, but ridiculously at variance with the estimate at which the princely chief of Tigre had placed the English
army.

With Snider rifles thrown over their shoulders, with a day's
rations in their haversacks, and with a military band preceding
them, playing some lively tune, the little army marched on.
The shining steel rifles glimmered against red uniforms and
sun helmets, and dazzled the eye; the steel scabbards rattled,
spurs jingled, as the blare of the trumpet gave the signal to
the Brigade to march. Behind these came the little gun
battery, the cannon looking demure enough on the backs of
the mules; and slouching and shuffling in the rear, waving
their trunks from side to side, were the elephants, with houdahs
on their backs, and mahouts snugly seated on the napes of their
necks.

The Commander-in-Chief, mounted on a magnificent charger,
with a brilliant staff, dressed in their best, towards making the
greatest impression on the susceptible minds of the barbarian,
cantered on ahead of the column.

In good order, as trim and neat and compact as when they
started, the little army arrived upon the field of the expected
durbar,[1] Mai-Debar. Happy place! most auspicious for the
interchange of courtesy and goodwill between man and man,
were they ever so much separated by race or religion! a fair
half mile in circumference, varied by green knolls and grassy
slopes, with a stream bubbling through its centre, was the
valley of Mai-Debar.

The muskets were stacked, the cavalry dismounted, and the
soldiers, after drinking the pure liquid from the stream hard by,
stretched themselves upon the sward, and awaited with admirable
unconcern the advent of the barbarian Prince.

The General and his staff sauntered up and stood upon the

---

[1] Durbar: Hindostanee term for levée, or council.

spire of the highest eminence, to watch the approach of the expected cavalcade.

Far away the rolling plain, dotted here and there with sharp peaks and clumps of kolquall, expanded into purplish infinitude. The mountainous Debra Demba pierced with his hoary head the storm-clouds! Ambitious ambas towered in every direction above plains and valleys and lesser mounts. Amba Shukulat in the rear, like a giant shorn of his locks, struck at a mass of cloud with his sublime head; Mount Masoba was in front, and Mount Arnedin disputing with his bristling sides the Dongola defile; and directly ahead was another shaped like a spur, with pensile puffs of white ether hanging near its base; no, it was not ether, it was dust. A squall of wind dissolved it, and the British videttes discovered that for which the galaxy of stars stood waiting near Mai-Debra — Prince Kussai's army.

'Every one to his place, clap your knapsacks upon your backs, snatch your rifles! fall in, dress up soldiers, and look your best!' In an instant of time the orders were obeyed. A solid line of stout armed, scarlet-dressed infantry; on their right picturesque squadrons of cavalry; to their left a battery of field-guns, manned by tall fellows with swab and rammer at their side, stand ready, drawn up on one side of the little stream in admirable order.

The English chief hurried to his tent and made his toilette deliberately. His staff surrounded the tent, and some mysterious conversation was held; after which three officers, attended by a detachment of the 3rd Light Cavalry in blue and silver uniform, suddenly dashed off towards the Prince, and informed him that the English General would advance simultaneously with him.

A sonorous rumbling of kettledrums and sharp words of command were heard on the heights opposite. The Abyssinian force deployed, and presented an extended line of cavalry and infantry. Two or three stalwart long-armed fellows were seen galloping about, aligning its front. A scarlet tent was pitched, and some one, who was probably the Prince, was seen entering.

Ten minutes of expectancy and patient waiting, and the Prince had mounted a gaily-caparisoned pony, and an English officer rode with hot haste to inform the General that the Prince was coming.

The General issued out of his tent, and a large elephant decked in holiday ribbons and trappings, that was standing near the door, was mounted by him.

On this elephant, followed by another, and surrounded by his staff and a select body of cavalry, General Napier proceeded towards the stream.

The Prince bore down the hill with 500 choice warriors, forming a wing on each side of him. Over his head was held the state umbrella, of maroon-coloured plush velvet, heavy with silver ornaments, by a very handsome man. At his right side was his spear and shield bearer; at his left his fusil bearer— the son of a Tigrean grandee. His generals preceded him on foot, being preceded in their turn by two of the English officers who went to meet him.

On the banks of the tiny stream Napier met his princely ally. Both dismounted, and a loving embrace followed; much after the style of American Indian sachems embracing peace commissioners. The Prince was apparently fervent; Napier exceedingly passive, almost ungraceful, in this new mode of friendship-acting.

Together, both Prince and Napier entered the durbar tent. Five head princes of Kussai's retinue, with shield-bearer and 'gold stick' privileged by their position, entered the council tent, with a dozen of Napier's principal officers. At the same time a salute was fired by the artillery, the noise of which caused visible commotion among the array of horses and men on the opposite side of the stream; and great uneasiness among the dusky warriors, who, with tilted lances, looked down upon the red-coated army below.

Sej Muroja, the consequential ambassador already mentioned, when all were seated in the tent, broke the silence first, by initiating a most formal *tête-à-tête* between the English General and Prince Kussai.

The Tigrean Prince, through whose dominions we were about travelling to the southward, was dressed in a silken robe with immense flowers, as variegated as the coat of Joseph, with a lion-skin cape, from which three pendicles hung down the back. His head was bare, being well supplied with the bountiful covering Nature has provided all Abyssinians with; and this was plaited into numerous ridges, well oiled, which hung over his neck. Over the Prince's head was borne the state umbrella, by a youth who, though black, was exceedingly well-favoured.

Kussai had a very gentle, amiable countenance, but appeared as if he might have been termed the 'Unready,' from the irresolution apparent on his features.

'Greet the Feringhee Negus for me,' said he to his satellite, Sej Muroja; and Muroja forthwith related to Napier, with some embellishment, the words of his master.

Sir Robert replied that he was very happy to see him, and said that it was with great pleasure a Christian nation like the English beheld people professing the same creed in Abyssinia.

In answer, Kussai said that he did not much like to see strangers in his country; but, since come strangers must, he preferred they should be Christians.

This certainly did not seem to portend very amiable feelings towards the English, while it was evident that, in spite of Major Grant's diplomacy, much suspicion was entertained of the crusaders' intentions. 'But,' Sir Robert said, 'if we came here, it was because bad men held our countrymen in captivity. When we have released them, we shall go back to our own country without disturbing your dominion in the least.'

'That's right,' replied Kussai. 'Theodore is a bad man; I hope sincerely you will punish him as he deserves.'

The General then kindly enquired the names of the Tigrean chiefs in his retinue, and it was ascertained that one of them was an elder brother, and two were his uncles. All the chiefs wore vantbraces of gold, and lion-skin tippets.

The presents which were intended for the Prince were then brought out and exhibited, to the undisguised admiration of all. These consisted of one double-barrelled rifle by Purdy of London, some Bohemian glass vases, and an Arab charger belonging to Sir Robert's excellent stud. They were all accepted by the Prince with profuse thanks.

At this juncture the meeting broke up; that is, all the chiefs and officers left the tent, excepting Sir Robert Napier, his military secretary, and the adjutant-general on one side; and Prince Kussai, his brother, and Sej Muroja on the other side.

What followed cannot be given in detail; but the result of
the secret conference was, that Prince Kussai agreed on his
part to allow free passage to the English troops through his
country; and, furthermore, promised to aid and abet them as
much as lay in his power with supplies of food, which the Commissariat Department was of course to purchase and pay for at
reasonable rates.   He furthermore expressed a wish that some
aid might be given to him after the result of Theodore's defeat,
as he would then need it.   To which Sir Robert promised everything, except in the matter of constituting himself the safe-
guard or an ally of the Prince, as he was bound but to obey the
Queen of England's behests; which were to release his countrymen and leave the country as he found it, without committing
himself in any such way as the Prince desired.   With this
Kussai was obliged to be content, satisfied that, if he could not
get the aid of the English, no other potentate or Ras could.

The interview then over, and good will and friendly understanding prevailing among the heads of the party, wine and
spirits were partaken of, and the parties hobnobbed with each
other, after the Saxon custom of drinking ' Waes Hael.'

The troops were then paraded and drilled for the edification
of the Prince and his army.

The British soldiers marched and countermarched, wheeled
to right and left, deployed, marched by columns of companies
on the double quick, all at the sound of drum and fife, and
other military music, to the intense admiration of the guest,
who critically viewed these proceedings.

In his turn Sir Robert Napier crossed the stream, accompanied by the king, to see the Abyssinian troops.

They were a fine body of men, well formed and athletic,
and certainly well adapted for irregular warfare.   There were
between five and six hundred of them, armed with muskets,
matchlocks, double-barrelled guns, Portuguese fusils, and guns
of all nations.   The infantry were simply armed with spear,
shield, and sword, according to their means.   They might make
as formidable antagonists in a mountainous country as any
trained troops would wish to encounter; but, with all their fine
appearance, they had never been able to stand before the
Emperor Theodore's troops.

The Prince, after a review of his troops, invited the English

General to enter his gorgeous tent, where every one was seated on the carpets, spread on the ground. Cakes, tej (fermented honey), herbs, and arrachi were produced, and all partook of them. Before leaving, Kussai declared Napier a great general and a great man, and, doffing his own lion-skin cape, clothed the General's shoulders with the hide of the forest king. He also presented him with his shield and spear. The shield was a masterpiece of Gondar genius. Its umbo was covered with solid gold, and the edges of the shield were decorated with tuberous masses of filagree gold and silver. From its centre drooped a strip of lion's mane.

The opinion which the Abyssinian soldiers had of the English army was, that on an open field English troops would be perfectly unattackable, but that on the mountains they could do nothing. The cannon were not so large as what they had been led to expect by their camp gossips. (Breech-loading Armstrongs with double shells they had no very high praise to bestow upon.) They had heard also, through the same dubious authority, that the English army had a species of weapon which could throw whizzing balls of fire through the air for the purpose of setting fire to houses two or three miles off. These (rockets) were weapons of enchanters. The English, according to their ideas, practised enchantment in many things; and though they had not such big cannon as the Emperor had, and could not fight very well on mountains, still, when they could not conquer by force of arms, they could have recourse to their enchantments, and were therefore invincible and entitled to respect.

Had Sir Robert Napier only expended half-a-dozen shells and rockets in exhibiting the real power of the contemned Armstrongs, much lasting benefit would have accrued therefrom; but little was known of what Theodore's power was, or of what sort of fortress he had at Magdala, and he cared not, in such a case, to waste his projectiles of war in a mere exhibition.

There was a general embracing before the head dignitaries departed, and each chief vowed to do his best in behalf of the other.

CHAPTER VI.

ANTALO—MODE OF MARKETING.

At Antalo, Smelfungus and myself found the pioneer force, under Lieutenant-Colonel Phayre, Quartermaster-General of the Expedition. We were no more to travel alone at will across the country; our future marches would be made with the army.

I have been minute in describing our personal experiences from Zoulla to Antalo, because we were detached from all other company on the route. We had not yet become initiated into the reality of the campaign.

Until the army appeared at Antalo, I enjoyed myself in chasing jackals and hyænas riding over the country at break-neck speed; visiting the towns in the vicinity; buying ghee (native lard), bread, milk, chickens, or in other words, foraging as I used to do during the American war.

A fair occurred every Wednesday at the town of Antalo, at which all the peasantry of the country assembled.

The first Wednesday that occurred after our arrival at this camp, I accompanied the Captain and some others to see one of these gatherings, where it was said the *élite* of Abyssinian society was to be found.

The ground marked out as the intended site of the camp was in the very centre of a plain which stretched from Antalo on the west to Afgool, 15 miles distant eastward, and was nearly equidistant between the former town and the ancient and interesting town of Chelicut. The stream Gundwa, or the remains of it during the dry season, ran round the western extremity of the camp. All this luxuriant country was covered with tall grass, which afforded admirable grazing for the animals of the Expedition.

Once over the Gundwa, we put our horses to a smart gallop, and in about an hour we found ourselves at the base of a series of mountains, which rose higher and higher, until they were crowned by the scarped peaks of Amba Antalo, where, under the shelter of an overhanging mass, the town was built.

The ascent was fully 2,000 feet from the plain. As we surmounted the level whereon Antalo was built we passed two turreted buildings, which crowned the summit of the plateau. These we found to be churches; on the top of each building was a cross formed of ostrich eggs, looking very like alabaster balls. Ascending still higher a steep hill, we entered a narrow street flanked on each side by adobe walls some 10 or 12 feet in height; and passing along this for a hundred paces or so we found ourselves in the market-place—a large open quadrangle, enclosed on all sides by the houses of the people, all of which were curtained from public view by mud walls. The square was filled to overflowing with people of all ages, who had come to the fair from the villages around. The men were dressed in robes of cotton cloth bordered by a deep band of crimson; the women were clothed in a cotton chemise, unzoned, hanging limp and loose about their persons, universally dingy with dirt. The young girls and boys were naked.

Near by was the straw-manufactured goods market, where umbrellas of straw, exquisitely plaited; baskets, cups, jars, cradles, plastered outside with bois-de-vache, baked hard. Near this, again, the market for the sale of tobacco was held. The weed had been compressed into a cake resembling a 'buffalo chip,' and emitted a horrible odour. Next to this was a miscellany of various commodities—iron, hair and shawl pins, silken neck strings, iron crosses, without which no Abyssinian Christian would be seen; strings of wax, jet, and amber beads; pumpkins, chillies, water-cresses, onions, honey, potatoes, garlic, leathern garments, goat-skin bags, water-sacks, and a long catalogue of other vendibles.

Everything that a European wanted was priced at a dollar, though the article might barely be worth twopence; were it a bundle of wood, a few cakes of bread, a jar of milk, a dozen of eggs, the invariable price would be a dollar for each. The Abyssinians are a very apt people. Ten weeks before, a stranger coming to Antalo might have had the whole lot for a dollar; but the English strewed their dollar coins so very extravagantly!

From the town we went to visit the churches, which looked picturesque, surrounded by the deep green of the sycamore. Like many other things which from a distance seem enchanting, the picturesqueness of the churches faded away as we approached. The outside resembled a cowshed, and the inside appeared not much better. In an adjacent house lived the priest. Seeing Europeans around, after showing us into the interior he begged us to enter his house, and after we had seated ourselves he proffered us horns of butter and goat's milk. We were rather astonished at this hospitality— unusual in our experiences from Zoulla to Antalo; but never- theless drank our fill of the delicious beverage. He then brought us some parchment books and manuscripts, which he offered to sell. Not the least disinclined to add to our stock of curiosities, my friend and I purchased some half-a-dozen. When we were about to leave, the priest asked us if we did not intend to pay for the ' haleeb ' (milk). We had not intended; but, since he had kindly hinted to us that the money for it would be acceptable, we paid him without much demurring, though we had given two dollars for the books.

Before entering the camp, my friend the Captain and I struck off over the mountains for a hunt. Hares were in abundance, quails were plentiful, guinea-fowls flew in every direction, antelopes sprung up all around us, and my com- panion made me almost believe he had seen a wild boar.

It was late that afternoon before we turned our faces home- wards. We had ranged over considerable ground, and had seen some very lovely dells and mountain scenery. The day was tending to a close; sweet evening was fast coming on; the sun was declining gloriously over the far extended landscape, gleaming with a mellow light over the plain, purpling distant mountains, and flinging long shadows upon the grass from the tall cliffs overhanging the road.

My dear friend the Captain tried to interest me in recapitu- lating his wonderful adventures, and from the manner in which he related them to me it was evident that he was a very vera- cious person. Indeed, his propensities to relate exhibited themselves at all hours and in all places. The knowledge of

the existence of genius in his person was mine by slow degrees;
for, whenever a thought passed across my mind in regard to my
friend, the memory of his high rank and his vast riches caused
obliviousness of what I was pleased to think was but a mere
aberration of mind.

We had arrived within five miles of camp. The footpath
to it was visible under the brightness of a full moon, except
when now and then some projecting peak or crag, around the
base of which we were riding, slightly shadowed the ground.
A ravine or gully lay before us; and in following the serpentine
path we were compelled to describe an arc, so as to avoid a
deep hole close by, which led us under the shadow of a high
rock. So soon as we neared the rock a deep growl was heard;
and singular to relate, my Nimrodian companion who had had
such terrible combats with a boa-constrictor, a tiger and tigress,
hastily backed his horse behind mine so as to more leisurely
reconnoitre before making any further advance.

'What is the matter?' I asked.

'Did you not hear the panther's growl?' he asked me in
return.

'Are you sure it is a panther?'

'Quite positive.'

'Well, there he is, then, right close to your horse's leg,' I
said.

'Where? where?' he asked, hastily pulling his horse backward.

'There he is in that bush, close to you on your right.
Look out with that gun; you will shoot me if you don't take
care. Why, man alive, can't you see him? He is going to
spring. Shoot him now, or you will be too late,' I yelled out
to him.

'I can't see him, I can't see him!' he shrieked out, despair-
ingly.

'Don't you see him crawling up to you?—about twenty
yards off, on your right?'

'Oh, yes!' and pop—bang went both barrels. 'Is he dead?'
he asked.

'He is down,' I replied.

'Ah ha! the right hand of the "old joker" hath not forgotten its cunning,' he said, with a burst of exultation. 'Did

I not tell you, my boy, I never failed when I once drew a bead ? ' ·

' Yes, often and often,' was my response ; ' but how was it you could not see the animal until he was close upon you? Are you afflicted with dysopsy ? '

' Sometimes, in moments of great excitement.'

' You don't mean to admit that you were excited just now, do you ? '

' Faugh ! me excited ? My dear fellow, I never was excited in my life.'

' Ah ! I beg your pardon, I thought you were ; yet it might have been an illusion of the senses.   But what sort of animal have you shot ? '

' Please see what it is, there's a good fellow, while I finish loading my gun.'

' Why, it's a hyæna,' said I, after looking upon the animal awhile, and internally wishing he had examined it himself.

' A hyæna, is it ? Oh ho ! very often the hyænas are more dangerous than panthers ; for instance, if they get a hold upon anything with their teeth, they never let go.'   And as he was talking thus, he drew closer, and just as he stood near its head the animal started up, and the Captain, though he had a double-barrelled gun in his hands, was indisputably though slightly startled.   It was only a death spasm in the case of the hyæna, and a hunter's spasm in the case of the Captain.   When the beast had ceased to breathe the ' old joker ' came up and deliberately shot the carcase in the head, ' to make sure pop of him, you know.'

It was late when we got back to the camp, having met with no adventure on the way save the shooting of a hyæna by my friend.

On March 3 the 2nd Brigade, General Napier, and General Sir Charles Staveley arrived at our camp at Antalo. It was a most exciting scene.   The news of the approach of the ' burra, burra Sahib ' (big, big master) with the Feringhee army had roused the whole province of Enderta, of which Antalo was the capital.   Thousands of scantily-costumed warriors, women with huge unzoned and ungarbed bodies, and brown-skinned little Cupids lying in litters on the roadside ; old age and youth, male and female, priests and laymen, had ranged themselves on both sides of the road for a mile or so, to await the advent of the Grand Sahib and his host.

Regimental bands had gone forth to salute their comrades with welcoming music.

General Collings and staff, Colonel Merewether, and other officers had ridden out to meet Sir Robert Napier.

The much-looked-for General and his army came at last, preceded by a host of musicians, who distracted the welkin with harmonious and inharmonious strains. As the first regiment appeared, the Antalo garrison vented their joy in loud cheers and greetings. Responsive to their voices rose the huzzas of the succeeding regiments, while the rearguards flung the echoes far over the plain ; and the natives, catching it up, changed the tones to the exhilarating Lululu in honour of the occasion.

First came an Irish regiment, each soldier bearded like a pard, and bronzed by the tropic sun, all weather-beaten veterans inured to life-long campaigns in India, to Himalaya snows, and fervid days in Scinde. Above their heads waved the regimental banner which was tossed several times on the deadly fields of the Iberian Peninsula, and had received reverence from Wellington.

Then came the 4th Regiment, the 'King's Own,' with a regimental flag tattered and torn by the gushing storms of fire and lead in the Crimean campaign. After these came the Native troops, the 27th Native Infantry, called the Punjaubees, tall, well-formed men from the Punjaub districts of India ; the Belooch Regiment — Beloochees with ample green turbans, and red fezes round their heads, dressed in green uniforms with red facings ; and the 10th Native Infantry, composed of sepoys, who were taken from particular districts and tribes. Each man of the latter regiment had thirty or forty kinsmen in the ranks.

After them, again, came the cavalry, the Scinde Horse, with plaited crimson cloth folded round their heads like turbans. They were dressed in green cloth uniforms, and their horses had green shabracks. Each man was armed with a short double-barrelled rifle and tulwar.[1]

[1] Indian sword.

The officers wore silver helmets on their heads. Behind the cavalry regiment came Sir Robert Napier and Sir Charles Staveley, attended by their respective staffs, well dressed and well mounted. A good deal of effeminacy was visible here which detracted much from their otherwise martial bearing. One young lordling wore kid gloves and a green veil.

The Head-quarter staff was followed by a company of the 3rd Light Cavalry, Native Indian, all smart, soldierly-looking men, though some of them, native officers, bobbed along in their saddles as if they were riding to rackets on a hard-mouthed native tattoo,[1] displaying immense calves through the lack of straps to their trousers. The uniform of this troop was a bluish grey, with white shoulder-straps and cuffs.

In the rear of these came the artillery, a battery of six Armstrong guns; the elephants and the transport train, the whole stretching over a distance of seven miles. The muleteers formed quite an army by themselves.

There were 7,000 mules in the transport train attached to the 2nd Brigade of the 1st Division, commanded by Sir C. Staveley, and for these were required at least 3,000 men; but including the camp followers, mahouts, elephant attendants, and camel drivers, this force was increased to nearly 5,000 men.

Besides these people may be added the native settlers, vendors of tobacco, ghee, haleeb, barley, bread, and herdsmen driving the beeves and goats for the commissariat. They were a prominent part of the procession, and figured conspicuously by their numbers and motley appearance. I must not forget, also, the multitude of dhoolie bearers,[2] dhobies,[3] and the bheasties,[4] who swelled to a large extent the strange multitude called together for the crusade. Such varied physiognomies were never seen in one modern army; while the colour of their uniform, clothes, rags, turbans, tarbooches, was of the most kaleidoscopic kind.

Poles from the Bastarnic Alps and bearded stalwart fellows from the Sarmatian wilds chatted gaily with voluble Frenchmen and staid-looking Germans. The turbaned Turk, with the passionate fires of his temperament somnolent under the

---

[1] Pony.   [2] Hand ambulances.   [3] Washer men.
[4] Drivers of oxen bearing water.

virtues of his inseparable narghileh, trudged along leisurely, with no desire on earth save that which he was then exercising; and swarthy Arabs inhaling fumes of a sweet morsel of Latakieh. The olive-cheeked Jew, remarkable for his mysterious limpid eyes of jet, stalked onward at the head of his little string of mules—for the nonce oblivious of the country, the crusade, and his occupation, indulging in day-dreams of some treasure he had hidden somewhere, or of his elevation into the business of a diamond-merchant on his return from African territory. Esurient-looking Parsees or Guebres from Persia ever on the look-out for a chance to turn an honest rupee into two, plodded onward industriously side by side with parlous Jewish mule-teers. Well-oiled Sikhs from the Mahrattas held animating conversations with gentlemanly Sepoy guards from the Deccan; and rolling on in never-ending succession were ferocious Soumalis from Soumali Land, ugly-faced Berbera men, Janus-hearted Bengalees, wild-looking Shohoes, athletic Nubians, and daring Hazortites.

The column, in spite of its martial air, had something of a piebald look. There were white men, red, tawny, yellow, and black men, all mingled up confusedly. Cockneys from London, Celts from Ireland, and sandy-haired Scotchmen marched side by side with Hindoos.

The commissariat mules were unloaded of their treasures at the park of the department; the baggage animals emptied their miscellaneous packs near their owners' tents, and were then marched off to their encampment near the river Gundwa.

The artillery wheeled leftward to their quarters on an eminence, and the furbishing of steel cannon and gear, the watering of horses and distributing of rations began, and the audible, contentious bustle and the droning hum of a vast camp commenced in this encampment on African soil, as it would in any other country.

Around the commissariat was shortly seen a strange sight, which I might have witnessed many a time on the march to Antalo had I accompanied the army thither.

The dollar bags were brought out by the Parsees who transacted business with the natives. Some order amid such a concourse was necessary when hundreds of Abyssinians, having heard of the liberal payments made by the English, contended

who should first dump his goat-skin bag full of barley into the
scales, and handle the bright new coins, their favourite money,
the sight of which always made their eyes sparkle with unusual
brightness.   Chuprasses [1] were already on the ground, arrang-
ing—with almost futile attempt at order—the eager, expectant
Abyssinians.   The natives brought barley, flour, oats, grain,
bread, honey, herbs, wood, cattle, goats, sheep, mules, mussucks,[2]
leather straps, &c.

There were about a dozen Parsees employed by the assistant
commissariat to purchase the provisions for the force.   To
avoid confusion, they were set apart ten paces from each other,
so that the bread should not be thrown with the flour, and the
chickens and ghee should not get mixed up with the bread; and
the wood should not be thrown among the chickens.

Each man was surrounded then by the market people, and
he and his bag of dollars was shrouded from all obtrusive observation. To men who from childhood are trained to deception and
lying, the tempting bag of dollars was irresistible.   Any person
watching the scene for a moment would have been able to detect
go-betweens—Parsee boys—mixing in with the crowd, and by-
and-bye receiving a certain quantity of dollars, and who, upon receiving the money, slowly edged out again to disappear among
some black tents, the homes of their employers, which were
situated near the commissariat park.   During one day these
by-plays would occur very often, and probably each of the
sellers of provisions cleared something between seventy-five
and a hundred dollars.

Strange to say, I never heard of a Parsee being brought up
on a charge of theft. Their influence was so great that I
suppose it would not have been politic to be very harsh to them.
There were but few officers who were not indebted to some one
or other of these clannish people.   Little bills could not be
cashed without their aid; and, as military officers are generally
freer with their money than any other people, in an extravagantly priced country like Abyssinia applications of that nature
were frequent.   To have applied to the military chest would
have been an abnormal proceeding; and they therefore chose to
pay high discount to those Indian usurers, rather than take a

---

[1] Policemen.   [2] Bags made of animal hides.

little trouble to check the nefarious proceedings of the Parsees.

I stated above that the transport-train, after unloading, had been withdrawn to their encampment near the river. For the better understanding of the term transport-train, I shall proceed to a description of it, as in the following pages the words will be frequently repeated.

The general transport-train was raised in Bombay for service in Abyssinia on February 14, 1868, but it may be said to have derived its origin from the two Punjaub trains specially organised in October of the preceding year for the campaign in Africa, to which it owes its foundation and organisation.

The Highland transport-train was organised for service in the Highlands of Abyssinia, a portion of which conveyed the baggage and provisions for the first and second Brigades of the First Division, under Sir Charles Staveley, now marching towards Magdala, while the lowland transport-train was reserved for service between Senafe and Zoulla.

The Highland transport-train consisted of four divisions of 2,000 mules, each subdivided into troops of 150, and the following was the scale of the establishment of each division :—

    1 Captain.
    3 Subalterns.
    4 European inspectors (elected from British regiments).
  20 Jemadars, or Troop Sergeant-Majors.
  80 Duffadars, or Serjeants.
667 Muleteers.
    2 First-class Nalbunds, or Farriers.
  10 Second-class Assistant Nalbunds, or Farriers.
    1 Head native Blacksmith.
  10 Second Assistant Blacksmiths.
  10 Bheasties, or ' Water Carriers.'
    2 Rope-makers.
    1 Salootrie, ' Native Veterinary Surgeon.'
    1 Moonshie, or Native Writer.

At the commencement of the campaign one man had charge of two mules only ; and they were as much as he could look after properly. Stable, watering, and all other line duties at fixed hours were conducted with the same routine as in the Bengal irregular cavalry regiments. The muleteers were trained and drilled to their work and kept in as strict a state of discipline as the circumstances of its being a newly-raised corps would admit of. The Jemadars kept in Persian the nominal roll of each man in their troop, also the pay account. The 'long roll' of every man was kept in Persian by the moonshie; the pay account, was kept in English by the captain.

The officers commanding the Punjaub trains had summary powers, and could sentence to two years' imprisonment in addition to corporal punishment. The whole was under the direction of a sub-director, who was one of the head-quarter staff, and reported direct to the assistant quartermaster-general, army head-quarters, assisted by a staff officer and officer inspector, the latter of whom constantly travelled from one station to another, and reported direct upon all matters that might come under his observation. The train worked regularly between Attigratt and Magdala, principally in carrying baggage, ammunition, hospital stores, and commissariat supplies for the force for one month. At intermediate large stations a staff officer had been permanently appointed to receive all reports from officers in charge of convoys, and he forwarded the same to head-quarters. At these posts several reserves were kept, from which sick and weakly mules were replaced. Mules purchased in Spain, Egypt, India, Persia, and Abyssinia were employed.

The Spanish mules were principally large and ill-bred; and, from my own observations, were naturally of a weak constitution, and unable to bear great changes of climate. They suffered much from a want of a sufficiency of food, especially of grass, of which they did not eat enough, even when it was procurable; for that which was obtained was dry and coarse, and they did not relish it. When once in low condition it was almost impossible to bring them round, even with rest and grazing, abundance of which they got at the sick depôts. The few well-bred Spanish mules there were answered well. The Persian, Indian, and Egyptian mules, on the contrary, were very much more enduring, required less grain, would eat any description of grass, and thrive upon it.

The Abyssinian mules were not very enduring; all those purchased were in good condition and fresh; still they broke down in larger numbers than any others. On March 1 there were

MAJOR-GENERAL SIR C. STAVELEY, K.C.B.

5,412 animals in the Highland train, and of these 1,002 were sick. On April 1 there were 6,691 and of these 1,002 were sick. On May 1 there were 7,690, and of these 1,944 were sick. Since March 1 there were purchased in the country on the Highlands 2,720, and 1,740 were transferred from the Lowlands. Since the same date there were lost on the road 449, and 1,257 were sent to the Lowland sick depôt at Koomaylee and struck off returns ; and 309 were destroyed or died at Highland sick depôts. 2,418 were discharged from hospital and sent to work. The mules had scarcely a fair start, for they were generally put to work the day of disembarkation.

A marked difference was apparent in the working of the men who were sent to follow the army. Those who had the advantage of the experience resulting from the long march from Rawul Pindee, in India, to Kurrachee, in the Persian Gulf, a distance of nearly 1,000 miles, kept their mules in admirable condition, taking into consideration the difficulties and annoyances enumerated by Dr. Hallen ; while those men selected from the low rabble of Cairo and Alexandria, in Egypt, and sent by the hundreds for actual and necessitous service, soon had their animals in a most woful plight. They were, besides being fractious and turbulent, entirely unequal to the task of climbing the rugged mountains, from their previous indolent habits.

The saddles in use were various ; the three principal were the Bombay pad, the cumbersome Turkish, an English saddle termed the ' Otago,' and the McMahon.

The Otago saddle in my opinion is the best that can be invented. It is made of good stout English leather. The sides are made of English leather sewn over two square pads which overlap the flanks of the animal. These are connected by strong curved iron bows enclosed with leather, which rise three inches above the spine of the animal. Two hooks are fixed on each side, to which stout straps of the necessary length, folding any kind of baggage, are slung. Its weight is but 28 lbs. The McMahon is also a very compact saddle, but its excessive weight of 66 lbs. precludes all possibility of its ever being adopted for general use.

The veterinary surgeon attached to the first division of the army of Abyssinia, whose report I have consulted concerning the cases of sickness and mortality so frequent among the animals, said that they were owing principally to ' hard work at great altitudes ; constant exposure, with want of rest ; insufficiency of food and bad quality of the grass, irregularity in feeding, galls owing to bad saddles, and want of time and means for repairing the same, and carelessness of muleteers in saddling and loading.   The muleteers were employed as ghora wallahs (grooms), in the position of public followers from the Punjaub, for the divisions raised in that district ; and men from the same part of India were subsequently substituted after their arrival at Antalo for the Arabs and Turks, as the latter were found to be indolent, apathetic, and totally unmanageable.

The Jemadars and Duffadars were selected from the most intelligent of the muleteers.

The officers of the transport-train were gentlemen who volunteered for the service out of regiments retained in India. They were a most abused set of men, by those who were not capable of appreciating the tremendous sacrifices they had made.

It will readily be perceived by the nature of their duty that they had nothing of that brittle stuff I found at Zoulla and Senafe, but real, manly officers, devoted to their duty, who had entered heart and soul into what they believed to be a good cause.   They were always in the saddle, riding hither and thither, from troop to troop, urging, coaxing, and threatening to keep the muleteers at their posts, and up to late hours of every night, as may be imagined from the condition of the country.

As soon as the road between Zoulla and Senafe had been made passable for wheeled carriages the train was strengthened by a supply of Bombay bullocks and Maltese mule-carts.   But for this timely assistance it would have been sorely straitened, as mule-carriage in the Lowlands began to fail, and the supply of camels collected in Aden and Persia, which had for a while revived in quality and numbers, also began to diminish.   The first passage of carts up the Senafe Ghaut, consisting of a convoy of eighty carts, in sole charge of a native military inspector, was so successful, that every effort was made to increase this kind of carriage.   Captain L. A. M. Graeme, 102nd Foot, was chosen to raise and organise the cart division, and his efforts were crowned with the utmost success.

As the supply of draught mules had ceased, and those in use beginning to fail from hard work, with the sanction of the controller of supply and transport, Captain Graeme commenced converting mule-carts with broken neaps into bullock-carts with the ordinary yoke. This was a great success, as those in charge were enabled to purchase a very fair supply of draught bullocks from the natives of the country, and thus they had as many as 600 carts at a time available.

The veterinary department was a source of great anxiety to head-quarters, for on leaving Bombay the establishment was miserably insufficient. However, among the French and Italian muleteers there were collected some eight or ten men who had a fair knowledge of the first principles of veterinary practice. This was the utmost help that could be rendered to the surgeons, many of whom were young men who had but recently entered the army, but who nevertheless by dint of energy and willingness did good service. These officers were under the immediate orders of Veterinary Surgeon Lamb, whose charge was an enormous one, consisting of upwards of 30,000 animals, even at a very early period, but which rose eventually to the number of 53,000 animals, with only ten surgeons to assist him. These gentlemen had to perform, under the existing state of things, extraordinary and unusual duties, having constantly to compound their own medicines, administer them, and attend individually to the washing and dressing of the animals in the sick depôts, for want of subordinates to assist them.

The efforts made to procure sufficient animals for the Expedition cost much labour and money. The English consuls at Malaga, Alicante, Cadiz, Barcelona, Genoa, Tripoli, Beyrout, Constantinople, Smyrna, Gibraltar, and Malta were specially directed to use all means to procure mules. Officers accompanied by veterinary surgeons were despatched to examine and select those fit for service as they were brought to the several depôts. Under the complicated system of English departmental government a larger number of officers was required for this work than were really needed, and as a consequence the whole became very expensive.

To expedite as much as possible the forwarding of stores to the front, and to aid the Lowland Transport Train, a short railway was laid, extending from Zoulla to Koomaylee. This prevented the possibility of the mules being kept on the Zoulla beach, which had proved so fatal to them when they first arrived; and obviated the necessity of sending the bullock-carts to travel over the hot sands.

As fast as each ship unloaded its cargo on the bunder [1] the coolie gangs loaded the railway cars, after which the locomotive rushed with its train to Koomaylee Station, a distance of twelve miles. Koomaylee was thus made a vast depôt, and the arrangements were such that two or three hundred carts could load at the same time; and from thence to Senafe and Attigratt was a good road. The rails used in the construction of the railway were very slight, weighing only 30lbs. per yard, or 25 tons per mile. The sleepers were six feet long, and three-fourths of a cubic foot in diameter. The locomotives, three in number, were poor, rickety things, but still well adapted to the task in hand. They managed to haul twelve or fifteen freighted cars from Zoulla to Koomaylee in an hour.

The telegraph proved most useful to everyone; to the general officers as well as to the various depôt superintendents. If there was a deficiency of provisions at one post, a telegraph message sent to the superintendent at Senafe or Koomaylee would bring a fresh supply in two or three days, or a news despatch could be sent to the mail steamer directly from the front, at the very last minute of her departure from Annesley Bay.

It was proposed, when the Abyssinian Expedition was first organised, to lay a Red Sea cable from Souakin, on the coast of Egypt, as far as Annesley Bay. From Souakin a telegraph could have been laid to Cairo, where it would meet with the Suez, Cairo, and Alexandria line, which connects with the London, Malta, and Alexandria cable. But the enormous outlay that the construction of the proposed line would demand deterred the English Government from pursuing this enterprise.

From Antalo to Magdala was but seventeen days' march, allowing ten miles for each day. General Napier, then, with some reason believed that thirty days would bring him before Magdala. For the transporting of the provisions for 10,000 men, inclusive of muleteers and camp followers, for thirty days,

[1] Pier.

allowing 150 lbs. for each animal, would require 4,000 mules ;
for conveying baggage, tents, rocket and mountain guns,
ammunition &c., rather more than that number.

It would not do by any means to trust entirely to the supplies procurable from the country.   With fifteen days' forage on
hand, we might well hope, from previous experience, that we
could keep starvation away.   Halts might be frequent, extending the seventeen days' march to two months', perhaps.   Some
flour and bread might also be expected from the country
people, when once they had seen the glimmer of real silver
coin.   From which the reader may perceive that against contingencies the army was very well guarded by the calm and
skilful foresight of the veteran, General Napier.   What did
happen on the eventful march will be seen in the following
pages.

How far wrong and how far right Napier was in his calcula-
tions will also be seen.   With one hundred thousand dollars in
his treasury, Sir Robert Napier could do wonders.   Backsheesh
to chiefs and presents to village magnates were necessary, and
would ensure, if it were possible, supplies of all kinds.   Dollars
were our only coin.

The currency in Abyssinia is coticular blocks of rock salt
eight inches long, one inch in width, and one in thickness.
These are dug out from the salt plains of Hashed.   No coin is
recognised except the Austrian dollar of the year 1780, introduced, since that year, to replace the native gold withdrawn.

General Napier tried hard to introduce rupees or shillings
in the country, and even wrote to Prince Kussai about it ; but
it soon proved impracticable.   Had he been successful, the cost
of articles would probably have been just one-half the prices that
were paid.

The natives, besides being very tenacious for the dollar,
were also very careful that the distinguishing marks were
visible on the coin ; otherwise they were refused altogether.
These are a portrait of the Empress Maria Theresa, with a
diadem of pearls around the head, a pearl brooch on the
shoulder, and the mint mark S.F.   It was very essential,
therefore, that all the dollars were of the true orthodox stamp.
As there were not enough dollars of that reign extant new
ones had to be coined, and it is a curious fact that 500,000

of this obsolete coin were struck at the Imperial Mint at Vienna expressly for the Abyssinian Expedition.

The first morning after the arrival of the army at Antalo, I determined to call upon the General-in-Chief, it being the custom in all countries for civilians to report themselves at the head-quarters of a military cantonment.   Hitherto he had been as invisible to me as the Mikado of Japan or the Grand Llama of Thibet, engirt as he was by the reverence due to his high position, and surrounded by an impenetrable cordon of guards, and a barrier of brilliant uniformed officers, with high-sounding titles.

'Can I see the General?' I asked Captain S——, who was acting aide-de-camp, a fine, handsome, sturdy gentleman.

'What General?' said he, enquiringly.

'Sir Robert Napier, sir,' I answered, blandly.

'Oh! you mean His Excellency Sir Robert Napier?' said he, slightly lifting his eyebrows.

'Certainly, sir,' I responded; 'I mean His Excellency Sir Robert Napier, Lieutenant-General, Commander-in-Chief of the Abyssinian Expeditionary Field Force, now in Abyssinia.'

'You cannot see him just now,' he replied, with the ghost of a smile upon his features.   'He is at breakfast.'

In the evening I called again, but His Excellency was busy with some native chiefs who had called to pay their respects.

The next morning I repeated my call, but unfortunately the Commander-in-Chief was again at breakfast.   In the afternoon I called once more, but he was taking his siesta.

The third morning I presented myself at the hour of nine, but he was asleep, having been very busy until midnight the night before.   In the afternoon, at two o'clock, I paid a visit at head-quarters, but he was at tiffin.

The fourth morning I called at the hour of eleven, and presented myself at Captain S——'s tent for the seventh time, with the view to obtain permission to call upon Sir Robert.

'Is His Excellency Sir Robert Napier engaged, sir?' I asked, very politely.

'Ah! may I ask your name, sir?' said Captain S——.

'There is my card, sir.'

'H'm! yes, I see.   Mr. Stanley, Correspondent of the "New

York Herald." If you will be kind enough to seat yourself, I will go and inform Sir Robert.'

I seated myself, and began to imagine what sort of a reception I should receive from the Commander-in-Chief.

'Sir Robert Napier will be happy to see you, sir,' broke in Captain S——.

Following this officer, I entered a commodious tent used as an ante-chamber, lined with yellow cotton and carpeted with Persian carpets, and partitioned from the interior by a silken curtain. Sitting at a table covered with newspapers and documents was a man in an undress uniform, with a slightly greyish head and moustache, looking as serene and contented as a farmer who had just been over his estates, and was satisfied with the expectation of a good harvest.

'Mr. Stanley—Sir Robert,' was the introduction of Captain S——.

At the sound of a voice the serene-looking gentleman rose and acknowledged the introduction with an affable inclination of the head, inviting me to a seat by a motion of his hand.

'Allow me, Sir Robert, to show you what introductory letters I may have for your examination.'

After reading them, he said that he was most happy to see me, and inquired whether I had been accorded every facility for accompanying the Expedition.

'No, sir; not quite,' said I. 'I was informed, by a telegram from the Secretary of the India Office, that all facilities would be afforded me, but I cannot say that I have experienced any yet. What I have is my own, save one baggage animal given me at Zoulla. I was misled when I arrived at Zoulla by persons who said that you were about to make a dash upon Magdala, and I refused a tent upon that account; but I soon found my error, and applied for one, but without success. If you can let me have a tent and a baggage animal, I should feel quite comfortable.'

'How did you come from Senafe?' he asked, in a kind tone.

'With Captain Smelfungus. He and I live together in a *tente d'abri.*'

'How! Is Captain Smelfungus in this camp?' he asked, in astonishment.

'Yes, sir; I have just left him.'

'H'm! I was not aware of that; but to recur to yourself. Really I do not believe that I can spare you a tent for yourself alone, as I intend to make forced marches towards Magdala. All officers are expected to live there in a tent; and, if you can content yourself with your friend's accommodation, it will be by far the best plan. However, if you want a tent very badly the quartermaster-general is authorised to sell to civilian gentlemen what they need. Horses, also, you can purchase from the natives about here. As for rations, I can give you an order upon the commissariat, and you can settle there every month, or whenever they call upon you. I must not show more partiality to you than I do to the correspondents of the English Press; but whatever privileges they have shall be extended to you. I would advise you to take as little baggage as possible, because to-morrow the Adjutant-General will issue an order to reduce all baggage to seventy-five pounds. In that case you have all you need in your two horses and the *tente d'abri.*'

'Thank you, General. I also wished to ask you if you will be kind enough to let me see any *précis* of intelligence concerning political matters you may think fit to send around to the other correspondents, as the "New York Herald" expects me to telegraph any news you may give me, or I may become acquainted with.'

'I shall be delighted. Certainly; you may see whatever the Adjutant-General, or Political Secretary, may give out to the other gentlemen.'

Before parting, the General kindly offered, if my ready funds failed, to supply me with any reasonable amount on a draft, and extended an invitation to me to dine with him that evening.

At seven in the evening precisely, the invited guests of Sir Robert Napier were assembled around the lengthy camp-table, heavy with plate, mugs and glasses, decanters and cruets. Smelfungus was of the party, profusely and wonderfully decorated with medals and orders of all nations. Before seating ourselves, Sir Robert made his appearance, dressed very unpretendingly in the dark undress uniform which he generally wore. Introductions followed his appearance. Colonel Penn,

of the Mountain Train Battery ; Colonel Frazer, of the 11th
Hussars ; Captain Speedy, from New Zealand, the interpreter
for the English army, and whilom commander of King Theodore's army; Mr. Richard Holmes, archæologist from the British
Museum ; Captain S——; Lord C—— H——, aide-de-camp;
Captain Harden ; Captain Smelfungus, of Universal Notoriety,
and myself were then seated. I had the good fortune to be
placed next to the gigantic Ajax—Speedy, and opposite to
General Napier.

Every one had provided himself with his own drinking-horn
or glass, knife and fork, spoon, camp-castor, and napkins. The
plate and other ware were the General's. His khansamah,[1]
khitmigar,[2] and hamili,[3] were very well-bred, quiet, and
attentive servants, who knew by intuition the wants of each
guest, and administered to them without any noise or disturb-
ance to break or distract the harmonious clatter of table
implements. The soup was brought in ; it proved to be an admirable and most nutritious dish, concocted out of compressed
vegetables from the commissariat. Scarcely a word was uttered
by anyone during the business of imbibing the soup.

The soup being despatched, the *entrées* were brought in by
the white-robed and cleanly servants, while the gentlemen leaned
back in their chairs to obtain fresh strength for a renewal of
their efforts.

The camp-manufactured wax candles diffused a soft light
over the table in harmony with the pale yellow of the tent
lining ; outside was heard the dull boom of bass drums, and
the shrill sound of the fifes ; and now and then the cry of
'Post—all is well!' prosaic sounds in comparison with the
romance of the scene enacted in General Napier's tent.

Indian patties, brain cutlets, and veal pies artistically
gotten up were laid before us. As there was nothing of that
chilliness apparent in set dinners, neighbours helped each
other freely with the various viands.

Conversation was not very lively, nor do I believe it is
good during dinner to speak much ; it generally cools the
dishes, engenders a little distaste for food, and appears very
much as if people contemned the culinary skill of the *chef*

[1] Butler. [2] Under butler.
[3] Bearer or keeper of bungalows in India.

de cuisine. All the gentlemen present were epicures in Indian cookery, and, like well-bred people, paid particular attention to the grateful dinner.

Then followed ragouts, potato cakes, vegetables and herbs of every description, cotelettes and fricassees, all of which received due attention. After these came beefsteak pie, boiled hump, roast fowls, and roast sirloin, cooked *à l'Anglaise.* As well as circumstances permitted these edibles were also excellent.

To close up the substantials came the *curry de poulet* and snowy rice, eliciting from the bearded gentlemen unqualified encomiums upon its delicate and piquant flavour. In my opinion the curry was abominable, giving me a foretaste of the liquid fire which is said to be the portion of all sinners in the Hereafter. When it was despatched it appeared to have unlocked the conversational powers of the whole party.

'The pleasure of a glass of Abyssinian tej [1] with you, Captain Smelfungus,' said Sir Robert Napier, blandly.

' With the greatest pleasure, Sir Robert,' replied the Captain, in blander accents; ' I am a great admirer of tej.'

Simultaneously we all inclined our glasses to the General and to each other, and each imbibed with satisfaction a hornful of the native brewed beer.

A slight sketch of each, as they were all most notable persons in the campaign, and will be heard of often as we proceed, will not, I hope, be out of place here.

The General-in-Chief was a man of fifty or fifty-five years of age, stout and well built. His face was remarkable for the kindliness of the blue eyes, the genuine gentleness of the countenance lit up by them, and the smile that continually played around his lips. To all Sir Robert was extremely bland, affable, and kind; sometimes there lurked in his tones something akin to a sarcastic *politesse,* and at such times he was more plausibly phrased than ever.

Colonel Frazer was the very picture of a military *chevalier,* grand, yet polite at all times; a steady, vigorous being, endowed with frankness, good nature, and redundant health.

Colonel Penn was a veteran in military matters, but not in

---

[1] Abyssinian beer.

age, being a youngish person. A frank and hearty cordiality and an irresistible energy were his principal characteristics.

Captain Charles Speedy was an Ajax in truth. His limbs might have challenged the envy of a Hercules; he stood six feet six inches in his stockings. The adjective 'long' could be affixed with justice to anything appertaining to him; he was long-shanked, long-handed, long-bearded, and long-wise in every way, shape, and form, except being long-sighted. He was, in fact, extremely short-sighted, and wore glasses at all times. The prominent features of his face were—high, salient cheek bones, a sloping brow, red bushy beard, and kindly blue eyes strangely contrasting with the determined expression of his mouth.

Lord C—— H—— was a manly-looking boy of twenty, with a full chest and a tall, robust figure.

Mr. Holmes was a specimen of the British volunteer, on all occasions ready to uphold British dignity. He was out in Abyssinia to collect manuscripts and antique articles for the British Museum.

'Ah! that's a jolly drink,' said Speedy, uttering a pathetic sigh, and wiping his beard after the tej; 'a very jolly drink, say I.'

'How do the natives make it, Mr. Speedy?' asked General Napier.

'Well, General, there are different ways of making tej in this country. In this province they put a quart of honey in a five-gallon jar, also a quart of strongest arracki; they then fill the jar with water, and let it stand for about five days, when the whole will be well fermented and ready for use. The Emperor Theodore is a famous man for the manufacture of good beer, but he seasons it pretty strongly with the arracki.'

The dessert was brought after this extraordinary revelation; and, while the delicious blancmange and custards were discussed, topics of conversation were skilfully introduced by the Chief.

With Mr. Holmes Sir Robert started a conversation about spiritualism, during which it transpired that the General knew much about that cabalistic science.

When the coffee was brought, and our Habanas (the gift of Sir Robert) were fairly lit, the General commenced on other subjects. It must not be imagined that he monopolised the conversation; quite the contrary. Nor was conversation confined to any one subject; it embraced literature, fine arts, poetry, sculpture, philology. From these subjects we drifted by easy stages to a learned dissertation upon Governments, where the living wrongs committed by dysnomic legislation were most graphically related; relative comparisons between Monarchical Governments and Republics, where the General most strenuously maintained his opinions by multifarious quotations from classical lawgivers and modern essayists, of course in favour of the supreme excellence of the Constitutional Monarchism of England.

CHAPTER VII.

SMELFUNGUS AND I PART—THE BRITISH PRESS—IRRESPONSIBLES AND THEIR
PROPHECIES—DISMAY OF THE ANGLO-INDIANS—UNHAPPY THEODORE!—
POOR CAPTIVES!—POLITICAL INTELLIGENCE—SIR ROBERT'S ABLE GENERAL-
SHIP AND DIPLOMATIC TALENTS—DELAY IS EXPLAINED—THE 'ELEPHANTS
OF ASIA' IN ABYSSINIA—THE TRANSPORT TRAIN.

IMMEDIATELY AFTER the events recorded in the last chapter I
found it expedient to dissolve my connection with my good
friend Smelfungus. Accordingly I purchased of him his tent, .
and, moving up to head-quarters, pitched it close to Captain
Speedy's, the place appointed me by the Quarter-Master-General.

After the dinner at the Commander-in-Chief's quarters,
during the two weeks the army remained at Antalo the invitations to mess came thick upon me.

The English Press was very ably represented in Abyssinia,
and the correspondents consisted of some remarkable literary
lights. I do not mean remarkable in any invidious sense; but
each one, really and truly, was remarkable, after his own kind.
They all messed together; and though there were contrasting
elements in their natures, yet they seldom disagreed; in fact
there was more harmony in the Press tent than in any other
bell-shaped domicile in the army. I hope I shall not be tra-
ducing these good gentlemen, as I bear them much love and
respect, if I give them the credit of being the most sociable
mess in the army, as well as the most lovable and good-
tempered.[1]

The wheels of the campaign plaustrum, in spite of the dark-

---

[1] The British Press was represented by my friends Dr. Charles Austin, D.C.L.,
Times; George A. Henty, Esq., *Standard*; W. Owen Whiteside, Esq., *Morning
Post*; Alex. Shepherd, Esq., *Daily News*; —— Adare, *Daily Telegraph*.

some prognostications of legions of irresponsibles, moved on with dignified pace, like the car of fate, slowly but surely.

Very few indeed had as yet fallen victims to the voracious carnivorous animals that were said to infest every mountain, forest, patch, and plain in Abyssinia.

Not one had, as yet, mistaken poison for honey, or died by eating the delicacy; neither had a soul suffered any great inconvenience from drinking water from old wells.

No Houssas had been found, neither had the Shohoes, the Agames, or the Hazortites proved unfriendly; rather the reverse. Whatever the commissariat chose to buy could be bought.

No hippopotami had darkened the sunlight of our paths, nor had any esurient lions and cheetahs frightened any one of us with their glaring eyeballs in the dark; no loathsome cerastes had nestled in our blankets; no disgusting yellow-spotted, pink-headed fly had been seen.

The murderous tetse had vanished out of sight and out of our minds; the tzaltzalya had spread his whirring wings and had hastily stolen away; the earth had become fair to look upon; the mountains looked down benignantly upon us; the plains laughed joyously for us; the water flowed abundantly from the spouts of Norton's pumps; even the air appeared more lucent after our arrival at Antalo.

Lo! it was miraculous! All England stood amazed; the false prophets, with their false vaticinations, must need hush their everlasting moan!

A general order was issued that a further advance would be made in two or three days, with the command that all soldiers should reduce their kits to seventy-five pounds, and that officers must positively limit their baggage to seventy-five pounds, and their studs to two horses. It was further decreed that twelve soldiers should club together in one tent, and that two officers must be contented to live together in a tent.

'Oh, Anglo-Indian officers! dainty, heaven-born children! scions of patrician parents! had it come to this? Uneasy and wrathful were they in this extremity! The prudent old Chief had reserved this painful determination until they had arrived at Antalo.'

'Seventy-five pounds—seventy-five; just think of it! It is a most preposterous order! I declare I will resign sooner than

go on with a miserable seventy-five pounds of baggage !' and
fluffy old majors and colonels, who had served twenty-five years
in India, vented their indignation in fine style. Sir Robert
was inexorable; but the Anglo-Indians were equally determined to take as much as possible—on the sly ! Numerous
were those who threatened to write to the War Office ; but in
the meantime the Expedition would be over.

The Pioneer Force, under Quarter-Master-General Phayre,
was ordered ahead to clear the road, remove obstructions, and
report on the prospects in the front. Two days passed by, and
the report was presented by a courier from Colonel Phayre.
It was admirably written : ponderous and learned, elaborate
also, though a little too florid, with a clause, however, that a
road could not be made across the Aleggie range in less than a
week.

' It cannot be accomplished sooner,' wrote Phayre ; ' the
country before us is nothing but a succession of ranges intersected by deep gaps and tangled ravines. The slopes of the
mountains are almost perpendicular ; a series of terraces, or
escarpments, must be cut through before the army can hope to
proceed further.'

' So, this is bad news,' said Napier to Speedy ; ' I wonder if
no other way can be found without going too far out of our way.'

Speedy asked of some of his spies the best way to Lake
Ashangi, our first objective point.

' Oh, the road was easy ; there was a path would take the
army clear to the Lake,' replied the spies.

These sanguine people were sent with Captain M'Gregor ;
and assuredly a path was found, both straight and clear, without any deviation from the intended route. Colonel Phayre
had relied too implicitly on the information of natives.

It is a noticeable fact that the literal truth was never told in
Abyssinia. All Abyssinians are the most stupendous liars, and
everything related of the country ahead was certain to be found
false as we journeyed along. Colonel Phayre's reports, when
based upon native information, became as incredible as the effu-
sions of the London seers ; and the rumours that were wafted
one day of the political affairs or the whereabouts of Theodore,
were sure to be belied by the official *précis* of the following day.

We were as ignorant of the next day's station as though a

Quarter-Master-General had not been sent before us to explore and lay down the line of march.

One day we received intelligence that the vineyards of Abyssinia had been found ; that the country presented the aspect of superlative civilisation ; that the maidens were very Cleopatras and Judiths in their nut-brown beauty and queenly graces ; that cultivated fields expanded as far as vision could penetrate—in short, that everything was diametrically opposite to what we discovered it to be when the army came upon the scene. All this was as wonderful as fable, as enigmatical as a rebus !

I am not singular in these opinions. All the correspondents remarked the same peculiarities ; and it was passing strange, in such a deceptive atmosphere and illusive mirage, how anybody could ever say again one word of truth.

'Theodore was coming.' 'He was not coming.' 'He was sending presents.' 'He was retreating to Kuara.' 'He was fortifying near Lake Haik.' 'He was going to dispute the Bechilo Crossing.' 'He was sending presents and praying for reconciliation.' And he was doing neither. Such a compound of inscrutable mysteries was never manufactured without some legerdemain of the Evil One.

Now that we have brought the Expeditionary Force to the point of moving forward, let us glance for a moment at the political affairs of the country, and see whether Theodore has been prospering since the date Britain declared war against him.

We left him in our introduction struggling manfully against a cursed fate ; beating back his Briarean-handed foes which beset him round about. His vast empire was dissolving like the waxen wings of Icarus before the fierce light of revolution. The lurid fires of towns and hamlets burning flashed their portentous blaze athwart the midnight sky ; the wails of widowed women and fatherless children rent the air ; the groans of dying warriors, murdered by cruel hands, called loudly to Heaven for vengeance !

It is said that 30,000 men, women, and children were destroyed by crucifixion, the relentless courbach,[1] or by shooting, stabbing, or decapitation, within three months. At such times

---

[1] Courbach : a whip, made out of hippopotamus hide.

he appeared like a demon. He was crazed with drunkenness
and despair. He slew his best friends and councillors, and
condemned to death tried and trusted warriors. Truly is it
'whom God purposeth to destroy, He taketh away his understanding.' Unhappy Theodore! None was more wretched
than he!

The captives were in chains. His hatred of them was increasing. Three or four of them had been condemned to death,
but the sentence had been commuted to imprisonment. The
batch of English and German prisoners lived on, having but a
precarious tenure of life at the best so long as they lay at the
mercy of the tyrant Emperor. Theodore but retained them
under the impression that his own ends were attainable only
through their safety.

The captives were supplied with money through Colonel
Merewether's indefatigable exertions in their behalf. Money
was powerful even in prison, and their otherwise painful condition
was greatly ameliorated through the luxuries which the money
purchased.

The British Head-quarters now and then received letters
from Consul Cameron, the Envoy Rassam, Lieutenant Prideaux,
Surgeon Blanc, the reverend gentlemen Stern, Flad, and
Waldmeier, all symphoniously choral in their lamentations,
teeming with an agony which seemed to have become
chronic.

Cameron, in his letters, indulged in harrowing phrases, and
was nearly certain that he would never experience the joys of
liberty again; but he was resigned, though melancholy. Poor
Cameron!

Rassam prayed earnestly, in the agony of his spirit, that
Sir Robert Napier would deal leniently with Theodore, lest
his vengeful temper should cause him to wreak murder on
them. Poor Rassam!

Mr. Stern relieved his misery in inditing florid accounts of
the country, and highly interesting descriptions of the cruelty
of the 'Wild Boar of Ethiopia,' upbraiding the British Government for its unaccountably apathetic movements. His letters
were truly pathetic, in composition. Unhappy Stern!

Messrs. Flad, Prideaux, Blanc, and Waldmeier wrote very
interesting epistles, wherein they endeavoured to enlighten

the British public as much as possible concerning the habits and customs of the people.

The Emperor Theodore received the war proclamation with well-acted calmness, and showed his contempt by not replying to it, or to any letter sent by Sir Robert Napier. For a time the man was a mystery; but information was smuggled into camp of his movements and his intentions through the aid of funds which Rassam possessed.

The following is a synopsis of the news received up to March 12, the date the army left Antalo for Magdala :—

December 9, 1867.—Theodore was at Debra Tabor, continuing his evil and imbecile course of alienating the affections of his men by ' killing them in great numbers.' He made incessant efforts to increase his army, but the largest number he could rally round him was 8,000 men.

December 15.— A messenger arrived from Mr. Rassam, bearing a letter which stated that a large force under Menilek, King of Shoa, was between Magdala, the prison fortress, and the Emperor's camp, and that there was an intention on Menilek's part to make a *coup d'état* in favour of the prisoners. Kussai, of Tigre, was professing friendship, but there were sinister reports that under the mask of friendship he intended some mischief.

December 19.—The reports about the Prince of Tigre were proved to be unfounded. A message came to camp from him wherein he asked, very naturally, why the ' Feringhees '[1] and he should not be friends? That his foes were Napier's foes, his interests British interests. ' Take therefore forage, and my prayers.'

December 20.—Kussai's ambassador arrived at Senafe, and was delighted with his reception. He returned home a sworn friend of the English.

December 22.—An important messenger from Mr. Flad arrived at Senafe. Theodore had swept down from the heights of Debra Tabor, and with one fell blow had annihilated unready Menilek, whose *coup d'état* was to liberate the prisoners. He had reached Magdala, and was then about to return to Debra Tabor with his captives.

---

[1] Ethiopian for ' English.'

December 25.—Dr. Krapf, the interpreter at that time for the army, returned from an exploring tour, and cross-questioned the messenger who brought the startling tidings on the 22nd inst. The man confessed it was all a falsehood, after it was proved to be so by the date of letters at hand. The attack upon Menilek had occurred in October according to the false tale-bearer; but the last letter received from Mr. Flad was dated November 7.

January 6, 1868.—Authentic information arrived through a letter received from the prisoners at Magdala, stating that Menilek had retired to his own country of Shoa without having accomplished any *coup d'état*. The country intervening between Magdala and the Emperor's camp at Debra Tabor was clear of all enemies.

January 15.—Information received that Theodore was moving upon Magdala with a force of 10,000 men and 20 cannon, and that Gobayze, Prince of Samen, was following cautiously up with the view to watch his movements.

January 20.—Major Grant, the Nile explorer and *compagnon de voyage* of Speke, started on his mission to the Prince of Tigre with presents. Major Grant did good service in Abyssinia.

January 21.—A letter arrived from Rassam, stating: 'Theodore is moving slowly upon Magdala. Wagshum Gobazye has run away. Theodore writes to me, " How are you? Are you very well? I am quite well. Fear not, I am coming to your assistance. Keep up your head; I shall soon be with you. I have one big mortar and plenty of long cannon. They make a terrible noise; but they are heavy to move." ' Dr. Blanc wrote in this wise: ' We are delighted to hear you are coming to liberate us. How it will end no one can say. We have all prepared for the worst; but we have at least the satisfaction of knowing that our deaths will be avenged.'

February 2.—Captain Charles Speedy arrived from New Zealand, to act as interpreter for the English army.

February 9.—*Précis* of intelligence furnished by officials read as follows: ' Letters were received to-day by Colonel Merewether, the political officer, from Mr. Rassam and Dr. Blanc, dated Magdala, January 17, with enclosures from Mr. and Mrs. Flad, dated King's Camp, *en route* to Magdala, January 9.

All the captives are reported well up to date. A detachment of troops, which had left Magdala January 8, had joined the King in his camp, and had received charge of a party of about four hundred prisoners from the camp to Magdala. The imprisoned Europeans were among the number. Their leg-fetters had been removed, and handcuffs substituted, so that they might march. It is said that Mr. Rosenthal would accompany them. The king was using every endeavour to have the road made—working with his own hands, and making the free Europeans (his gun manufacturers) help. He had made some slight progress, and had arrived at the bottom of the valley of the Jeddah river. Mr. Rassam calculates he would reach Magdala about the end of February with his camp, though by abandoning the latter he could in one day arrive there. The people of Dalanta continued submissive, but others had rebelled again. His soldiers had suffered from the scarcity of provisions and transport. It was reported at Magdala that Menilek, King of Shoa, had again set out for Magdala, better prepared to act against Theodore than on his former visit. A detailed communication from one of the captives sent to one of his friends in England, and there published, has by some means reached the King's camp, and is in the hands of Mr. Bardel. Apprehensions are entertained that it may do injury.'

February 12.—Ambassador of Tigrean Prince arrived. Loud-sounding blare of trumpets and thunderous welcome to the dusky representative of His Sable Highness by the English army. Friendly sentiments interchanged.

February 25.—Prince Tigre arrived at the English camp at Houssein, and confirmed with his own lips friendly intentions.

February 27.—Another *précis* issued out, stating that Wagshum Gobazye and King Menilek were both near the Emperor Theodore ; so near that the camp fires of Gobazye could be seen from Theodore's camp. The Wagshum had sent him a message which was deemed impertinent, and Theodore therefore put the herald to death.'

March 6.—Monsieur Munzinger, French consul at Massowah, accompanying the field force as political adviser and interpreter, started on a mission to Gobazye's camp to endeavour to ally him with the English.

As we have traced in a compact manner the political

HOUSE AT MAGDALA WHERE CONSUL CAMERON AND THE REV. MR. STERN WERE FIRST IMPRISONED.

features, I may be allowed to pass a few comments upon the position in which Sir Robert Napier found himself in the midst of these little by-plays of Theodore, Wagshum Gobazye, King Menilek, and the Prince of Tigre ; and, as the condition of the prisoners is improved, we'll let them rest until we come before Magdala.

Sir Robert Napier's aptness for command was established no less by his prudent advance into the interior of an apparently unknown country than in his skilful diplomacy with the Ethiopian potentates.

Theodore, the Emperor, by his misgovernment and unfitness to rule over Abyssinia through his unbridled passions, alienated from him every prince and ras, or chief, of any consequence in what was once his empire.

It therefore remained for Sir Robert to take advantage of this fact, and endeavour to retain that which Theodore had lost —the friendship of the natives. His well-balanced mind was admirably adapted for this task.

The chiefs gradually became stipendiaries of Sir Robert, and exercised but little their own wills, save in what was acceptable to his policy. Prince Kussai had been already secured. Wagshum Gobazye, Prince of Samen, was said to be inimical to Theodore, and it was reported that he had an army of 40,000 men prepared to fight him at any moment. Now that the army was approaching his country, attention must be paid to him, and his services, alliance, and good will secured.

Monsieur Munzinger was despatched with presents and peaceable overtures to him.

Between Lasta and Enderta, in which latter province Antalo lay, another warlike chief was encamped on a precipitous mount thirty miles distant from Antalo. This man's name was Welda Yasous-Welda, son of Jacob—who was more like a robber chief than a recognised power. To this person Napier also sent invitations to visit him, with proposals for alliance and neutrality. Warlike Welda gave equivocal replies; and Wagshum Gobazye was inclined to be more of a braggart than ever. With 40,000 men, thought he, what could he not do ?

Still his answers were not literally menacing to the English ; they contained, however, unsatisfactory bombast as to his power, and his intentions towards Theodore.

He was more inclined to neutrality than alliance.

'But would he promise to permit his people to sell provisions to us if they were adequately remunerated?' asked Sir Robert.

'Yes; the people could sell what they wished; he would not prevent them. Was not Theodore his deadly enemy, as well as Napier's?'

Sir Robert was too patient a man to be flustered. There was a perpetual sending of despatches, and an endless series of complimentary letters issuing from Head-quarters to him; presents were sent, with innumerable kind greetings and well-wishes. No prince could well withstand such forethought of his success and such magnificent presents without feeling that it was incumbent upon him to reciprocate. The vanity of the Wagshum was flattered by the attentions and kindness of the British Negus, and he succumbed as well as a man of his suspicious mould could well do.

He wrote back a most gracious reply, declaring his intention to aid Napier as much as lay in his power, and his determination to attack Theodore as soon as circumstances permitted.

There was another person in the Trans-Magdala country whom it was wise to conciliate; this was Menilek, King of Shoa. He had also an army of 40,000 men ready to attack Theodore. But he was such a peripatetic and fabulous potentate, believed to be in existence somewhere, but not a settled entity like Gobazye and the Tigrean Prince. It was doubtful whether communication could be had with him.

In the Cis-Magdala country, to the left of the route to Theodore's prison fortress, ranged the ubiquitous Gallas under their respective queens—Ambo Gallas, under Queen Walkeit; the Wollo Gallas, governed by Queen Regent Musteval. These professed Mahomedanism, while we were Christians.

Sir Robert was prepared for all sorts of castes, nationalities, and religious faiths. In his train was a Mahomedan prince, Meer Akbar Ali, who had requested permission to follow the General. He was taken under the supposition that he would be useful. This Hindoo prince was a most proper person to enter into negotiations with Mussulman governors. His dark skin, his faith, his native love of finery and gorgeousness, was in his favour, and he was selected for the business, to depart only when we should have arrived in the neighbourhood of the Galla lands.

Major Grant was also accredited to Welda Yasous; and so well did he succeed in his mission, that Welda sent his brother to see the Feringhee General and accept proposals for peace.

Captain Moore, Arabian and Persian traveller and Arabic interpreter, was despatched with letters and presents to the local chiefs, and to assure them that there was no intention on the part of the English General to molest them in any way. His efforts were signally successful. The chiefs were unanimous in expressing a hope that the English General would conquer and occupy the country, as the liberal policy which guided him throughout in his intercourse with the natives was so completely at variance with their preconceived notions of his object, that out of their very hearts they desired that they could always experience the same.

Thus the seeming dilatoriness of the Old Soldier of China is explained. Besides the prosaic tactics of the march, there were delicate diplomatic tasks for him to perform before he could advance with confidence into the far interior. An audacious, self-reliant general like Sherman would very probably have marched directly on to Magdala long before Napier would; but whether the happy consummation which finally rewarded the effort of the cunctative English General would have been shared by vigorous Sherman is extremely doubtful.

When the report of the Quarter-Master-General with the Pioneer Force arrived, the army broke camp and marched for Musgee, eight miles distant from Antalo.

On account of the stupendous barriers said to be ahead, much anxiety had prevailed among us as to the nature of the country through which our route lay. From the summit of the Amba of Antalo an extensive view of the country was obtainable.

From all we could see from that elevated position, it seemed a very region of precipitous mountains. Huge reddish pinnacles shot upward against the pale blue sky; slate-coloured reefs of granite rocks lay piled one upon another; hills upon hills were tumbled into platoons, or were rolled into abrupt serrated ridges. Through a ravine plunged the whitened road that led to

the dark defiles from Antalo plain, which could be detected like
a milky way through the opacity of its surface.

The march to Musgee did not warrant the word 'arduous.'
It ran over a strip of the plain whereon our camp was
situated near Antalo. Musgee was situated at its extremity,
on a triangular area of ground washed by two streams of purest
water.

As I said before, the officers managed to elude the strict
letter of the order relating to baggage. Out of six servants the
major part of the officers had at Antalo, only two, perhaps, were
actually left behind ; and out of five baggage animals three had
been retained, exclusive of their chargers ; so that an immense
train of followers and animals, extending over the whole distance
between the two camps, accompanied the army, and by its
exceedingly encumbering length retarded any rapid advance.

Marching with the army proved almost intolerable. The
hardships of those whose business it was to keep up a lengthy
correspondence after arrival at camp were but now commencing.
After being in the saddle close to our baggage, to protect and
escort it, for twelve or fifteen hours, as the case would be sometimes, it was no light task to sit down and hammer away at a
letter, jotting down the incidents of the route, and raking up
latest political intelligence. Then to snatch a sleep of three
hours, hastily swallow a cup of sugarless tea and dry azinous
bread, and mount saddle again at five o'clock in the morning to
undergo the same experience, was no joyous picnicking. All
this might have been easily remedied.

One can perceive that an army of 5,000 men and a train of
10,000 animals, travelling along a footpath by single file,
would necessarily occupy the day in journeying ten miles, even
if nothing went wrong and no baggage got displaced. Then,
if the steady marching of such an army and its baggage train
would occupy a day, what must it be when there is an incessant halting to remedy the packs of two or three hundred
mules, which flounder and sprawl at every third step, causing
displacement of baggage and unmitigable wrath?

The general mode of watching was similar to that followed
in all expeditions. The reveille sounded at daybreak ; the
vanguard bestirred itself, and in half an hour would be *en route,*
accompanied by an officer of the Quarter-Master-General's de-

partment, who went ahead to select and mark out camps near the place designated as a station by Lieut.-Col. Phayre, in charge of the pioneers two days in advance.

The transport train, through assiduous pricking and slashing by the officers, was drawn out before the respective regiments; the baggage was packed and slung on the mules, after which it would form line along the road until the column should have passed.

By six o'clock all would be ready; the military band struck up a merry tune; the infantry, cavalry, artillery, naval brigade, the elephants, and the transport train, were then fairly moving onward.

By noon the head of the column generally arrived at their new camp. Two hours afterwards the head of the baggage train appeared, and until midnight, and even next morning frequently, laggard animals would be arriving. Sometimes we were compelled to halt one day, that the animals might recruit their strength. When a short march of five or six miles was made—unless the road was very steep—the last of the baggage train would manage to reach camp before sunset.

At the end of every day's march the natives would begin to pour in with their supplies of grain and straw, meal and bread. Visits of chiefs were made; and on such occasions the General-in-Chief would hold a levee, whereat officers were expected to be present.

On steep marches the elephants toiled up laboriously with their ponderous burdens, and the other baggage animals invariably passed them; but on moderate roads these gigantic animals with their 1,800 pound loads were masters of the situation, and made excellent time. They were at once the terror and delight of the natives who crowded around them, who seemed never tired of watching their uncouth and unwieldy forms ascending or descending mountains. An ascent of 1,500 feet told seriously upon them, and their hard puffing and loud trumpeting were eloquent of their sufferings. They were invaluable on the campaign. They were fed with about thirty-five pounds of bread and forty pounds of straw or the coarse yellow grass of the country per diem. There were some a great deal larger than the renowned 'Hannibal' of Barnum's, but the majority of them were comparatively young, and stood the fatigue a great deal better than the old patriarchs.

Upon first arriving in a camp, while the regimental cooks prepared the meals, the soldiers, after their ablutions, busied themselves in various ways. They were very fond of gymnastic exercises, and the Campus Martius at such periods presented a very curious and animating scene.

About sunset martial strains of music invariably burst forth from the brazen instruments of the military bands. For the first time the circumambient mountains echoed to the wondrous sounds; and each native, leaning on his spear, wrapt in silence, would drink in, with undisguised pleasure and intense delight, the harmonious sounds which soothed his soul. Twilight was a time which imparted pleasure to all. The music from a distance sounded indescribably sweet, and infused a delicious melancholy into the soul, causing our minds to wander over rugged ranges of Abyssinian mountains, across the blue seas, into our own far-off homes, where we imagined, in fancy free, the social family circles arrayed around the cheerful fire, discoursing of the absent ones.

We left Antalo on March 14. Mesheck lay eight miles above Musgee. From the latter camp the road wound upward along a ravine which was flanked by towering ranges of hills clothed with the most verdant vegetation.

Every kind of wild scenery was around us as we journeyed. On the conical hills blazed the bale-fires, to warn the master of Aleggie, who sat on his knoll of strength like a caged tiger, of our coming. From their summits mounted smoky columns— lurid and high—like smoke from burning sacrifices to appease avenging deities. Dark, suspicious faces, looked down upon us from behind bristling palisades, wrathy and eager for prey and plunder; but, like crafty wolves, they bided but auspicious opportunities to pounce ruthlessly downward from their giddy fortresses to kill and devour.

The people dwelling in hamlets on the lesser hills did not seem to have taken umbrage at the overflow of pale-faced strangers, if we may judge by the appearance of the apparently enchanted village throngs who lined the slopes, dizened out in their grotesque finery.

When within two miles of camp we were obliged to cross

the meandering brook every few minutes because of its exceeding deviousness. The ravine, also, at this place widened into a valley of respectable width, wherein vivid green patches of corn were visible, interspersed with noble groves and low-lying meadow-land, studded with tussocks teeming with flowers of varied hue in full blossom. The mountains still towered higher, fully 1,500 feet above us, on each side. Their uneven sides were hung with an abundance of tamarisk trees, now and then varied with a clump of kolquall or fir. Immense boulders at times projected boldly above the tree tops, and numerous ravines chequered the slopes.

After travelling eight miles the camp was seen beyond, and at the extremity of a noble avenue of trees. It was called Mesheck, from a fortified village of that name which crowned a hill hard by. It was a picturesque spot, like a small park, begirt with high mountains. A running stream, a level lawn, and patriarchal sycamores, were its principal features.

The 1st Brigade arrived at it on the 15th, while the 2nd Brigade rested at Musgee, and did not arrive at Mesheck until the following day.

The next march was that to Atzala, past the robber roosts on Amba Aleggie, inhabited by the vulpine Welda Yasous, *alias* Welda son of Jacob, to the certain detriment of travellers in general. The chief possessed at his beck and call, in the beehive dwellings on the hills, some 1,000 stout-armed spearmen and 500 musketeers. This was the capital of Woggerat, in which province we were now travelling. The amba [1] was a rhomboidal mass of grey granite and limestone, crowned by an elliptic cone several hundred feet high ; each side of the rhomb rose to a perpendicular height of 100 feet above the spine of the mountain on which it rested. The base of the peak covered an area of perhaps 200 square yards. Half way up was a ledge or terrace on which the dwellings of the garrison were built, appearing at their altitude like eyries.

Through a field-glass several swart faces could be detected glowering upon us as we passed. I opine that the chief, Welda Yassous, himself looked down from behind some friendly coverture, though it was believed that he had deported himself

---

[1] Amba is a native term for Mount.

away from his home, suspicious of some sinister design on Napier's part in passing directly by the base of his fortress.

For the first six miles the road led through scenes of forest shades and intervening spots of quaking quagmire, by mantling stream and by patches of cultivated land that lay within the environed valley. Then suddenly it inclined upward, and a long pull of two miles up the slope of the amba already mentioned was gotten over before we finally surmounted the spire of the ridge directly underneath the mount, which was fully 1,500 feet above our camp at Mesheck.

From this high eminence we could discover our new camp at Atzala, situated on an oval-shaped plain, 2,500 feet below us, looking like an emerald set in wood, so the comparison suggested itself by the striking contrast between the verdant plain and the engirdling hills, each of a height with that upon which we stood looking downward with astonished countenances at the extraordinary freaks of nature.

Without climbing the opposite heights, which appeared to be of equal altitude with that of the amba, further progress from that valley seemed to be impossible.

The prospect of the morrow's march could not therefore be very captivating. Captain J. Sidney had reported ' One hundred and sixty-six animals belonging to the Abyssinian Transport Train died on the last march from Mesheck to Atzala!'

The last of the transport train did not arrive until next morning after that fatal march. One or two ' specials ' passed the night on the top of the hill.

' Welda,' son of Jacob Chevalier de Furioso, who kept airy lodgings on the lofty amba of Aleggie, at an altitude of 10,000 feet above the level of the sea, after much coaxing condescended to come down and visit the chief of the crusade, not in an Icarian manner, but in Jupiterian state, surrounded by tawny blackguardism, as exhibited in the polished faces of some three hundred musketeers, who travelled to head-quarters as if they were imbued with the idea that they were just as good as any of the English soldiers. Oh, the unbaptised canaille!

Welda sauntered into Sir Robert's marquee in true lordly style, and sat on a camp-stool in a dignified manner.

He was accompanied by his brother, who looked a worthy scion of the same parent—a stately-browed, frizzly-haired, restless, rat-eyed, high cheek-boned, heavily-auricled, thin-lipped and oval-chinned rascal.

Welda himself was an intelligent looking man enough, of medium height, robust form, muscular limbs, and not of a bad countenance by any means. But his conduct betrayed an insufferable pomposity. His love of state and barbarous magnificence was indicated by the golden armlets which he wore, and the collar of silver filigree work enriching his tower-like neck; in the splendour of his shield, draped with the lion's mane; and the circumstance of state with which he had surrounded himself. His musketeers were armed with the Portuguese fusils of the sixteenth century, and double-barrelled shot-guns.

A long conference then ensued. Welda promised to allow his people to sell provisions to the army; and the Commander-in-Chief presented him with a double-barrelled gun with ball and caps; besides thirty new dollars for himself and twenty for his brother. When questioned about Theodore, Welda said at last that he did not think Theodore would fight; but Welda's brother, having received the presents, spoke differently. He said, ' I shall now advise you differently from what my brother Welda said just now. Todoros will fight, fight as long as he has breath in his body; he is cunning as a fox, quick-sighted as a lynx, brave as a lion. He will sweep down upon you in the dead of night. Beware of him; he never shuts his eyes, neither does he dream.'

Napier. 'Your brother Welda said that he would not fight.'

Welda. ' I said it only to try you. It is the custom before we open our hearts to strangers. My brother has said well. Todoros is brave; he will meet you; he will not run away. Beware of him in the night before the mist vanishes, or dawn of day.'

Napier. ' I am obliged to you. But come with me and you shall see our elephants and big guns.'

They went out together, and the elephants, in obedience to the mahouts, performed many manœuvres, such as lying down, kneeling, trumpeting and charging, which amazed the two chiefs greatly. The military bands discoursed liveliest strains of music, and the 33rd Regiment performed many evolutions for their benefit. Two chargers were then brought up and presented one to each of the brothers, after which they departed to their amba, protesting their sincere friendship. They were preceded by their own martial band blowing through cow-horns, playing on pandean-pipes and flutes, and twanging three-stringed guitars. Their presents to Sir Robert consisted of two shields, two fine spears, pots of honey, and two dozen oxen.

Mukhan, the next position to Atzala, and distant from it sixteen miles, was occupied on March 16 by the 1st Brigade.

As Amba Aleggie towered upward to a height of 10,000 feet above the level of the sea, and as the valley of Atzala was 2,500 feet below the summit of the amba, the camp at Atzala was then just 7,500 feet above the sea. The valley was about two miles broad, bounded on the opposite side by the range of Debra Musa, whose sloping rugged sides looked formidable indeed, when we arrived at the base, and scanned its ambitious height. It was much steeper than Amba Aleggie, and the road had therefore to be made to run in zigzag fashion before the army could attempt to scale it.

Within two hours after it had started the head of the column appeared upon the brow of the mountain, and below them on the southern side (for we were marching due south) another plain, but infinitely larger and still deeper than Atzala valley.

Half way downward we passed the village of Pilago, and directly above the village appeared one of the most extraordinary natural formations I ever saw. It was a polygonal-shaped mountain, upright on each side nearly 1,000 feet, the summit of which covered an area of three square miles. It was named by the natives Mount Debar, and it appeared to be of equal height with that of Amba Aleggie. Why Theodore should have neglected such an impregnable position was a question only to be answered by the supposition that Magdala was still more impregnable.

After reaching the plain on the southern side of Debra Musa, and on the left of Mount Debar, we crossed a stream, and our route lay over undulating ground covered with luxuriant grass. Gradually we left the open plain and travelled through avenues of shrubbery. Again and again the scene changed to a more or less wildly romantic one, past amphitheatres, through sylvan groves, over tiny jewelled rivulets, across open fields, downward through dells embosomed by tall firs and oaks, and up to the broad sunlight again. Refreshingly picturesque ravines and nullahs, teeming with undergrowth, were passed in quick succession, until we had journeyed sixteen miles, when the camp of Mukhan appeared upon an open ground, just as day deepened into twilight.

CHAPTER VII.

ABYSSINIAN SCENERY—LAKE ASHANGI IN THE LAP OF THE MOUNTAINS !—
A HOPEFUL HEIR-APPARENT—OUR GROG IS STOPPED.

As might be expected, after such a long march, the troops were obliged to halt at Mukhan the following day.

The district around the new camp was a very populous one, and the natives came pouring in in great numbers to see the strangers, bringing with them at the same time various commodities to sell. The people hereabouts were slightly paler in complexion than the Tigreans and the inhabitants of Attigratt. They were more muscularly formed and of a wilder appearance than any we had yet seen.

Presently there appeared three turbaned Mussulmen from the Galla lands, blowing their prodigiously long cow-horns. They were avant-couriers of a Galla chief, who came in the name of his people to offer peace and friendship to the British Negus, Sir Robert Napier, which I need not aver was gratefully accepted by the ' old man,' who reciprocated the goodwill of the Galla chief and bestowed upon him pecuniary rewards for his friendly promises.

Antelopes, hares, and other varieties of game were abundant around Mukhan, and the English officers turned out *en masse* as amateur sportsmen, and great was the devastation they committed upon the preserves of nature. Some one (name unknown) shot a wild elephant in the neighbourhood, and left the huge carcase where it fell to sicken the country. There were some graphic accounts rendered of successful hunts, and every one had a trophy or two of the chase.

On the morning of the 18th the army broke camp and marched for Lake Ashangi.

For the first six miles our route lay across an undulating country bordered by groves and airy heights, until we came to the base of Mount Mosobo, where shortly overlapping ranges enclosed us as we laboured up. It was a region like unto what I have seen travelling a-foot through the southern Pyrenees—a region of tropical beauty, opposite in character to any we had as yet beheld in Abyssinia. Upward we toiled, around the base of successive slanting gradines covered with tall pineries. Virgin forests sprung up on all sides, feathering upward to an infinity of height spreading out to right and left, and intervened o'erhead into pleasantest of frescades. In their shades sprouted flowers of rarest hue.

The road up the mountain was very bad, and in some places highly dangerous.

After four miles of an ascent we emerged out of the shade of the grove, and saw the sky above us, and another slight acclivity clothed with myrtle and sweet-briar, and presently we stood upon the topmost height of Mount Mosobo.

There, beneath us 3,000 feet, slumbered Lake Ashangi, enclosed in the lap of mountains.

An army is of too prosaic and commonplace a temperament to stand long idly looking down upon scenery; and so, though we all should have dearly loved to have stopped hours longer upon Mount Mosobo, and enjoy the tropic beauty of the scene, up-pouring troops upon troops swarmed around us; pressed onward, and in the irresistible current swept us with them.

Soon began a series of convolutions around bases, of projections, past segregated masses of quartz and rock, whose varied colours became all intensated by the vivid brightness of the sunlight or the deep shadows of overhanging cliffs, and pendent boughs of pine.

Round and round we felt our way by the side of looming walls of gneiss, flanking all wayward mounts, and coursing in a serpentine stream over bridal paths running along giddy precipices, until at last we arrived upon the plain of Ashangi.

We camped close to the lake on a park-like plain, after our march of fourteen miles. The nature of the route prevented an early arrival of the baggage trains. In the meantime the smooth placid water tempted us to plunge in and cool our bodies after the fatigues of the journey. There were not the slightest indications of any hippopotami ever having been there ; or of any other animal more ferocious than man, which proved the instability of all the rumours that had prevailed.

It is true there were two or three sedgy pools near the lake, but the piscatorial and amphibious tribes were limited to a few small fish. The whole surface of the lake itself was dotted with innumerable wagtails, ducks and wild geese, the shores teemed with numerous species of horn-bills, herons, and several kinds of large rostral birds—ibis, snipe, pelicans and toucans.

Across on the Galla side of the lake there were plenty of antelope in the abundant woods, and panthers, hyæna and jackal ; but that was a region where the Afrit [1] of the Mussulman held sway ; at least so the natives endeavoured to make us comprehend as they appeared very anxious that no Feringhee should go there lest bad afrits should cut our—ugh ! (they drew their fingers across their throats, an intelligible sign enough).

Had it not been for the refreshing bathe and swim we had enjoyed, very probably we should have felt too languid to dispute the existence of the evil spirit ; but aroused from slothful torpor by a healthy ablution, we disputed the point with as much ardour as any ' Jack Easy,' and in the very headstrongness of our natures rode around the lake. The Galla's side was undoubtedly mysterious and sombre-looking enough to have been the home of all the demons known in mythology, and we for that reason, bearing also in our memory the significant sign across the throat, did not think it worth while to explore its penetralia.

Several specimens of petrified wood were found along the eastern shore of the lake. The northern side of it, a little to the east of where our camp was situated, and covering an area of nearly five miles, was a most treacherous piece of ground. It was one vast quagmire,[2] a stagnant marsh of Cocytus. It was cut up by numerous fossulates, full of Stygian slime. My Arab horse sank to his middle before I was aware of the

---

[1] The Afrit of the Mussulman is analogous to the Ahriman of the Persians, and the Loke of the Scandinavians—an evil spirit, a bad geni.

[2] There is a tradition among the natives, that three hundred Galla horsemen in the heat of a charge upon the Abyssinians were caught in this quagmire, that they sank quickly out of sight, not one of whom escaped.

treacherous nature of the ground, and a wide circuit had to be made before I was enabled to reach camp. The circumference of the lake was about ten miles.

It was fully midnight before the last of the transport train arrived at camp ; my baggage was unfortunately one of the hindmost, and I had therefore to wait about ten hours before my tent was pitched.

The next day we were obliged to halt ; the animals were unable to travel. Our forage was giving out, and a fresh supply had to be procured before we could move. The rations were therefore reduced one half.

About midnight an envoy arrived from the paramount chieftain, the Ulema—priest and king, spiritual and temporal ruler, over the nomadic Mussulmen tribes in the immediate district of Ashangi. This envoy was the hopeful son and heir of the Ulema, a vacuous-eyed individual—hare-brained too, if I am any judge of physiognomy.

He bore a letter to Sir Robert Napier, wherein, after the usual circumlocutory addresses and well wishes—a peculiar custom of the country—the Ulema stated his desire to have peace with the Feringhees, and he asserted his theological opinion, how that the Old, the New Testament, and the Koran, were part and parcel of the religious belief held in common between us Christians and themselves Mussulmen ; he also promised to remember us in his prayers ; commended us to the care and watchfulness of Allah, and warned us to watch carefully against our enemies, as they were many and strong. In accordance with the custom of his country he sent presents in the shape of one pot of honey equivalent in value to a silver dollar.

Sir Robert, having read this effusion of the Galla priest, presented the envoy with robes, such as should adorn the person of one in his high station. Not a quiver of a muscle, or the wink of an eyelid, or the dawning of a smile betrayed the pleasure he must have felt, as he was being enrobed with a gilt stole and a gorgeous robe of coloured muslin, but remained as passionless and as immobile as a waxen doll.

In the afternoon another Galla chief, accompanied by warriors carrying headless lances in token of amity and good-will to the army, arrived at the camp. The chief was a perfect specimen of a coloured 'John Bull.' This god-like son of

Mars also stated his wish to be on amicable terms with our Negus, and he in like manner was treated with the greatest deference, as was due to his Royal Heaviness.

Towards evening a different scene was enacted. A native— a Galla—was detected pilfering by a keen-eyed Hindostanee camp chuprassee, and he was forthwith dragged before his Provost-Marshalship, Colonel Fraser, to be tried for theft. A great wow-wow was instantly created. His countrymen—decidedly of clannish proclivities—hastened to surround him, and endeavoured to bear him off. One young man, supposed to be his brother, made the greatest outcry. The culprit had his arms pinioned behind him, and the formula of a drum-head court-martial was gone through. Colonel Fraser wished to give him three dozen on his bare back, well laid on by his satellites. Captain Speedy, hurrying there from a durbar, suggested that it would be better to permit him to be tried by a jury of his own countrymen. Fraser demurred, arguing for the force of example ; Speedy maintained his suggestions on the ground that peace had just been made with their chief, and that it would be highly derogatory to it and impolitic to be too harsh at the offset.

' But what am I to do? ' urged Fraser.

' Let his countrymen decide ; they will punish him worse than you will,' said Speedy.

' Very well then, let it be as you say,' replied the Provost-Marshal. Accordingly, a jury of the culprit's countrymen was convened, and at it they went, argufying and speechifying with ten times more *empressement* than the same number of New York lawyers would have done. Such fervour ! such style ! and such action !

The gist of the address of a grey-bearded elder was that nobody could prove that the man meant to steal it, and that he might have taken it in his hands merely to look at it. ' Is that a crime? ' he inquired. ' Yet,' he added, ' must we punish him, lest the Feringhees say—Ah ! these people will not punish their countryman, our trust in them has been misplaced. Therefore, I adjudge him to receive six blows with a stick, and to pay one-fourth of the article he intended to steal.' And according to the letter of the sentence was it carried out, to the great disgust of certain martinets.

On the 20th the English army occupied a plain, which was but a continuation of the low land around Lake Ashangi, near Muzzageeta, at a point seven miles from our camp on the northern side of the lake.   The route to this camp lay along the western side of the lake, and about four miles to the south of it was Muzzageeta.

From Muzzageeta, the English army journeyed to Lat, a distance of eleven miles.

For the description of this route, to avoid a tiresome repetition, let me say at once that it has a certain resemblance, on account of the many hills we passed (to use a low comparison, but yet a very apt one), to the teeth of a saw, or a succession of traverses, or marching up and down a line of inverted stalactites.   Had we been condemned to undergo Sisyphean tortures we could not have fared worse; it was very like a series of Sisyphean labours—the going up a steep hill and descending it a score of times on a day's march.

To those mounted upon ambling nags the journey was not so very bad, but to those travelling on foot, with fifty or sixty pounds of a load, it was indescribable torment; more comfortable to imagine such a thing than to experience it.   The heat was intense; that means that it was on an average of 98° Fahrenheit.   From Zoulla to Lat during the day it had never been less than 84°; at night there was often a decline of from 30° to 40°.

Rich compensation it was to cavalry men and foot soldiers when, after surmounting a mountain top which had tasked their energies for three or four hours, they were allowed to breathe the pure air of upper regions, and to rest for a while.

No sooner had we planted our tents on an uneven, slanting terrace of ground, where we were about to enjoy ourselves after our sorrowful toil, than another blow was given to our equanimity, by the threatened " order " of the Adjutant-General, Colonel Thesiger, relative to the abandonment of rum— vulgarized " grog "—with a codiciliary " order," announcing the intention of the Commander-in-Chief to abandon his tents and baggage at Lat, our present camp, and decreeing that twenty men should be apportioned to each tent, and twelve officers should batch together.

The order was a second edition of the one issued at Antalo,

except that it was sterner and more peremptory in its tone. Sir Robert Napier was aware the first one had been evaded by the Anglo-Indian officers. Moreover, no baggage was to be permitted to go to the front. Every mule being required to convey commissariat stores, unmounted officers were to be allowed to carry only a greatcoat, blanket, and an indiarubber sheet.

Staff officers and état-major of cavalry could take what they pleased on their chargers, but then—as Sir Robert Napier suggested—'When sent upon duty requiring speed they must make no complaints to him of their horses being wearied out.'

With such an understanding they might take their tents and baggage along. Sugar, rum, coffee, tea, potatoes, onions, and such luxuries, must also be left behind, as 'a forced march is to be made upon Magdala!'

The next morning, with a slightly diminished train and a singular paucity of tents and baggage, we broke camp, and started for Marowar, stated to be ten miles, according to Colonel Phayre's computation, who said he measured it.

The order issued at Antalo was but a tentative order to fathom the feelings of the officers; but the last was peremptory, stern and inevitable. Hence there was discontent, and an undercurrent of querulousness among the soldiers, who were beginning to grumble with a significance which, unless checked, might prove subversive of discipline.

The long marches in prospect seemed to make matters worse. Men talked, and there were others to carry the tale to head-quarters. Emboldened by the continued silence in regard to their murmurings, the soldiers spoke fearlessly in presence of the commissioned officers, even before those who were on Sir Robert's staff. Colonel McLeod, Assistant Adjutant-General, passed some soldiers on the route, and was not recognized any more than if his ghost had hovered invisibly near. He became enraged, and sternly questioned them as to where their respect for officers of the staff had fled to? The Adjutant-General reported the facts to Sir Robert Napier.

We halted at Marowar that night after a march of thirteen miles. Next morning we started for Dildee. Higher and higher rose the mountains, wilder grew the region! Whither, and in what way, was the march to end?

What could Magdala be like if Theodore had passed by

these legions of peaks, each one of which would have been unscalable? Yet the soldiers but grumbled—those of the 33rd, however, in a noticeable way. Col. Phayre had reported this march to be ten miles only. It turned out to be more than eighteen. The soldiers, muleteers, and animals, fell down by the way in dozens, in scores; yea, in fifties, out of sheer exhaustion.

Towards the latter part of the march we passed several picturesque ravines, through all of which meandered streams. Under the banks the tired people rested, almost despairing of ever being able to make camp.

At five o'clock in the evening only the skeleton of an army had arrived at camp. A thousand men dashing from behind the hills would have annihilated the whole force by attacking them in detail as they came straggling along in weary groups. It was midnight when the soldiers finally reached their new rendezvous at Dildee. The baggage trains did not come in until ten o'clock next morning, thoroughly worn out with hunger and travelling.

As a consequence of the protracted march, we were obliged to halt again for one day at Dildee, the army sadly in need of it.

In the morning soldiers of the 33rd Regiment of Foot were drawn up before Sir Robert Napier, who reproached them for their conduct—saying that they were a disgrace to the army of Abyssinia; that he had heard through officers' reports of their murmurings and grumblings on the last march, and that since they had been pleased to conduct themselves in such unseemly manner he could not permit them to have the honour of being the Advance of the column, and he would therefore reduce them to Rear Guard under Sir Charles Staveley, and the 4th 'King's Own' should replace them. Anyone would have pitied them, as they stood with unmoved countenances, but with hearts almost bursting with sorrow, listening to their General's reproaches.

The 33rd Regiment is one of the finest bodies of men in the British army, and it was a great blow to its reputation to be thus summarily disgraced. But even the Duke of Wellington, who formerly belonged to it, had occasion in the Peninsular campaigns to punish this regiment on account of the reckless spirits in it.

The camp at Dildee was a very beautiful and picturesque spot. It was in a narrow valley enclosed between two high hills covered with wood, through which ran a small river—the Mya. This little river seemed like a river of Eden to us after yesterday's fatigues. We all disported in it, with the freedom and enjoyment of boyhood. It was the first real river we had seen in Abyssinia, and we grew suddenly poetical about it. There was another feature about the river which made it interesting. It had worn itself a channel deep into the earth, and its banks were from fifty to two hundred feet in upright height. These banks were clothed luxuriantly with rarest vegetation.

General Field, in command of the pioneer force, had waited two days for our arrival at Dildee. The day we rested at this last camp he moved on with Quartermaster-General Phayre to the next rendezvous, which was said to be a place called Wandach, so as to be one day ahead of the 1st Brigade.

The next morning the army continued its march along the slope of a mountain called Mezgar Amba. It ascended in a consecutive succession of slanting ridges, until we were led to believe that there was actually no end to its height.

About mid-day we surmounted Mezgar Amba, which rose 10,700 feet above the sea. The change of temperature from summer heat to subarctic cold was experienced at this high altitude. Cold drizzly sleet and rain, and gusts of icy wind, which met us, made us shiver in our summer linen.

Misery loves companionship it is said. To impart our miserable feelings one to another we left our tents, and in the cold wet mud such converse as utter wretchedness evokes was entered into. Those who by a wise foresight had provided themselves with wood, while in Dildee Valley, and had secured it in their tents during the rain-storm, immediately had hosts of friends who warmed themselves by the friendly fire-blaze, and allowed their hearts to expand into intense admiration of the owner of the camp-fire.

The slopes of the hills were bare of wood. The only kind of shrubs, or plants bearing any resemblance to wood, were several green palmated-leaf trees, which could never be used for firewood unless well baked first. The natives were asked to bring firewood to the Commissary; and in about an hour, directly after sunset, a bevy of zoneless damsels, bearing on their heads huge loads of wood, arrived at the camp, causing all men's hearts to rejoice. Late in the night the 4th ' King's Own,' reached camp to replace the 33rd.

About ten o'clock the army struck camp, and still following the slanting declivity of Mezgar Amba, marched to Muja, a distance of six miles. The road ran over the same slope, gradually descending, above the Takazze, or the Blue Nile. On the left of the road the Takazze flowed, of course, only a small stream at its source, but a hundred miles to the westward it was a considerable river.

The camp of Muja was situated between the Takazze and another deep ravine whose precipitous banks prevented us from getting any very large supply of water.

From Muja we travelled next morning to Santarai, a camp named after a village which crowned one of the grassy dunes on the Wadela plateau.

It was one of the brightest days of the spring time in Abyssinia, the day we marched from Muja. For the first seven miles the route led over the continuation of the slope of Mezgar Amba, gradually descending to the Takazze. A most gradual descent it was as may be imagined from the fact of the slope extending over a distance of fourteen miles. It may be likened to a sectar, the two radii being the ravine on our left and the one on our right, through which the Takazze ran. Between the two radii our route lay. After travelling fourteen miles from the summit of Mezgar Amba, or seven miles from the camp at Muja, the Takazze suddenly made a bold curve.

Upon arriving at the Takazze we found we were hemmed in unless we retreated backward from whence we came, or followed the zigzag windings of the Blue Nile.

'We neither intend to go back nor follow the Takazze,' said Captain Speedy ; ' but we intend to scale that apparently interminable wall you see before you !'

' Great Cæsar ! '

In very truth most men would have uttered the same exclamation, upon seeing the lengthy sky-wrapt walls of granite. But what has been done, will be done by Napier and his army and the horde of followers ; even if it takes two days to do it.

Already the iron souls of the 4th Foot are half way up the perpendicular walls, serpentining at right and acute angles with each other, and round and round like the windings to the summit of a cathedral spire. The staff is also determined to try it, and spurring our horses we fly tugging up the slope. A short half-hour, and we have surmounted the first terrace. A gradual acclivity is in our front, for three or four hundred yards, and another terrace of equal height and formation is to be scaled, and having accomplished this we stop to breathe, and get a glimpse at the bewildering variety of nature.

The higher we ascend the grander the scene! Under a sky of gorgeous sunshine, as blue as ever vaunted Italy boasted, spreads the wildest land, growing each day wilder and more rugged as we advanced upon Magdala. The mountain tops are tinged with the brightest colours of the solar light!

Upward unceasingly we toil, but often pause to rest. Fleetest and most hardy of Arab horses become blown; hardiest of iron frames relax the firm tension of their nerves; the stoutest ruttiers succumb, and nought but good long rests will save them!

Each highest angle or curve of road formed a perfect fortress. The slightest effort would overturn a ton weight of rock. Through the excessive boulderiness of the slope, ten stout-hearted men could have defied a thousand, and added another worthy Thermopylæ to history!

The English army shall profit by it, and all things that end well must be well. Uttering Napier's pious ejaculation, ' God with us,' we will even ascend higher, and view the upper world once more, and inhale the breezes from the topmost elevation of a height of 11,000 feet above the level of the ocean. A plateau, like another Laramie region, undulating as far as vision of man can reach, in nude dunes, bereft of the least signs of shrub, rolling like a vitrified sea—such was the Wadela plateau rising above the nether world: strangest freak of nature this! A country combining all temperatures!

Having, like heroic souls, surmounted fairly the Wadela plateau, Colonel Cameron requested of his regiment, the 4th Foot, to give three cheers, saying that it was a worthier feat to have marched up this height of 3,000 feet than to have conquered Theodore; but the men, though willing, were too much wearied to give more than the feeblest dithyrambic expressions, and, from the state of their feet, it was cruelty to prevent them from lying down the moment they arrived upon the soft sward of the plateau. Far into the night the brow of Wadela, at the entrance of the exit from the Pass, was covered with the forms of tired men.

Two miles from the edge of the plateau our camp was pitched, near the village of Santarai, on an amphitheatrical-looking spot, just twelve miles distant from Muja.

At this place we found the Pioneer Brigade, under General Field and Colonel M——, than whom none was oftener mentioned by the gossips of the press—never in a flattering sense, which was a great pity and a cruel omission on the part of those sent to report upon the events of the march to Magdala.

At this camp of Santarai, within the bosom of an amphitheatre of gentle hills, there was not a particle of wood to be procured, as the whole plateau of Wadela, a hundred miles in circumference, has not above two or three dozen trees upon its surface.

But the dollar in Abyssinia is an omnipotent genius; it causes water to flow from solid rocks; it brings forth food in abundance for 20,000 men and 60,000 animals; it makes the plains to overflow with goats and bleating kids; it causes the inhabitants to pull down their houses, and give their rafters to the Commissariat Department, that the food which they had brought for the soldiers should also be cooked! This latter was the case in Santarai.

Two rafters—that might support a light roof—about ten feet long, were sold for a dollar. Those who had a sufficiency of dollars, purchased as much as they needed; but one day's fuel cost a whole dollar. The rations were decreasing rapidly. Our animals were on half rations; very often on quarter.

Private gentlemen wishing to keep their animals in a saleable condition, were compelled to buy from natives such quantities of grain as were procurable. Then there were the servants, who required a dollar's worth of bread extra of the rations received from the Commissary. A quart of milk or a little honey would be another dollar; so that it may be safely concluded that our expenses, including Commissariat bills, since leaving Antalo, averaged seven dollars per diem. A bag of two hundred dollars seemed to last but a short time.

This high plateau of Wadela has seen stormy times. Very many times did Theodore ravage it when his power was in its zenith. It now belonged to Wagshum Gobazye, and Dajaz Mashesha, his pot-valorous uncle and commander-in-chief, the day we arrived at Santarai was making an expedition a few miles to the eastward with a force of 1,500 men to punish a rebellious village which had demurred paying the heavy taxes imposed upon them by His Highness the Prince of Lasta and Samen.

His arrival at the village to the eastward occurred while we were arriving at Santarai, and friendly messages were at once interchanged between the respective commanders-in-chief.

Munzinger, long since reported missing by rumour, is again in the land of the living, feasting upon Habesh[1] cabobs, and imbibing His Excellency Dajaz Mashesha's tej safe in a friendly camp. The French Consul has been playing the polite diplomat in a satisfactory way: he has won the heart of the suspicious Habesh General; and, in accordance with an invitation from Sir Robert Napier, Mashesha promises to pay a visit on the morrow.

We, safely housed on the other hand in our camp, on the bare plateau snugly embosomed by bare hills, fare as well as might be expected with people suffering from half rations, no grog, and bad bread. The mountain air of Wadela is chilling, and acts like a refrigerator upon frames acclimated to the rigour of oven-like valleys. Hoar-frost crowned every slope and eminence on the morning following our arrival. By reason of our blankets left behind, we suffered very much from the sudden change of temperature, and six blankets were required before anything like comfort was obtained. These we were obliged to purchase from the natives at a dollar each.

The Quarter-Master-General announced that we were but seven days' march from Magdala. At Lat it was only six days' march to Magdala, so now, after marching five days, we found we were seven according to Phayre; but Charles Speedy, interpreter-in-chief, declared it as his opinion that we

---

[1] Arabic term for Abyssinia.

were twelve days, if we allowed margin days, to guard against accidents adverse to rapid travelling.

The intelligence and political departments busied themselves in feeing chiefs—in bribing them to good conduct; the Commissiariat were always busy in purchasing what was brought. Abyssinian chiefs, proud in their cloud-begirt ambas, were as slippery as eels; no sooner had the army left the depôt than they became clamorous for more gifts, though they were already paid in cleanest brand-newest coin, their specialty. The supernumaries were immediately sent back to satisfy rapacity and rectify errors, or restrain with strong hand what couldn't be appeased by soft words.

Major Grant and Captain Moore were sent to Dildee because of some complications occurring consecutively there—nothing in themselves, but very evil unchecked.

Colonel Merewether, political officer, in his head-piece of silver, and clothed with due authority, obtained permission, or rather was commanded, to satisfy pugnacious Welda Yassous, who in spite of his earnest promises and solemn covenant with Napier, ratified by breaking of bread and eating of salt, had departed from the straight path, and ever since our absence had been diverging wider and further from his plighted faith, until he had broken out into downright hostility.

It was even said that the couriers of Sir Robert Napier had been captured and ill-treated; that if they made stout determined resistance, they were emasculated. That Welda Yassous and his brother had peremptorily forbidden their people from selling their provisions was indubitable. That they had prohibited native carraiges or country people with provender to sell, from passing through their territory of Woggerat, between Mesheck and Atzala, was equally indisputable.

Since Sir Robert Napier had determined to trust to native carriage for the daily subsistence of his troops and immense trains, such checks and hindrances were replete with serious consequences to the advance columns. Harsh treatment of every petty chief would not avail us; rather the reverse. Each Ras on his own peaky amba imagined himself as good as the next one, or the High Wagshum, or the kingly Prince of Tigre.

Plaints of this sort were clearly not inaudible when such a personage as Col. Wm. L. Merewether was sent to relieve them.

On the second day of our stay at Santarai, at eleven of the forenoon, the pickets, standing like shadows on the highest points of the embosoming hills, shivering amid the drizzle and mist of a foggy morning, discovered an armed body of men approaching from the westward, and at once fired their muskets to signal the camp. Busy rumour immediately announced it as the approach of Theodore with 50,000 men! Rumour had been very uneasy ever since our arrival at Santarai.

At the first signal of the picket, Colonel Fraser, then officer of the day, galloped up to the outposts and scanned the strangers closely, after which he informed the outposts not to fire, as they were friends who were approaching. Head-quarters had known since the previous evening that these strangers, who were no other than the troops under Dajaz Mashesha, Commander-in-chief of Wagshum Gobazye's army, were to come into camp on the morrow to pay a visit to Sir Robert Napier; for Munzinger, the envoy, had sent several couriers to inform Sir Robert of Mashesha's intention.

When therefore the vedettes of the Scinde horse and infantry pickets had fired their guns to warn the camp, the infantry had thrown themselves forward, in skirmishing order, and were advancing upon the supposed enemy; and had not Colonel Fraser arrived at the most opportune moment, the consequences would not only have been disastrous to the native cavalry, who were marching with careless confidence towards the friendly camp, but it would have arrayed against the English Wagshum Gobazye and his army of 40,000 men, who, believing that he had been treacherously dealt with, would no doubt have eagerly sought alliance with Theodore to avenge the murder of his men. The British army would then have had Theodore, with his 30,000 men in front, and Wagshum Gobazye, with 40,000 men, in flank and rear.

When it was known at head-quarters that the Abyssinian General-in-chief was coming, the various regiments were called to quarters instantly by beat of drum and fife. The soldiers put on their cleanest coverings for their sun helmets and white cotton gloves. The 3rd Light Cavalry advanced towards Sir Robert's tent, turned, and faced about to front. They were all in spick and span silvered and belaced uniforms. Their officers, mounted on their finest chargers, drew near, and, selecting a small troop of the cavalry to accompany them, at the word of command dashed off to meet Mashesha. In the meantime four regiments under arms were marched up to the western extremity of the camp, where they were drawn up in lines. All looked expectant to the summits of the bluffs on our left, for from that direction they were to come.

At the end of about half an hour a long line of cavalry crested the ridge. After a pause, during which it may be supposed that they examined the strange camp below, with its thousands of men and animals, the whole of the cavalry marched down the steep sides of the bluff as coolly and as deliberately as if they were on a level plain. There was no appearance of commotion or jostling, though the precipitous faces of the elevation were uneven enough with rough boulders and toppling stones. No European cavalry could have done so without being huddled together in groups, and in stumbling and sprawling masses.

The appearance of Mashesha's force was very imposing, though the unevenness of the ground caused one end of the advancing line to be fifty feet below the other; but it kept admirable order and ranks. As they reached the base of the bluffs they halted and formed column. The English officers and their body-guards surrounded the Abyssinian general; and as he crossed a little stream which separated the camp ground from the heights, the artillery commenced thundering a salute which caused the native ponies to leap with affright, and disarranged the column.

The regiments presented arms to the princely General, and, forming column, preceded in advance of the native army. The military bands struck up 'Garryowen,' which delighted the ears of the dusky soldiers with its martial strains. In this manner they marched up to where Napier's tent stood, when the English formed open square, through which Mashesha was conducted towards Sir Robert's marquee. The English General came to the door, and immediately both generals embraced each other.

General Mashesha was a medium-sized man, about five-and-thirty years old, with a kindly good-humoured face, full of lively intelligence. He was dressed in a flowing syma of silk. His vest and wide trousers were of richly-coloured silk brocade, a necklace of amber beads was around his neck, and an elegant torque made of filigree gold set with vari-coloured glass was worn round the head. He had just come from a successful expedition against the rebels in western Wadela, and wore the torque with a horn in the centre, as a sign that he was exalted on high. This wearing of the horn after victory is an allusion to the scriptural description of the horn of the righteous, and to the 'lifting up of the horn on high, while the wicked man speaketh with a stiff neck.'

The Abyssinian force numbered probably 800 mounted men, composed of the *élite* of the army of the Wagshum. They were all fine stalwart natives, dressed uniformly in robes, with a broad scarlet band around the hem. Gaudy bracelets were around their arms, necklaces of amber and beads about their necks, and anklets adorned their ankles. Each soldier carried a couple of spears and a shield of ox-hide, while about half of them were armed with matchlock muskets. Not one of the horses was shod.

It is not known what took place between the English and Abyssinian generals after they entered the tent, nor is it very material. The substance of the conversation was doubtless that Napier wished well to Gobayze, and Gobayze wished well to Napier; Napier hoped that Mashesha was well, and Mashesha expressed his hope that Napier was well. Furthermore, Mashesha expressed his undoubted belief that Theodore would fight—a cause for infinite gratification to the British general. He (Mashesha) said he was very much afraid of Theodore— which we had all guessed long before he told Sir Robert. Then Munzinger, Speedy, Thesiger, and Mashesha were invited to drink a little rum with Sir Robert.

The interview ended, Mashesha was presented with a horse and a double-barrelled gun, of which latter he seemed to be very proud, and the horse attracted the general admiration of the native soldiers.

Dajaz Mashesha had his shield borne by a youth of high degree—a perfect young Adonis; as handsome in his dusky beauty and aspect as any page who carried the shield and spur of the steel-clad barons of old. The shield was similar to that owned by Prince Kussai.

From these people we were able to form some idea of the

strength and power of the Emperor Theodore. If his troops were armed only a little better than Mashesha's, Magdala was certainly a place that would task the strength of the British force to the utmost to capture it. There was reason to believe that they were, as Theodore was always victorious in combat with native chiefs and princes. All about Theodore was, however, involved in mystery. Whenever spies came into camp with reports of him, they were so exaggerated that it was impossible to arrive at the truth. The prisoners' letters were also as vague productions as would task the keenest diplomat to understand anything about Magdala and its surroundings. In such a case there was nothing left for General Napier but to tramp onward, with the hope that before many days the whole problem would be resolved, and all the stories, and doubts, and surmises ended for ever.

On the evening of the same day General Sir Charles Staveley, with the 33rd Foot; Punjaub Pioneers, six companies; Twiss' Steel Battery; a company of the 3rd Native Cavalry; the Naval Brigade, in charge of the rockets; another regiment of European Infantry, the 46th Foot; the 3rd Dragoon Guards (Europeans), and the left wing of the Belooches, arrived upon Wadela.

On the third morning after our arrival at Santarai, the 1st Brigade broke camp, and started for Gazoo, sixteen miles distant.

The whole surface of Wadela plateau resembled the same amount of land on the Illinois or Nebraska prairie, except that rocks protruded a little oftener above the soil and grass on the Abyssinian plain. Several fields of green wheat were passed, while adjacent to them were patches of vetches and teff ready for the sickle. Small villages rose here and there, surrounded by an abattis of wattled fence-work. There was always a pleasant breeze, which, with solar topees on the head, obviated the effects of the extreme heat.

Charles Speedy, our interpreter, the quondam General-in-Chief of the Emperor Theodore, was commanded by Sir Robert Napier to go to the different villages on the plain to buy grain and flour. Mr. Speedy proposed to me to accompany him, that I might see something of the people untainted by British dollars. I very gladly acquiesced in the proposal.

The first village Captain Speedy and I entered was Wota, on the left hand side of the road over which the army travelled, and about three miles from it.

When the inhabitants saw us approach, two or three of the elders came forward and bowed themselves to the ground before us. Speedy delivered his 'Dïdwal achoo,' which means 'Health to you,' or 'How are you,' the customary morning greeting of the people. They prostrated themselves with their hands to their heads three times, and Speedy salaamed likewise in true Ethiopian style.

By their gestures, which I interpreted to an invitation according to the Scriptural texts:

' "Come, Sooltan,"[1] if now we have found favour in your sight; pass not away, I pray ye, from your servants.[2]

' Turn in to us. Here is a pleasant shade under our village trees, " rest yourselves,"[3] and we will fetch a morsel of bread, and comfort ye your hearts; after that ye shall pass on.'

My friend consented to the arrangement, saying at the same time that it was all very pleasant.

We sat ourselves under a beautiful green kousso tree, and ' directly they set before' us bread, butter, and milk, and the hind quarter of a kid. They stood by us ' under the tree,' holding their robes before us, so that evil eyes should not rest upon us while we partook of their hospitality, and we ' did eat.' The milk was delicious, the bread was excellent, and the kid was rich and tender.

After the extempore meal had been despatched, damsels arrived, bringing their corn and flour for sale, while we spread our silver dollars before us in tempting array. The first quantity of flour we bought barely weighed six pounds, for which we paid one dollar, but from this minimum we gradually rose to the maximum of thirteen pounds for a dollar. The heaps of flour and grain became larger, slowly, until they formed a miniature illustration of the great pyramids of Ghizeh in Egypt. We also purchased goatskin bags, thongs, and mules.

Before we departed from Wota, Captain Speedy and I had 4,000 lbs. of flour, and 3,500 lbs. of grain. When we left

---

[1] My lord.　　　[2] Genesis xviii. 3.　　　[3] Genesis xviii. 4.

we could hear the ardent blessings as they were showered upon us by the kindly villagers.

From Wota we went to Mai-Ayni, where the extraordinary stature of Captain Speedy excited unusual attention and respect. When he moved among them, they all moved away precipitately from before him; but, jocularly disposed, this son of Anak would swing one of the young men aloft with his powerful arms, just to show them what 'Basha Felecca' could do.

The Syndic at Mai-Ayni was a singular being. He had been suffering a long time from a cachectic disease, probably the tabes—until now he seemed to be nothing more than an automaton skeleton. He welcomed us with many salaams. Around him skipped and danced a dozen laughing girls. Merry boys and youths sauntered out of the village at the sound of our voices, leaning on each others' arms, like so many Davids and Jonathans. Such interesting tableaux vivants were not witnessed by me in Abyssinia before this day. Pitchers and jars full of richest cream were brought before us, and one young damsel, as rich in graces as Cleopatra could boast, out of honour to me—so said Speedy—insisted on my drinking cream while she held her pitcher to my lips.

We managed to secure at this village 40 mules, 30 horses, 1,700 lbs. of flour, 3,000 lbs. of grain, 300 cakes, bread, 75 goatskin mussucks, and about 100 fathoms of hide rope. Late in the evening we arrived at Gazoo, the camp of the English army.[1]

From Gazoo the army moved on to Abdecoom, distant fifteen miles.

The aspect of the country was precisely similar to that between Gazoo and Santarai; if there was the slightest difference it was in being a little more level.

Accounts of starvation in store for us between Abdecoom and Magdala were rife. Theodore, it was said, had burnt the

---

[1] This was very important service which I assisted Speedy to perform for the British Government. Yet I regret to say the magnanimous British Government have never even thanked me for it, least of all have they given me a medal, such as all Abyssinian heroes obtained, which I consider to be a strange oversight on the part of the British Government, and deserving of gravest reproof from the author of this history.

country throughout, and taken every pound of grain away, so that all Napier's finely-wrought plans seemed to be destined to sudden misfortune. But then so very many unfounded stories had been in circulation that people would not place any great confidence in anything said in Abyssinia.

Still rations were scarce, our supplies had long since been devoured. For a week before we had had no coffee, sugar, potatoes, rum, vegetables, or ghee. We were living on the toughest of beef—one pound per day, sugarless tea, ten ounces of flour, four ounces of rice, and half an ounce of rock salt, while our servants got but half of the soldiers' ration. Our horses were living on eight pounds of grass, and four pounds of barley grain, and mules on four pounds of grass and three pounds of grain. Ahead the prospect was still darker. Two days at the furthest would consume all our stores. If our Commissariat Department was not replenished daily, starvation would certainly ensue, unless we killed our animals. Yes, by that method we might be able to live a month, until further supplies could be brought from Antalo. The idea was, however, anything but pleasant.

From Abdecoom the army on the next day moved on to Sindhe; only three miles distant from the former. Sir Charles Staveley's second brigade reached that same evening Abdecoom; thus the two brigades were but three miles apart.

General Mashesha, uncle to Gobayze, in the afternoon called upon Sir Robert Napier. He was accompanied by about three hundred cavalry. He came to inform Sir Robert that he had purchased 6,000 lbs. of flour and grain, and would send it to his camp directly. This was very satisfactory news.

It being late in the day, Mashesha proceeded to a village on an eminence not very far from camp, and quartered some of his men among the villagers. Leaving that village, he was proceeding with the remnant of his band to another about three miles north, when he accidentally came upon an outpost of General Staveley's camp, consisting of a corporal and four men of the 3rd Light Cavalry.

These soldiers did not know Mashesha; it is doubtful whether they ever heard of him; but greatly exercised in their minds by the many rumours they heard each day about the dash and enterprise of Theodore, they believed

that this strange party approaching them was either the
Emperor himself, or else one of his reconnoitring parties, and
when they saw the cavalry advancing they shouted to warn
them off. But with that obstinacy which, I believe, all
sorts and conditions of men feel when thus warned peremptorily by sentries, they still advanced. This seemed to confirm
them in their suspicions. The corporal ordered a picket to
fire. The sentry's shot was instantly answered by two from
Mashesha's party. The corporal ordered his men to discharge
a volley at them, and then gave the word to charge. These
five men charged upon nearly fifty Abyssinians, sword in hand.
Mashesha perceiving the error, gave the order to retreat, but
the pickets were upon the rearmost, cutting and stabbing
away like heroes. They pursued the rest for some distance,
and seeing that they could not overhaul them, the little outpost squad returned to their post, having killed two and
wounded two others.

As might be expected, the noise of the firing caused
great excitement among the respective camps of Napier
and Staveley. The second brigade was instantly under arms
and drawn up in line outside of their camp, while the 3rd
Light Cavalry scoured the adjacent hills in pursuit. The
4th European Foot, and the Beloochees, were marched out of
the camp of the 1st Brigade, to take post on an elevation
that commanded a large surface of ground near it. Every one
believed for a short time that the firing had been between the
pickets and the enemy.

Mr. Munzinger, the interpreter, sallied out, however, with
a portion of the Scinde Horse, to enquire into it. The village,
whither Mashesha and his men had fled for refuge, was fortified, and its walls were manned by an infuriated body of men,
who believed that the Feringhees were treacherous people who
had professed friendship for their chief, that they might murder him, and it was with some difficulty that they could be
persuaded from firing upon the Scinde Horse.

After many expostulations on the part of patient Munzinger, Mashesha was induced to come to the gate to listen
to him, and after a fund of logic had been exhausted, the
mistake was forgiven by the Abyssinian chief. Three hundred
dollars were given to the families of the deceased as blood-

money, which satisfied them, as the dead men were no chiefs,
and common peasants' lives are held but lightly in Ethiopia.
The wounded were taken to the British hospital tent and tenderly cared for.

From Sindhe, the army marched the next day down the
Jeddah ravine, and up again on to the summit of the opposite
plateau of Dahonte Dalanta, which was also 11,000 feet above
the level of the ocean.

From Sindhe was six miles to the edge of the ravine. A
ravine is defined as a long, deep and narrow hollow. It is a
good definition, but were I to describe the Jeddah as a
long, deep and narrow hollow between two plateaus of equal
height, few would understand the true depth and extent of the
Jeddah ravine; were I to say—as I must say—that this ravine
that yawned beneath our feet as we stood on the extreme edge
of Wadela was thirty miles long, 3,800 feet deep, and two miles
across from Wadela to Dalanta, I should convey to the mind of
the reader the true proportions of the Jeddah.

To many even this exact description of it would not enable
them to realize its awful depth and appearance; and as I believe
it incumbent on me to endeavour to make it comprehensive, I
shall proceed to details.

Two miles from the camp we left the grassy district, and
an expanse of scorched plain lay in front of us—the handiwork
of Theodore's soldiers. As we surmounted a gentle rolling
ridge, a huge and gloomy wall of rocks looming up on the
opposite side of Wadela was seen. On approaching the brow
we discovered the remains of Theodore's camp. This was the
Bet Hor, so often heard of in the newspapers, where his camp
stood while the road down the deep Jeddah ravine was constructed.

Around the old camp was the blackness of desolation,
which invariably follows in the train of ruthless war. There
were still to be seen the booths, or arboured branches in-
terweaved with the yellow grass, which served as tents for
the soldiers. The bleached bones of sheep and oxen covered
the ground. Here and there were scattered pieces of the
black native blankets; kneading troughs of wood; horn spoons
and bottomless horn cups; head-rests for girls, and innumerable
pieces of leathern aprons or loin-cloth. Here were remnants

of quern and other stones, by which the camp women ground their corn; and broken Ethiopian distaffs strewed the whole place.   A significant sign to us of the cruelty with which Theodore governed his people was a human arm in the last stages of corruption, and a scull with a great hack across it, which was found in the very centre of the camp.

Upon arriving at the brow of the plateau, the road which Theodore had caused to be made through the solid syenite and basalt rock opened before us, through which the head of the column at once passed.   For the first hundred yards the declivity was very steep ; when measured, it was found to be 1 foot at the base to 3 feet in height.   We could perceive the marks of Theodore's drills in the rock throughout the whole length of this pass.   It must have been tedious work to him, working as he did with imperfect tools, an enemy hovering about him, in the form of Wagshum Gobazye, and another one advancing upon him in his rear, in the shape of the British force.   The natives say that he worked as hard as any of his men ; that he was constantly riding about on a white mule, to observe his men while at work.   Now he would encourage and praise, anon would he threaten ; at times would he flog ; and at another place he would order a man out to instant execution ; then again he would dismount, and proceed to show what real earnest work could do.   After five weeks' hard work, he had constructed from the height of the Wadela to the hollow of the Jeddah, and up to Dalanta plateau, a road 8 miles long and 30 feet wide.   Over this road we travelled, from terrace to terrace, down slanting gradines, and down the side of ledges, which in some places were almost perpendicular, until we arrived at the bottom of the ravine.   The bottom was covered with pebbles and round boulders.   In deep hollows alone could be discovered any water.

About two P.M. the head of the column was once more put in motion, and it was almost piteous to watch the faces of the soldiers as they tugged manfully up, with a certain air of half despair and resignation depicted on their features.   It is of no use to follow them up step by step ; suffice it that they halted every fifty paces, and about four P.M. they surmounted the plateau of Dalanta.

From here we could trace the army baggage trains along

the whole length of the ascent and descent from Wadela downward and upward to Dalanta. We were slightly astonished to find that, though we had started from Sindhe camp as early as six o'clock in the morning, the last of the baggage train had not yet, at four o'clock in the afternoon, commenced the descent of Wadela. Those who had their baggage so far behind might at once make up their minds that they must pass the night without their tents, blankets, or suppers, unless good Samaritan souls could be found to feed them for that one night.

At six o'clock Ali came up, triumphant, with my baggage. Captain Sweeny had stopped him twice, but on the road so numerous were the mishaps that occurred that the Arab boy brought his two animals clear of them and without hardly a pause, except to breathe.

Not one-half of the baggage trains arrived at our new camp on Dalanta that evening. When once night fell, those who were still toiling among the convoluting defiles became enwrapped in thick gloom, and each man had to make the best of what he could in the premises. They unstrapped the saddles, and dumping the packs down on the ground, they tied their mules or horses to some bush close by, threw themselves down under the shelter of a tree or shrub, and slept their fatigue off as best they might.

It was very late even next day before the over-strained transport train arrived upon Dalanta plateau. Animals had died by scores in the ravine.

Further movement for two or three days was impossible. There were no rations on hand. According to a Commissary's report there were only 4,000 lbs. of grain! This amount divided amongst the 5,000 animals of the 1st Brigade would give each one about six and a half ounces of grain, after nearly thirty-six hours fasting! Sir Charles Staveley's Brigade, with 5,000 more, was to come up this day, and there would be none left for it. Captain Speedy was therefore despatched with haste, to see the Chief of Dahonte Dalanta, who had always refused Theodore supplies, and had been the most rebellious of all rebels, to enquire whether provisions could be bought. His mission was very successful. He returned with one hundred

mules loaded with grain, which amounted to the gross weight
of 10,000 lbs., and a promise from him of 100,000 lbs. on the
following day.

This was indeed joyful news for a portion of country that
was said to be ravaged by Theodore. The chief showed unmis-
takable enthusiasm for the British cause, and all hearts were
elated. The next day, which was April 7, witnessed an abundance pouring in, from all sections of the province of Dalanta,
of grain, flour, bread, honey, chickens, grass, chopped straw,
oxen, and goats.

On the evening of April 7 the long-expected 45th Regiment
and 3rd Dragoon Guards arrived at our camp upon Dalanta.

The army was now concentrated, and numbered over 5,500
fighting men. The quick march performed by these two
bodies of men from Antalo to Dalanta showed what the two
brigades could have done had they been divided into regiments on the route.

About ten o'clock on the morning of April 8 the
Commander-in-Chief, accompanied by his staff and an escort of
cavalry, proceeded to the southern side of Dalanta plateau to
reconnoitre Magdala, which was plainly seen from the extreme
height of the plateau. He reported that he saw the tents of
Theodore's army on a plain below Magdala, and many guns in
position on the fortress. There were also many thousands of
people seen moving about there, as well as smoke curling
upwards from the camp fires.

Sir Robert called it a very strong place, stronger than any
amba we had as yet seen ; which confirmed everybody in his
impression that the Emperor would never have neglected the
strong places we had passed on our road, had he not held a
much stronger position. The great desideratum was satisfac-
torily answered. Theodore was in Magdala !

When Sir Robert came back from his reconnaissance, the
Engineers were set to work to make escalading ladders and
torpedoes, in readiness for an assault up the cliffs of their
numerous strongholds.

Our minds were at ease. Plenty reigned in the camp.
Theodore was known to be in his fortress. To-morrow we
should see Magdala from our camp on the southern brow of

Dalanta. All felt happy at the thought, and such merry laughter and ringing heartiness was never heard in Abyssinia as was heard in the English camp upon Dalanta.

The rival bands of the different regiments played their most enlivening music. The Naval Brigade had invited the Punjaubee band. The members of it squatted themselves in two decorous segments of circles, between which the lively 'Jack Tars' tripped the light fantastic to the really charming music the band discoursed.

The natives left the camp bazaar and their trading to gaze open-eyed at the strange actions of the sailors, which they could not understand, believing them to be part of a worship to their divinity.

By and by the music ceased, and all retired to their quarters. The last tap of the tattoo was sounded, lights were put out; but in the effulgence of the summer night who could sleep?

The gentle winds came whispering over the upper regions of the earth. The Southern Cross could be traced, with its base resting upon the Bechilo confines, the Pleiades shone out clear. A profound repose brooded over the camp; the very angel of rest seemed to have laid his hand upon the sleeping hosts!

CHAPTER VIII.

POLITICAL INTELLIGENCE—MAGDALA—THE LAST ACT OF THE DRAMA—THEO-
DORE OPENS FIRE—CHARGE OF ABYSSINIANS—COLONEL PENN'S LAST
SALUTE—THE JACKALS AND HYÆNAS AT WORK.

BEFORE proceeding with the description of the march from the
northern brow of Dalanta to its southern side, let us glance at
the political tidings which have reached camp since leaving
Antalo. The last news I have quoted is dated March 6, 1868.

March 16.—A letter came from Mr. Rassam, stating that
Theodore had at last informed his soldiers of our coming,
adding that the English were a nation of traders, and that
although they could manufacture good guns, and could use
them at a great distance, they were wanting in courage, and
no match for Abyssinians in a hand-to-hand fight. He promised
that his soldiers should defeat the English with great slaughter,
and that all their rich treasure should be divided among them.

March 19.—We heard from Mr. Munzinger that he had
arrived at Wagshum Gobazye's camp, and had been received in
a very gracious manner by that prince. He also stated that
Gobazye extended the hand of friendship to Sir Robert Napier ;
and that whatever he could do to forward his expedition, he
had promised would be willingly done.

Theodore was at Bet Hor, working as hard as possible to get
his guns across the Jeddah ravine. He had burnt the country
and ravaged it round about. The Chief of Dahonte Dalanta,
had also rebelled, and a few skirmishes had taken place
between the Emperor's forces and the men of Dalanta.

March 20.—Another despatch came. Theodore had arrived
upon Dalanta plateau, and was marching on to Magdala.

March 29.—Colonel Phayre, with the Pioneer Force,
despatched a note to General Sir Robert Napier, informing

him of the capture of Munzinger by some of Theodore's scouts, and that Theodore himself was recrossing the Bechilo River, which lay between his fortress and Dalanta plateau, with the intention of fighting the English.

March 31.—Munzinger sent a note, written by his own hand, from Dajaz Mashesha's camp, that Gobazye's General would visit Sir Robert Napier, upon his arrival at Santarai. Theodore, it was said, was still at Magdala; had never recrossed the Bechilo since his arrival at his fortress.

April 4.—News came that the people of Dalanta had submitted to Theodore.

April 8.—Rassam sent a letter to Colonel W. L. Merewether, who had come up with the army again from his successful mission to Welda Yassous, saying: 'Beware! the Emperor is moving. Be very careful; he may attack you in the night. Deal leniently with him for our sakes. The least burst of temper that he may be in, may end our lives. Have you any good Champagne and Havannas in the camp? Keep some for me, dear Colonel, by the time I am with you. God bless you.'

On the morning of April 9 both brigades moved from our camp above the Jeddah ravine to the southern extremity of the plateau overlooking Magdala. The plateau being but six miles across, and smooth and level as a billiard-table almost, we reached the southern brow of Dalanta in a couple of hours, and pitched our camps just half a mile from the road leading downward to the Bechilo.

Join me, reader, on the extreme brink of Dalanta plateau, and let me show you, to the best of my ability, the famous Magdala and its surroundings. Open your eyes and behold the scene!

From the edge of the plateau on which I stand I look down directly below my feet and see a wall of sheer rock about fifty feet in depth, then a sloping terrace running forward a hundred feet or so, abruptly terminated by another precipice of a like depth, along the base of which winds a well-made road for a hundred yards, when it turns and descends another terrace, and so on from ledge to ledge it winds through its tortuous convolutions until the eyes rest on a river—the Bechilo, 4,000 feet below!

Across this turbid stream the vision traces another road, whitened by travel, inclining up another ravine—the Aroje, for about five miles ; when it is lost from the view by a jutting abutment of an aslanting hill, until we find the road, fainter than before, ascending at a sharper incline, a high hill topped by a small plateau. From this plateau arises apparently perpendicular to us, at eight or ten miles distance, a frowning mass of rocks, divided into two differently shaped mountains. The one to the right is Fahla ; the other is Selasse ; a low ridge connects them. Behind Selasse I am told is Magdala. Not a particle of it is visible. The obtruding proportions of Selasse prevent us from seeing it. Move a mile leftward of where the camp is, and the massive outline of the royal fortress is distinctly seen. These triple scarped heights contain Theodore, his army, and the captives. Theodore has certainly selected the most impregnable heights for his eyrie !

The Bechilo valley must be seen to be realised. Away to the right and left horizons of the sky trend indistinct and interminable groups of peaks, which have their numerous points buried in the blue firmament. Mountains crossing mountains, hills set upon hills, shoot upward as if purposely placed one upon another by a race of Titans in a vain endeavour to pierce the boundless bulwark of ether which God erected between the immortal realms and the globe. Stream-girdled ranges appear like vaporous masses in the dim far-away distance, aching the eyes in the attempt to define their extent and reach. From these giant heights decline in gradation the lesser heights, until they reach the bottom of the valley, in the centre of which rises isolated the colossal triple fortresses of Theodore.

Night's shadows sink at last, and gradually envelope the yawning gulf, until its gloomy profundity becomes as dark as the fabled realm of Pluto. In the centre of the circle, environed as it is by the deep valley on all sides, and by the circumambient mountains, which recede from all points to a distance of ten miles, is dimly seen through the sabling twilight the lofty head of Selasse, illumined by thousands of twinkling fires, and with all its precipitous escarpments made visible by the many fires ranged around it.

While standing on Dalanta, we almost imagined we could

hear the sounds of revelry in which the enemy's host indulged. With a powerful night-glass wild fantastic figures could be made out, capering round the flickering lights, and we could distinctly distinguish groups squatting around them, as if they were discussing the events which the morrow foreshadowed. As the altitude of Magdala and its sister fortresses of Selasse and Fahla was the same as that of Dalanta, had the night been clear the different positions of the camps on their summits might have been easily distinguishable.

> I am a soldier, and my craft demands,
> That whereso duty calls within earth's compass,
> Or the unmeasured scope of fathomless ocean,
> I do forthwith obey.—HYDASPES.

So might all British soldiers have said, as they were paraded for roll-call before their tents on the morning of the 10th April, which was Good Friday, in 1868.

Breakfast was despatched early that morning, and at the sharp blare of a bugle the tents were struck. Colonel Phayre, with his six companies of Bombay and Madras Sappers and Miners, were sent forward long before daylight, to remove obstructions and prepare the road, where the late rains had destroyed it.

Another sudden startling blare was sounded, and the head of the 2nd Brigade commenced the long descent to the Bechilo.

Lo! now the last act of the drama was about to be enacted! In spite of the little blunders and impediments of the march from Zoulla to the Bechilo; in spite of the mysterious warning conveyed to the Political Officer—Colonel Merewether—by a messenger from Rassam the captive envoy, 'Beware! the King is moving'; in spite of the rumours of the sudden sweep down meditated by Theodore; the army had survived the march, the alarms and terrors of the night, and the head of the column had actually started on the descent in full view of Selasse and Fahla, whereon Theodore stood, with his eye-glass, surveying the scene!

Sir Charles Staveley had fully recovered from his rheumatism, and imbued with energy and spirit, led the 1st Brigade, which consisted of the 4th European Foot (King's Own), Colonel Cameron; a company of the Royal Engineers, commanded by Major Pritchard; two regiments of native Indian Infantry—the Belooches and 27th Punjaub Pioneers; two companies 10th Native Indian Infantry. They were all dressed in their newest and best uniforms, and presented a very gay and animated appearance.

The tactics and plans of the battle and the storming had been discussed till late on the previous evening, and all the heads of departments and commands fully understood what was expected of them. England and Sir Robert Napier expected every man to do his duty! The programme seems to have been as follows :—

Colonel Phayre was to take his sappers and miners to repair roads and clear obstructions, to occupy and to hold the head of the Aroje ravine directly beneath Selasse.

General Sir Charles Staveley, commanding the division, was to march the 1st Brigade across the Bechilo River, then to make a détour to the right, march up a steep hill intervening, and occupy the little plateau below Fahla.

The 2nd Brigade, consisting of the celebrated Irish regiment, the 33rd, or the 'Duke of Wellington's Own,' 600 strong; the 45th Regiment Infantry (Europeans), 500 men; Colonel Penn's six-gun battery of Mountain Train Artillery, 100 men; the Naval Brigade in charge of the Rocket Battery, 100 men; the Armstrong Battery of six 12-pounders, two 8-inch mortars, manned by about 200 men, were to proceed up the Aroje Valley, the Mountain Train Artillery in advance, and endeavour to secure a small knoll and a narrow ridge directly under Selasse, at the distance of 1,500 yards to the left of the hill known as the objective point of General Snyder's (First) Brigade.

The after conduct of General Napier was to be ruled by circumstances. When he should arrive in camp he intended to write another appeal to Theodore to consider well his position, and the dread consequences of refusal to deliver up his captives safe and unharmed.

Thus man proposes always, but the Almighty disposes, and our wisest intentions are very often frustrated by chances inimical to them. Sir Robert, with the calmness of an educated General, had theorized much upon the glorious science of war the evening previous at the Council Board, and had supported his tactics by choicest logic, proving thereby his perfected attainments in the abstruse details of military engineering, and through the whole there glimmered enough light to demonstrate to us that he intended at the first opportunity to transform theory into practice, and become master of Magdala!

At ten o'clock in the morning we who stayed behind to eat at our leisure comfortable breakfasts, and smoke the very last of our cheroots, had our tents struck and rolled up. The last of the 1st Brigade had disappeared beyond the brow. The English Commander-in-Chief mounted his charger, the Staff sprang nimbly to saddle, the " specials " *en masse* order their horses up, and, booted and spurred, mount also.

Under the warm sun-blaze the 2nd Brigade, led by the Chief himself, followed by his brilliant well-dressed retinue, defiled statelily down the slope. In all the gorgeous panoply and glory of numbers, the living stream of warriors commenced their descent for the Bechilo River. Regiments well-known to the world were in this compact army. This might not be a great battle, but the victory would be the same, and the glory of it appreciated as much.

Down, down, and still down the lengthy slope they descended, slipping, stumbling, sliding, amid dense clouds of dust. Round and round in a hundred convolutions an hour, from terrace to terrace, with a train of elephants, camels, mules, horses, asses, and camp followers, extending far away out of sight over the summit of Dalanta. Strangely bewildering was the continuity of the descent, and exceedingly provocative of thirst was the fierce sunshine, which increased its heat as we neared the Bechilo from the incandescent rocks and fervid dust; but the soldiers heeded it but little.

After two hours we found ourselves at the bottom of the ravine close to the Bechilo—a river, muddy and swift, about fifty yards wide, and four feet deep. The head of Snyder's column had already forded the river, and was at that time debouching to the right. In a short half hour it had crested the first hill.

The scene of confusion in and on both sides of the river was somewhat alarming. Three hundred men with a field-piece or two appearing suddenly above this arena of strife might have ensured a decided victory for that one day at least, against the crusaders. The soldiers still waiting for their turn to ford the Bechilo were squatted, or were recumbent upon the river banks, thickly covering acres, while the 2nd Brigade, following in the rear of General Napier, only added to the tumult, hurrying as they did to the stream to assuage their thirst. The animals of the Rocket Battery, and those conveying the mountain artillery of Penn's Company, plunged eagerly into the water for the same purpose, and when restrained, kicked and plunged, and made the confusion worse confounded.

On arriving at the river, the 2nd Brigade were ordered to cross. They accordingly took off their shoes and stockings, rolled up their trousers, and walked boldly in, getting a very efficient wetting up to the midriff, which slightly diminished their ardour. After crossing they formed column again, and preceded by the Naval Brigade and Penn's Battery they marched slowly up the Aroje ravine, at the head of which it was supposed the Bombay and Madras Sappers and Miners, and the companies of Punjaub Pioneers, stood guard.

There were no outlying pickets or scouts posted on the flanking ranges of heights. Seemingly the English army feared no enemy lying in the rear of those frowning crests, that rose up on each side of the ravine like walls. The regiments' careless mood and bearing at this time were very noticeable.

Slowly the 1st and 2nd Brigades wended their respective ways, marching on two radii, forming a V, the base of the figure being the embouchure to the ravine close to the river. The 1st Brigade was even then swarming over the brow of a steep hill. If all went well, both brigades were to join below Selasse and Fahla.

The Commander-in-Chief, staff, and followers, lunched on the river banks. After resting an hour, they struck off to the right, joining Snyder's column, then labouring up the slope. At 2 p.m. both brigades arrived at the bases of their objective points, very much fatigued. There was still a tremendous mountain—whose slope was nearly a mile in length—to climb, before they could surmount their positions.

I was riding with Colonel Milward—the senior officer of the artillery—in advance of the 2nd Brigade, marching up the Aroje ravine, when I saw him beckon to Lieutenant Nolan of the Armstrong Battery, and heard him whisper to him to go forward, and ascertain whether there were any signs of the enemy in front. We were at that time three miles up the ravine, and had as yet seen or heard nothing from Colonel Phayre or his Pioneers, though he had reported to Sir Robert Napier that he held the 'head of the ravine.'

A dead silence brooded over the hills and narrow valley. It was a suspicious silence ; so it seemed to us, knowing as we did that a wily enemy with a large force was in front. An ambush might have been planted somewhere on our flanks. There were many available places where the enemy could have hidden and taken us at disadvantage. The hills in the hands of a civilised force defended by modern engineering would have rendered the ravine as fatal to a dense column as the Afghan Passes were to the British forces. From the prominent points in front, in rear, and flanks, we might have been subjected to enfilade, slant, and reverse fires. In the way we were proceeding, it appeared to me we were going blindfold, and I suppose the idea struck Colonel Milward.

Lieutenant Nolan, after buckling his waist-belt tighter round his loins and preparing his sabre and pistols, spurred forward on his errand. I requested to be allowed to accompany him, which request Colonel Milward politely granted.

For another couple of miles we travelled along the ravine at a sharp pace, looking keenly to the right and left, until we arrived at the bottom of a hill which terminated the great ravine of Aroje, where a narrow tangled gorge deviated in an oblique direction to the left and beyond the eastern extremity of Selasse.

As the king's road ran up this jutting abutment of a hill, we spurred upward without hesitation, Lieutenant Nolan taking the lead, pistols in hand. For a few hundred feet the rise of the slope was one foot in two, and it was with great difficulty we clambered up. Upon surmounting its summit we discerned Selasse looming up high and vast o'erhead in dangerous contiguity, if a few expert sharpshooters were on the watch above.

There appeared, singular enough, no indications of an enemy on the heights. That there were inhabitants was obvious from the many columns of smoke observable ; but why they permitted the strangers to mount the plain below them was a problem. Judging from their perpendicular escarpments, they appeared unscalable to us. Tough work we imagined was in store for the British army when they should attempt to storm Selasse and Fahla—those twin giants that guarded the entrance to Magdala—if they were in any way protected as they ought to be.

To our surprise we discovered a group of officers reposing on an open green sward, in full view and range of both hills. This group of recumbent forms we found directly were Colonel Phayre and a few young officers.

When Lieutenant Nolan arrived at the scene, I heard M'G——, 'the best sabreur in the army,' utter the opinion that he for one did not believe there was anyone in Magdala.

'By George, neither do I!' exclaimed another officer. 'Nor I,' was echoed around.

'O, pshaw!' said Colonel Phayre; 'even last night there came a messenger from Rassam bearing a letter reiterating his warning, and stating his belief that Theodore intended to fight.'

'But I have ridden up that road that you see there within pistol-shot of the summit, and don't you think,' expostulated warmly Captain M'G——, 'that if people were there they would have fired upon us?'

'That proves nothing,' responded the sceptical Quarter-Master-General. Theodore is cautious and sly as a fox. He was not going to sacrifice any good chance which he may think he has, for the sake of bagging a couple of men. He may have decided upon giving up the prisoners, and, of course, in that event he would not fire; and, stop! by Jove, there is the Chief and his Staff!'

Looking in the direction indicated by Colonel Phayre, we saw the Chief surveying with his field-glass the precipitous faces of Selasse and Fahla; and at the same time the head of Penn's Battery appeared in our rear, panting and breathless after their vigorous clamber up the slope of the steep mount which Nolan and myself had come up but a short time previous.

These two positions, intended to be occupied respectively by the two brigades, were connected with each other by an almost level ridge, along which ran the road constructed by the Emperor on his march from the Bechilo to his fortress.

The site of the camp of the 2nd Brigade was the round summit of the hill at the termination of the Aroje; from it ran the road, undulating and dipping into slight hollows, until it reached the plateau directly at the base of Fahla, where it was protected by scant copses of underbrush and scrub oak, when it suddenly emerged again into view on the surface of the plateau.

This level ground was but a narrow neck connecting the base of Fahla with a group of termite hills whose cones had been eroded, and were unevenly connected together by the débris which almost filled the different depressions that existed formerly between them, until now they represented but a series of undulations dotted at intervals by clumps of low bushes. This ground had been chosen by Sir Robert for the camp of the 1st Brigade while viewing it eight miles off from the heights of Dalanta. It was a happy selection, protected as it was by the steep slopes in the rear, and commanding as it did all approaches to it while still well out of range of any artillery that the Emperor might have upon Fahla. Ten 24-pounders (Armstrong's) mounted upon Selasse and Fahla would undoubtedly have rendered both positions completely untenable; but then Sir Robert was but measuring himself against Theodore, an African potentate.

Scattered over the ground between the two positions were the remains of the Emperor's old camp of wicker-booths still standing, and broken native crockery lying about, with the filth and offal emitting a fearful smell.

Colonels Milward and Penn busied themselves in looking for the cleanest place to pitch the tents that were expected to appear during the night, and the Artillery sergeants followed them about, to measure the camp when selected. The two colonels then rode to the edge of the eminence nearest the fortress, and began to comment leisurely upon its appearance, and the probability of battle.

It was 3.30 P.M., my attention being called to that fact by hearing Colonel Penn utter the remark aloud after looking at his pocket chronometer.

' I wonder if that fellow Theodore intends to fight at all,' said Colonel Milward.

' Not he,' replied Colonel Penn, bluntly.

' There are men on top of the hill, sir !' exclaimed a sergeant, who had been examining the stupendous mountains with great interest.

'What?' bawled out Colonel Penn; 'aye, by Jove! I see them.'

'Egad, they line the whole summit from one end to the other!' said Colonel Milward, aroused now to activity.

'And by Jove,' added Penn, 'Theodore has opened the ball! See you the puff? D'ye hear the music of chain-shot?'

Two seconds of expectancy, ' boom!' came the loud report, and half a second later a huge chain-shot flew over our heads, burying itself fifty feet in our rear.

'They are coming down, sir!' roared the sharp-eyed sergeant.

'Where? Where? Ah! I see them,' responded each of the colonels.

'Bring up those guns, boys! Get ready for action!' shouted Colonel Penn, bounding towards his company, waving his sword and uttering various commands in true syntonic military style. 'Look smart, boys! Bear a hand quickly! Keep your eyes open, my lads!'

'Boom! boom! boom!' went three monstrous guns in quick succession from the most salient angle of Fahla, discharging 68-pound chain-shot over towards where stood Napier and his staff. Again the huge cannon vomited their thunders and iron balls, and one continued blaze of fire and globular wreaths of white smoke announced to us below that he had ten guns at work in real earnest.

One more glance at the charging squadrons of the enemy told too plainly that there was no time to lose, if a reception à l'Anglaise were intended to be given them.

'Down with those guns,' commanded Colonel Penn, swinging his sabre ominously above some laggards' heads. 'Lead the mules to the rear. Man guns. Here, you sir, what are you driving at? keep your wits bright, or by Jove I'll spur you to it.' A very prompt, forceful man was Colonel Penn, though he did use energetic language; but what of that? it was life and death with them all now. Over 3,500 of the enemy were galloping—animated by wild fury—down the hill.

At this crisis an aide-de-camp arrived upon the scene, with

instructions for the Naval Brigade, which had just come up, to hurry up to where Sir Robert Napier was.

Captain Fellowes of the Royal Navy, commander of the frigate ' Octavia's ' crew which formed the brigade, touched his hat upon hearing the order, and giving the word to his men, drew his sword, and led them at a quick pace to occupy the position designated. In a short time they emerged upon the crest of the plateau, whereon Sir Robert Napier stood surveying the advance of the enemy.

All this differed *toto cœlo* from the well concocted plans of the last evening. Not thus had Sir Robert intended to act. By a wayward fate his schemes were suddenly quashed, and we may well imagine the confusing consequences. Happy is he, however, who is always prepared for emergencies!

Following the route taken by the 1st Brigade from the Bechilo Valley, Sir Robert had overtaken Sir Charles Staveley, the divisional commander, with whom he entered into conversation, and in company they had continued their march up the consecutive gradines of the interminable slope. After an hour they had surmounted the eminence already mentioned, whence they obtained a good view of the extreme edges of both fortresses—Selasse and Fahla. Near this elevation the six companies of Sappers and Miners were found.

Immediately after his arrival upon the summit, General Staveley decided that, as the steepness of the road up which he had climbed presented so many obstacles to the rapid advance of the baggage trains of his column, it would be wise for them to take the more accessible route up the Aroje Valley, and with that order an aide-de-camp was despatched to the conductors and superintendents of the trains.

For a thorough conception of what these two fortresses represented, I must state that they appeared in the form of a line, or an arc of a circle, with 3,000 yards for a radius. The centre of this line was a low gorge, with a depression of 100 feet between Fahla and Selasse, by which easy communication was maintained between the two fortresses. The salients of Selasse and Fahla were about 1,000 yards apart. That of Selasse commanded all approaches from the east and the Aroje Valley, and its fire was easily convergible upon the road which led up to the gorge which divided it from its counterpart Fahla. The salient point of Fahla commanded the western approaches, and the exposed summits of the undulating ground—upon which General Napier stood—so could it sweep the road which led to Magdala. The camp site of the 1st Brigade was distant 1,500 yards from Fahla, and 2,500 yards from Selasse. The rounded plateau whereon Penn's battery was posted was distant 900 yards from the most salient point of Selasse, and from Fahla 1,900 yards. So that if the two fortresses were defended by guns of modern construction, the sites chosen for the camps of the brigades would have been within range of the guns on the heights; in fact, they could not be tenable for half a minute.

The prolongations of the natural lunar formation of the heights commanded so well the road leading to the intermediate gorge, that it would be simply an impossibility for an army, except at the risk of annihilation, to have marched up; and on all other sides the slopes were almost perpendicular, so that, were the enemy but armed with rocks, barely one of a storming party could live to reach the summit. There were no prominent hills in the neighbourhood from which an efficient slant or reverse fire could be maintained by the assaulting column. A traverse, again, planted along the prolongation of the intermediate gorge, would have perfected one of the most defensible places in the world; and a concentric fire would thus have been brought to bear upon the only entrance to Magdala.

About 3.30 P.M. two men were seen going from gun to gun on the salient of Fahla—the summit of which rose to the perpendicular height of 1,000 feet above the ground on which Napier and his Staff stood. It was supposed that they were loading them; and a critical survey, made through a field-glass, verified the supposition. Still the group of English gentlemen did nothing but stare, even when they saw that six cannon which they had counted were being got ready.

Supposing a dash was made from the heights upon the group! Their rapiers would have availed them nought. Inevitably they would have been captured—Napier and his gold-epauletted Staff—unless they would ingloriously fly! The 1st Brigade had not yet come up. The hills were very steep; the soldiers were very thirsty and tired.

Suppose an ambuscade among the underbrush on the slope to their right, which, indeed, was well adapted for such a sinister purpose! What would have become of the General-in-Chief, Sir Charles Staveley, General Milby, and their respective staffs? Certain capture, so long as they were armed with dress swords only!

Was there anything to prevent such a catastrophe? No! certainly not! The body of the column was still half a mile behind, clambering, almost in despair, the rugged steeps!

The grove and underbrush had not been examined by anybody. There was no telling what might happen. No keen-eyed and cautious scouts had gone over the ground. Those numberless little gullies, covered with tangled bushes, had not been penetrated. The Sappers and Miners, fatigued with hard climbing, had as soon as they reached the top of the plain below Fahla thrown themselves on the ground inert and apparently lifeless.

With the exception of the Sappers and Miners, Sir Robert Napier and other chiefs of the crusade had no help at hand. And, 'Just God!' the enemy were seen pouring down Fahla slope *en masse*; and a pearly wreath of smoke, a thundering report, and a chain-shot, shrieked the Emperor's defiance!

In quick succession flashed the fire-flames from the rude mouths of his cannon—in quick succession rolled the white smoke—in quick succession a series of hideous wailing sounds was heard in the air; indubitably, Theodore was in earnest!

His sullen, defiant silence so long kept in reserve had burst its bonds, and lo! the concentrated thunders of his determination shook the heavy mountains round about, and the little upper world around Magdala gave tongue now for action!

'Away, you sir,' Sir Robert commanded in sharp tones; 'bring up the "King's Own" on the double quick; and you, sir,' to another aide-de-camp, 'order the Naval Brigade here instantly; and you, Sir Charles Staveley, let the Punjaubees deploy across that narrow plateau in front, but do not fire until the enemy are within two hundred yards of you.'

'Ah, is that you, Colonel Phayre?' Colonel Phayre had now come up from the hill at the head of Aroje. 'Have you examined those ravines?' Colonel Phayre responded indistinctly.

Nearer and nearer was the advent of the enemy, 3,500 strong. They all appeared confident of the issue. Their war songs came pealing towards us. We could see their cavalry caracoling and bounding joyously along ; the foot soldiers leaping and brandishing long spears and swinging their black shields. With loud chorus all sang the death-doom of the invader.

Onward, still onward they came, horsemen and foot soldiers vieing with each other. They flung their flowing symas, their bezans, and many flung their loin clouts away, and with lances and shields in rest they bore down the hill, reached the plateau, and inundated it with their dusky bodies. A clear open plain was before them, over which they rolled like a huge wave !

Where was the opposing force to sunder and dash this furious wave backward ? Sir Robert Napier sat on his charger serene and impassible, surrounded by a group of men who were not nearly so unaffected as their chief. There was but a thin line of men ; and in these confidence could not be placed. Each and all might be brave, but had there been six companies of the Irish regiment with their utter *abandon* and quick scientific manipulation of cartridge and bayonet, there was room to hope a check at least until their comrades could come to their relief—but six companies of Bombay and Madras Indians armed with the ancient muzzle-loading Brown Bess ; ' Rather dubious' was everybody's internal comment. A crisis stared the General in the face !

Closer the Abyssinians drew, until we momentarily expected to see them launch their spears, and annihilate the Sappers and Miners.

' Here it is, General; the Naval Brigade has arrived !' said a smartish aide-de camp.

' Very good,' responded Napier ; ' let Captain Fellowes take position on that little knoll in front.'

Not a minute too soon did the little band of sailors appear on the scene. Quick as lightning and prompt as powder are sailors when they hear the well-known voice of their commander ! No useless time was wasted here !

' Action, Front !' shouted the naval captain. ' Action, Front !' repeated the lieutenant and boatswain ; and hardly had the words died away from their lips before the sailors had unstrapped rocket tubes and carriages, and had them arrayed on the knolls; muleteers in the rear with their animals; rocket carriers with their ammunition; rocket men ready with their pry poles.

'Fire!' and, even in the act of launching their spears, a stream of fire darted along the enemy's ranks, ploughing its fiery way through their swaying masses. Another, and another, rushed through them; and cheer after cheer issued from the lips of the sailors and marines.

The battle had begun! The cheers of the Naval Brigade were echoed fiercely behind; and as the General turned his head, he saw the 'King's Own' coming up on the double quick, with cartridge-boxes rattling on their hips, and men's fingers manipulating cartridges and fixing their Sniders for the strife as they ran.

When they first heard the loud boom of Theodore's cannon, these soldiers were a mile to the rear, reclining on the slope— peevish and fretful—disposed to curse themselves and the world. But no sooner had the sound of cannon vibrated on their ears, than with a bound they sprang to their feet, expectant as a hare when listening to danger. An aide-de-camp galloped up, and delivered his orders. The effect was magical! The limping gait was dropped, and the woe-begone features became lit up with an intense pleasurable excitement. In an instant of time they were on the double quick like hounds in full cry. There were only 300 of the 'King's Own' together; the others of the regiment were on duty as baggage guards, but they were pure Britons!

The enemy, from very astonishment at the novel sound caused by the rockets, halted and cast inquiring glances at each other, as if to ask 'What manner of things are these?' but urged by their chiefs they made another desperate essay to advance.

A low ridge of ground rising but a few feet above the narrow plain, and a hollow, divided the enemy from the Rocket Battery. The Sappers and Miners had been withdrawn for the support of its flanks, and thus a clear space was kept for the rocket guns to do good execution, and incessantly they vomited their fiery darts at the enemy, now but fifty paces from the battery.

Without pause or hesitation the 'King's Own' kept on

their way, forming line the while. Into the little hollow in front of the battery they shook themselves, with their arms. Into, and through the scrub oak, and underbrush, and in a second almost, the head of the 4th Foot crested the slope and confronted the enemy, a few of whom were on the rise on the other side.

'Commence firing from both flanks,' rang out clear as a silver bell from Colonel Cameron ; and, instantaneously, two quick volleys of musketry were flashed in the faces of the dusky foe, and like a stream of fire volleys ran from side to side without a pause, raining such a storm of leaden hail, that for the second time the enemy halted from sheer astonishment. It was as if they were paralysed at the very moment they intended to launch out their spears, and one could almost fancy that these weapons vibrated in their hands, from the impetus they were about to give them. Slowly they seemed to regain consciousness, and horrified, they gazed upon the awful result. Strangest sight was it to them, who had ever been victorious in the field of battle, to see their own men tumble by the dozen, by scores, by fifties, into the embrace of death.

'Retreat!' cried the chiefs. The enemy did retreat, but not fast enough. They broke out *en tiralleur*, and endeavoured to take vantage of boulders to escape the whizzing bullets; but the bullets found them out, searched out each bush and mound and rock, and stretched the men behind dead upon the ground.

Here was one running for dear life for a copse; but suddenly you saw him leap into the air and fall on his face, clutching the ground savagely. Here was another one with head bent low, in the vain thought that if his head escaped he would be safe, making all haste to get into a hollow, out of reach of the leaden storm ; but even as the haven dawned upon his frenzied eyes, a whirring pellet caught him, and sent him rolling down the incline. There was another one, just about to dodge behind a massive boulder, from where he could take slight revenge, but before he could ensconce himself, the unerring ball went crashing through his brain; and there was another one about to plunge in hot haste down a ravine to the left, who had his skull shattered by a rocket, and with a dull sound the body fell down the precipice.

E E

Some chiefs there were who turned round to take a parting shot, and some who, not entirely panic-struck, strove to re-form the natives. They were partly successful, and under their leadership a thousand of them precipitated themselves down the steep sides of the narrow plain, and seemed determined to capture Penn's Battery, still isolated on their little knoll below Selasse.

Leaving the ' King's Own,' the Sappers and Miners, and the Rocket Battery, to work their will upon the few left on the plain, I took advantage of the lull of the skirmish to hasten down along the road towards gallant Colonel Penn, to see how that officer was prepared to meet the impetuous onset of the Abyssinians.

Not so much confused as they appeared to be were the men of the Battery. The guns of the Mountain Train Artillery were ranged in a semicircle, and when the enemy were within 500 yards of them, Colonel Penn smilingly gave the word to ' Fire.' A sharp yelp-like report, and six shells flew through the clear air with a strange diapason, and directly burst among the advancing masses. Another astonished pause !

Here were other strange things for the Abyssinians. Music like that of distant harps ; while missiles were tearing and rending men to pieces. Simultaneously the enemy, instead of advancing, turned their faces upward to listen to the novel music, and seek the solution of what was a dark enigma to them ; and while they were halting, and listening, and dubitating, they heard a concatenation of sharp cracks above their heads, and immediately afterwards a thousand pieces of iron were flying amongst them, laying whole groups of them level with the ground.

They ascertained at last that the horrible hubbub proceeded from the little knot of men on the knoll. ' Forward ! ' yelled a bull-hearted chief, Dajatch Deris, using his spear freely among the most craven. Coerced into activity by gesture and example, they leaped downward like tigers, mad rage in each heart, up and across knolls and curves, and down again into a ravine choked with wild olive and tamarisk, until they were at the base of the hill whereon the Battery was posted.

When first this demonstration on the part of the enemy was observed, the Punjaub Pioneers, who had come up soon after the first volleys were fired by the ' King's Own,' were ordered to its support, and just as the Abyssinians were coming up towards the battery, they showed themselves to the enemy on each flank of the guns.

'Commence firing' was the command; and again rattling volleys were discharged in the faces of the sorely harassed natives—who had almost made sure of capturing the cannon—dashing them backward, and downward into the bottom of the ravine many times quicker than their advance. A moment's consultation was enough to convince them that here there was no hope. Against shell-vomiting cannon, and against a very wall of fire, discharging bullets by the hundred to their one, what could matchlocks and spears effect? So, in spite of strong-lunged Dajatch Deris, the Abyssinians retired precipitately, firing as they went, and sad to relate, the poor Dajatch,[1] before he reached the bottom, felt a sharp twinge, and incontinently tripped and rolled helpless down the slope, with a broken leg.

Round the base of the battery knoll the ravine ran a serpentine course, emptying itself a few hundred yards below into the Aroje. It was overgrown with tangled brake and dense jungle. Along this ravine the baffled enemy crawled.

'They are going after our baggage, sir,' said an expostulating voice.

'Ah, are they indeed? so they are!' said Colonel Penn after an examination.

'Right about! Left oblique! Forward, march!' were the sharp, firm, composed orders given at once, and understood. The Punjaubees went sweeping across the knoll in an oblique direction towards the brow overlooking the Aroje.

Up the Aroje were advancing long trains of baggage, ammunition, and commissariat stores; pell-mell; in confused masses and in straggling lines. Our involuntary thought at the sight was, 'What if Theodore had only 200 men with muskets in their hands, on some of those hills overlooking the Aroje valley!'

Warned by the thousand-fold echoes, that the embosoming hills flung far and near, Captain Roberts, who was at the time commander of the baggage-guard, mustered a few companies of the 'Duke of Wellington's Own,' and two companies of the 4th Foot, and stationed them at the head of the valley, which—

---

[1] Dajatch is an Abyssinian term for General.

it was supposed a few hours previous—Colonel Phayre and his men were guarding. No sooner had they done so than a confused noise was heard a little above, and presently a large body of men issued out of a narrow gloomy gully; and, as Captain Roberts said, 'By Jove !' the enemy was upon them.

'About face ; fire!' and along the line of soldiers drawn across the Aroje ravine there ran a rattle of muskets; a clicking of triggers ; and a roar of sharp musketry ;—steady, deeptoned, like the thunder rush of an express train through a tunnel. Practised men were at work with the Snider rifle. The latest invented breech-loader was in expert hands—one could tell the difference by the mere sound; there was no break, no pause, no hesitation in it; it was continuously rolling.

The Punjaubees came directly upon the scene, looking down from the summit of the knoll, with their dusky faces as dark as the Ethiops'. They saw the enemy, and again the dreadful word for slaughter was given—a word which will be remembered by Abyssinians, and handed down to their posterity !

Again and again the loud-sounding muskets boomed loudly on the ear, and there was sure 'Death in the front,' and 'certain Destruction in the rear !'

The English had caught the enemy, where, had they known anything about military science, the Abyssinians should have caught the English. The enemy dropped dead on all sides. Their experience here was worse than any they had undergone as yet. Had they stopped ten minutes longer, not a man would have been left alive to tell the tale of the grievous disaster that met them. Here, as elsewhere, they seemed to be too much astonished to fire in return. I did not hear a single Abyssinian musket fired ; they seemed to wish to fight hand to hand, but the rapidity of the Sniders gave them no chance.

They wanted the baggage, it seems, but the Irishmen of the 33rd stood between, and the passion fit of ungovernable fury, which racked the enemy previously, gave way now to the sentiment of dull despair. Some six-hundred—all that seemed left of the thousand—turned swiftly about, when they found that no impression could be made. They dived back into the jungles from whence they came ; keen-eyed riflemen following them up—'potting' the fugitives unerringly.

The Punjaubees, fleet of foot and prompt at command,

swept to their old position near the battery, and deploying along the prolongation of the slope, calmly waited the flying foe to emerge out of the bushes. Not long had they to wait; as the dark forms bounded out of the recesses, the Punjaubees commenced their withering fire upon them once more, descending the slope as they fired. The position in which the Abyssinians now found themselves was a perilous one indeed. It was an open hollow with clear slopes rising abruptly about a hundred feet from the bottom. On one side were the Punjaubees, 600 in number—up the opposite side some fifty paces across scrambled the Abyssinians, with the main desire now simply to get away as quickly as possible from the dangerous place. How easy to imagine the result of the unequal contest, where slope lined with cool riflemen fronted counter slope clear and open as an artificial glacis!

The fight became a battue—a massacre! Down the slope rushed the Sepoys, with bayonets fixed to their guns, and fresher than the tired natives, they soon came up with them as breathless they panted up the deadly steep. Out of that very despair which the most craven heart feels when hard driven they mustered new courage, and determinedly they turned round, at bay.

The fiery, hot-blooded, impulsive Sikhs came hand to hand with Ethiopian mountaineers, fierce and as impulsive as they. Now came the tug of a genuine contest! The Abyssinians launched their spears, drew their curved shotels, and charged down with loud cries. The Sikhs, undismayed, rushed up to meet them with their bayonets, and deftly crossed weapons with them. Blows were nimbly warded, stroke was met by counter-stroke, and murderous thrusts skilfully parried.

Two companies of the 10th Native Indian Infantry rushed down from the battery knoll to the support of the Sikhs. No mercy was asked; no high-toned sentiment found utterance; no puny blows were dealt; heads were chopped off, arms and limbs severed from trunks, and dead men lay stark and stiff plentifully. But they were all Abyssinians; very few of the Punjaubees were wounded.

From the summits of Fahla and Selasse, enraged Theodore belched his war thunder, and in active reply might be heard the rockets—their magic tails of fire singing shrilly their death errands—the shells with their harp-like music and sequent fatal boom, the cracking, ripping volleys of musketry which mowed men down in heaps.

While the spine and rim of Selasse and Fahla were yet fitfully blazing with fire from Theodore's large guns, I turned away from the battery hill after victory had declared itself in favour of the Punjaubees, and the poor remnants of the thousands were sullenly retiring from before them, to witness the action on the plain below Fahla, which had again burst forth as loudly as ever. I noticed that the sky had become suddenly overcast, murky masses of cloud scudding wildly across it, and that there was every appearance of a tempest being at hand. As I reached the plain, the thunder rattled over the whole wide concave, the lightning lighted up ghastlily the upper world of rocks and mountains, and as it disappeared, left a deep gloom in the valley. Shortly the rain fell in torrents, and a gust of wind sent it pelting in our faces.

However, it was not for long; the thunder of heaven's artillery soon ceased; bright sunblinks pierced the clouds, illumining for a short time the Bechilo valley, until the gorgeous tints of the rainbow were revealed when the sun came in his setting glory to take a parting look and dye with hues of gold the sad scenes visible in this rugged spot.

The 4th Foot during my absence had been reinforced by the Beloochees, 600 in number, and at the time I arrived upon the scene they were about to move closer to Fahla, as some of the enemy were ensconced behind some of the rocky projections of the slope, maintaining a scattering fire upon their ranks, and one or two had been wounded. When they had advanced to within 300 yards of the base of the fortress, they opened fire once more upon every suspicious rock and bush clump until they searched every foot of its vast surface.

Several of the enemy stood boldly forward and used their matchlocks and double-barrelled guns bravely, but to little purpose. Horses and riders were seen rolling down hill from the effects of the British fire. Of little avail were bamboo bushes, or tamarisk and kantuffa shrubs, to stay the searching bullet.

Three hundred Snider rifles, six hundred Enfields, and a dozen rocket guns plied fearfully rapid, raked the whole length and breadth of the grim steep with their murderous fire; even upon the very height of Fahla itself the rockets did considerable damage, destroying several of the cannoneers.

Theodore narrowly escaped death from one of these missiles, as he was superintending the gunners on Fahla.

The Abyssinian musketry was pretty loud and brisk at this time, and was readily distinguishable.

Amid his gay staff, Sir Robert Napier's snowy sun-jacket was seen. Riding near to the group, I found him bending over the pommel of his saddle encouraging the soldiers. His usual placid smile had returned. Taking up his field-glass, he noted carefully the most prominent points of interest. The obstinate native combatants still firing away from behind but-ments of rock and mounds of earth; Battery Hill and the numerous ravines lay between the plain and it. The defeated enemy, sullenly retiring from before the Punjaubees, attracted his attention, and he saw groups of them still hesitating as if undecided whether to retreat up Fahla or make another attempt at the Battery. Shutting his glass, he beckoned to him an aide-de-camp, and said, 'Ride over to Colonel Penn, and tell him to play his Battery well upon those fellows below.'

Upon receipt of the order Colonel Penn directed his 'Steel Pens' (as they were facetiously termed) upon the enemy, and his ricochetting balls quickly dissipated the last remnant of hope that the Abyssinians might have entertained of capturing his Battery.

It was 5.30 P.M. Theodore's guns had ceased their fire; because, as he said, 'the English are not afraid of my chain shot; they march up in spite of my big balls;' and because he could not depress his guns enough to bear upon the soldiers as they were directly under him, 1,000 feet below; besides he was so furious (so we learned subsequently) at the terrible overwhelming disaster that he met, that he could not reason or decide upon anything.

Quietness was restored as the twilight deepened. The death storm abated after lasting two hours. Slowly we returned to our camps, or rather where we intended to make our camps, to Battery Knoll, and the patch of table-land beyond the plain.

We stumbled over the thickened dead, who, lying stripped, were of the colour of the soil. Among the thickets the slain warriors, enrobed in their cotton togas, appeared like white splotches upon the dim and indistinct ground, and in crossing the sinuosities of the ground we were often startled by the spectral gleam of the linen and silken kirtles with which many of the dead were enwrapped.

We reached our positions and found that the tents had not come up ; no provisions had arrived, save those belonging to the staff. But the camp followers had arrived by the hundreds, and were now swearing discordant symphonies of wicked invectives and uproarious jabble. Hungry mules brayed most dolefully, and horses neighed impatient for their rations of grain, and there was neither grain for animals nor provisions for men. The tired soldiers came and questioned one another eagerly about food. They were answered in the negative. They would have to wait until midnight, they were told ; their tents would not come until morning. Then there rose a fierce murmur of discontent, not loud but very deep.

'There had been no obstacles,' said they, ' to prevent them from coming, and why did they not come ? ' Why, oh people, are blunders committed everywhere ? why is it that there is such a thing as a blunder ? why can't all the world be right at all times ?

As to the officers, nature's cordial abhorrence of a vacuum was mitigated by a supply, which their servants brought in their haversacks, specially for their consumption. When stomachs are filled and appetites appeased, people are disposed then, if ever, to be loquacious, and so was it here. A social circle was formed round each camp-fire, and the novel battle and its scenes were hotly discussed.

About midnight most of the baggage was up, but the tents could not be pitched, and the only resource left for us was to bivouac close to the smoky and uncomfortable fires.

Sentries were posted plentifully around camp, and soldiers slept with their arms ready for immediate use.

Before rolling ourselves up in our rugs, and while thinking of the events that marked the day, our ears caught the sounds that betokened the presence of the beasts of prey. In ravenous packs, the jackals and hyænas had come to devour the abundant feast spread out by the ruthless hand of war.

Stranger lullaby than that which lulled us to sleep that night man never heard, and the last sounds our dulled ears caught were the jackal's shrill whelp, the hyæna's sonorous bay, mingling with the lichowl's mournful ' tu-whit-tu-whoo!'

CHAPTER IX.

CASUALTIES AFTER THE BATTLE—A WELL-DRESSED CAPTIVE—THEODORE IN
DESPAIR—RASSAM—STRANGE RECEPTION OF THE CAPTIVES—THE FINAL
SUMMONS TO SURRENDER.

ABOVE the stormy field of yesterday; above the strange valley
with its dead and dying; above the English camp; and above
the rocky citadels, the sun shone with its usual tropic warmth
and brightness, the morning after that fatal Good Friday.

The clock-bird loudly sang in the groves, the swallow
chirped merrily in the gladsome light, and the lark soared high
in the diaphonous air of those high altitudes trilling the
sweetest notes, which strikingly contrasted with the hoarse cries
of the owls that lulled us to dreams. Peaceful heights! happy
valley, was it not?

In the camp the soldiers lolled about as usual, and interchanged their salutations and petty gossip.

'Chokra idher ao,'[1] I heard General Napier say to a Hindostanee servant. 'Compliments to Surgeon Sahib, and tell him I
should like to see him.'

'Surgeon Sahib come,' reported the chokra.

'Ah, all right. Good morning, Doctor. Beautiful day, is it
not? Have you your report ready?'

'Yes, Sir Robert,' answered the surgeon, drawing a paper
out from his pocket. 'Here it is, sir; one officer—Captain
Roberts—and thirty-one privates wounded; Captain Roberts
severely, eight privates severely also.'

Curiosity satisfied on this point, we congratulated mutually.
'Only think of it, my dear old fellow; not one man killed, and
but thirty-two wounded!' Let us now ascertain the casualties
on the other side.

Captain Sweeny, of the 4th Foot, was sent with a detachment

[1] 'Boy, come here.'

to bring in to the camp hospital all wounded Abyssinians that might be found on the battle-ground. I went with him. A frightful scene was presented to our eyes. The significant sounds of the past night had prepared me in a measure for some horrors, but reality exceeded my conception tenfold. The beasts of prey had been at work upon the bodies. They had revelled in the unusual abundance of flesh. It is said there never was a famine, that plenty for carnivorous animals reigned round Magdala always, that these useful scavengers saved the country from a pestilence. But the night following the battle of Aroje there was a rare feast for panthers, jackals, and hyænas.

Carefully the dead were counted and buried where they fell, and most carefully the wounded fellows, who had lain in torture and pain the livelong night and had feebly warded off savage jaws, were conveyed in dhoolies and hammocks to the hospital, and after three hours of such work the following report was handed in at head-quarters :—

' Seventy-five wounded Abyssinians carried to hospital, 560 dead of the enemy buried by detailed party.' Dajatch Deris, the stout chief who had been so persistent in his attacks, was carried to the hospital with his leg broken by a bullet, just above the knee. The body of Ras Ourary Guvrie, supposed to have been shot by Captain Speedy, was sent to the fortress during the night by General Napier that his own relations might bury him, and that Theodore might see the end of one of the best of his counsellors.'

After breakfast I strolled to where the banner of St. George proudly waved above Sir Robert Napier's tent.

' What is in the wind, captain ? ' I asked Captain Holland, the Assistant Quarter-Master-General.

' Nothing but rumours and flighty twaddle flying loose around camp. I know nothing myself; but I believe Sir Robert will send a demand for a surrender to Theodore, and after that I dare say he will move upon the enemy,' answered Captain Holland readily enough.

Captain Holland guessed rightly. Sir Robert Napier did intend to send a peremptory demand to Theodore, and decide what was best to be done after the receipt of his answer. What had been accomplished had differed *in toto* from what he had conceived would happen, and he was not going to act hastily.

If Theodore was frightened by the exhibition of his prowess yesterday, then he was in a fit mood to listen to proposals. 'Agitel-el-Shaitan'—'hurry is the devil'—is the Arab proverb.

Upon Napier's forbearance depended the lives of the captives, and the signal success of the campaign. A precipitate movement upon either Fahla, Selasse, or Magdala would have deprived the desperate Emperor of every ray of hope that his present tristful mind might have clung to, which, no doubt, would have caused him to order the immediate execution of the captives. Besides, it were perhaps easy for him to deport himself to the inaccessible mountains of Kuara, hundreds of miles north-westward of Magdala, in which case the objects and aims of the crusade would have been irretrievably nullified.

Having ended these few remarks, let us trace the bearer of the surrender summons, sent by Sir Robert. The key of all doubts and surmises must be held by that messenger.

Upon arriving half way up the road that led to the gorge between Fahla and Selasse, the messsenger to his great surprise discovered two strangers coming down the hill, who in a short time proved to be L—— P—— and the Reverend Mr. Flad—missionary—who were on their way to Sir Robert, with a message from the Emperor. The messenger from the camp returned without performing his errand with the prisoners.

The camp, at the sight of the strangers, was instantly in a state of commotion; the social coteries dissolved, the soldiers hurried in tumultuous crowds; but suddenly, the whole relapsed into astonished silence at the apparition of a cocked chapeau, new red uniform, and a gold-rimmed eye-glass and cane.

'Who is he?' said one.

'Where the deuce is he from?' said another.

'Egad, this fellah must have dropped from the clouds.'

P—— was indeed apparelled grandly—a very military Beau Brummel.

But it was to pay proper respect to the illustrious General that he dressed himself so! Was it very wrong then, that when Mr. Rassam notified to him that he must go to the English camp and see the Chief, that he bedecked himself with all the bravery and paraphernalia of a British officer; that he shaved himself carefully; that he anointed his head and whiskers with the very last drops of perfumery and Rowland's macassar that he had in his dressing-case; that he blacked his boots until they shone like a mirror; polished his spurs; brushed up his uniform until it was spotless; cleaned his eyeglass so long unused in his trunk; and, to cap the *tout ensemble*, unveiled his chapeau, and, after furnishing himself with a dainty cane, without which no British officer's outfit is complete, sallied forth to see him in whose hands his life was?

With elegant *insouciance* and eye-glass fixed, young P—— sauntered up to the General-in-Chief's tent twirling his cane, when he was met by the General himself, who had already been apprised of the captives' coming.

L—— made a bow and introduced himself as Mr. P—— of the English Mission to Theodore, and his *compagnon de prison* as Mr. Flad. It was evident that during his incarceration he had not forgotten the rules of propriety.

In the presence of the General an evanescent bashfulness was apparent, but it soon disappeared; Napier's proverbial suavity being a wonderful panacea against such youthful indisposition. He was invited inside the tent, and after a few preliminary observations about the weather—which were perhaps entirely unnecessary—the object of the visit was broached.

Began L——: 'I have been sent down to you, Sir Robert, by the Emperor, to ask why it is you have come to this country? What it is you want? Whether you will return to your own country if the captives are released?'

'Thank you,' said Sir Robert quietly; 'my message you shall carry with you on your return; but tell me of the result upon Theodore of yesterday's battle.'

Then L—— P—— told of how Theodore had seen the gigantic elephants, with their ponderous loads upon their backs, coming down from Dahonte Dalanta, besides a long picturesque line of men in red, men in white, men in all colours of dress, followed by an interminable line of animals, winning admiration from him as he gazed upon the proud martial scene, causing him to exclaim that the great wish of a lifetime was about to be gratified. He was proudly elated at the thought that he should see how soldiers, real soldiers, would conduct themselves in battle.

When the General and his staff had appeared upon the

plateau below, he was instantly beset by his chiefs and warriors, who persuaded him to let them go and fight the Feringhees. After hesitating a long time he had seen the Battery of Mountain Train Artillery coming up, and thinking the ammunition-boxes on the mules contained the specie, he said : ' Go, my children ; capture the treasure, and drive the Feringhees away.' He had intended to lead them himself, but was dissuaded by his faithful chiefs.

He then had made the German engineers and gun-manufacturers in his employ fire the first guns. One of the big cannon had burst at the first discharge, killing a great many of his men who had crowded around it. The Germans, unwilling cannoneers, fired widely from the mark, which enraged him so much that he pointed the guns himself. But when he saw that English soldiers advanced upward steadily in spite of his chain-shot and 200-pound balls, he fell to weeping violently, gnashing his teeth and stamping upon the ground ; and his rage was increased when he saw his best army melt away before the withering volleys of the English.

He then began to threaten that the captives should pay for his defeat with their heads. When night came and the battle had ended, he took to drinking arachi to drown his bitter agony of spirit.

Oblivion would not come, as his sufferings of mind were so acute. He had thrice attempted suicide, but his adherents kept good watch and prevented him. He moaned in his deep distress, calling for his favourite Generals ; for Ras Feet Ourary Guvrie, Dajatch Deris, Dajatch Workee ; but not a word could his attendants answer, contenting themselves with gazing upon the wreck of the once mighty ruler of Ethiopia.

Then he would querulously say that he had an army of brave men once, but had only women now, and a paroxysm of hysterics would seize upon him as he thought of the fatal battle ; and much more to the same effect did Messrs. Flad and P—— repeat, to the wonder of Sir Robert.

Well might he pause, with starting eyeballs, to listen to the agonising shrieks of widows bereft of husbands.

' Stop those women from crying ! ' he yelled furiously.

So crushing the defeat, so sudden and overwhelming was it, that he could barely realise its extent or depth.

'Go and kill those Feringhee captives, the cause of this ruin!' he shouted to his soldiers.

'No,' whispered a grey-headed counsellor, 'Preserve them. Send two of them down to the English general, and find out what he has to say.'

'Ah! yes,' said he musingly, 'that is best. Send Sooltan Rassam to me.'

This took place about midnight of that fatal day of Aroje, the night of Good Friday. Rassam, as I have said in the Introduction, was one well fitted to soothe a monarch's soul.

Mr. Rassam appeared on the scene and made his low reverential obeisance before his sable majesty. Humble were his salaams. Sincere was the hope he expressed that the 'Illustrious Potentate,' 'The Ruler of the Universe,' 'Cherisher of the People,' 'The Shadow of the Omnipotent,' 'Defender of Christendom,' 'King of Kings,' and 'Just one of the Earth;' would live for ever.

Repressing all traces of the emotions that filled his being but lately, with kingly courtesy Theodore rose, and proffered him a seat on his own silken couch, besides native wine, arachi, and tej. With prodigious suavity and wonderful heartiness, Rassam drank the Emperor's health.

After complimenting him upon the bravery of his (adopted) countrymen, of which he was a representative, Theodore proposed the question to him, 'What shall I do?'

Various things Rassam urged upon him, but the Emperor could not make up his mind to do any of them, but he thought it would be best to send two of the most trustworthy of the captives to the Feringhee Chief, and ascertain his intentions.

Rassam came to think so also, and suggested that L—— P—— and Mr. Flad, as suitable and trustworthy men he knew.

'Very well,' answered Theodore, his mind greatly at ease. 'Tell them to prepare themselves to convey my compliments and my message to the General.'

A fraternal embrace took place between the two, and many loveable phrases were interchanged, before they parted for the night.

So much I learned from the other captives when they came to camp.

L—— P—— and Missionary Flad lunched with Sir Robert

Napier, during which most of the above story was repeated to the General and the few of his staff who lunched with him. Never meal was dispatched with more infinite relish than that was by the two captives.

After P—— had communicated with the officers of the army, both envoys were instructed with the message they were to convey to Theodore. 'Tell him from me,' said Sir Robert, 'that I require an instant surrender of the prisoners with their property, of himself and the fortresses of Selasse, Fahla, Magdala and all therein. He may rest assured that honourable treatment shall be accorded him.'

About 3 P.M. the two captives before-mentioned returned from the presence of Theodore (who had been infuriated for a short time at the answer they had brought him from Napier), with the prayer that the English commander should give him better terms, as he was a king and he could not surrender himself to any chief who served a woman. 'Rather,' stated Theodore in his letter, 'than surrender, I would fight to the death. Can you not be satisfied with the possession of those you came for, and leave me alone in peace?'

They were sent back again, on their return to the fortress, about five o'clock, with the same reply as before. 'You must surrender yourself unconditionally to the Queen of England. Be assured that honourable treatment shall be accorded to you.'

The whole English army looked upon them with compassionate tenderness, anticipating for them the very worst of calamities that Theodore's vengeful rage would pour on their heads.

When P—— and Flad had commenced their return to camp with the second prayer for favourable terms, Rassam's greatness of soul shone out through the mist of doubt and uncertainty which environed him.

'Trust to me, your Majesty. Grant that all the captives go free to the English camp, and I will guarantee that the English chief shall return to his own country.'

My candid opinion is that Mr. Rassam in his heart of hearts believed in a great measure what he said, and unreasonably assumed that he, and he alone, was the great object of the expedition; the release of his contemporaries and the punishment of the Emperor being matters of secondary importance. Suffice it that his Majesty believed in Rassam, and trusted to his postulated influence and persuasive powers to reconcile him with the Commander-in-Chief of the Crusade, and gave him orders to assemble immediately all the European captives, with every article of property that belonged to them, at the Thafurbar Gate of the main fortress, and when all would be ready, he would be there to bid them good-bye.

In an incredible short space of time the captives had grouped themselves *en masse* at the gate with their servants and their servants' servants, and their slaves.

Strange hubbub the piebald group kept up.   White and black faces commingled together, and a very Babel of languages was used in this confusion, consequent upon their intense hurry to get away.   There was no order in the mass.   They had been told by Rassam that they were to get ready within half-an-hour to leave Magdala, with its horrible memories and bloody associations for ever!   At once danced before their vision—HOME, no matter where it lay!

They had not long to wait.   Theodore arrived at the gate, clad in his proudest robes, surrounded by his principal men. All the captives salaamed to him except one.   His Majesty then gave the orders to open the gate, and the prisoners defiled before him, bowing low as they passed him.

'Adieu Cameron.' Theodore addressed the Consul.   'We part friends, I hope?'

It was on the Consul's tongue to return cordially the parting salute; the knowledge of the man's coming doom awakened a kind of repentant pity in him for the fallen potentate; but the word could not issue smoothly.   Like a phantom rose before him the wasted years of his life; he remembered his former condition, enwrapped in a dingy gray blanket chained firmly to the wall; the fiendish tortures practised upon him; the cutting whip; the fate which impended over him for so long; and oh, the live-long days of mental agony and utter misery which he had passed as he witnessed the days flitting by, one by one, an interminable round of punishments; apparently uncared for, by his vacillating and unready government.   What wonder, then, that the spirit of the Saxon shrank from forgiving the wretched tyrant.   But his position was still hazardous in the

extreme, and policy compelled him to acknowledge the conciliatory farewell to Theodore.

'Adieu, your Majesty;' and statelily salaaming, Consul
Cameron passed on.

'Fare you well, Mr. Stern. Forgive me for what I did to
you.'

Whether it was the Christian spirit that urged him to forgive, or whether it was the deadly fear he entertained of Theodore that caused him to respond to the salutation is not
known; sure fact is it, that Mr. Stern salaamed low, saying,
'Farewell, your Majesty.'

'Ah, Kerens, is that you? Good-bye; I shall not see you
any more.'

Young Kerens, with his fresh youthful impulses, with all
his sufferings still strong in his mind, went up to Theodore,
and clasping his hand, bent his head in obeisance, and said,
'Good-bye, your Majesty; I wish you well.'

'Ah, Sooltan Rassam, Good-bye. I depend upon you to
reconcile me with the English Chief. Be sure to come to me
to-morrow with good news.'

'Rest assured, your Majesty. I shall be certain to come
to-morrow with good news. A good night to you, my Emperor;' saying which, he walked through the gate, passed the
barriers, passed the lowest barbican, and then breathed freer
and healthier, he and the rest of them, from Magdala along
Islangi, rounding Selasse, close to Fahla, and then in a long
deflux until the twinkling lights of the British camp awoke
fondest hopes within them, and in hearty unison rose on the
clear night air the thanks from their relieved hearts.

Instantly the news flashed through the army, and soldiers
hurried to head-quarters by the hundreds to wait for the released captives.

About 7 P.M. the first of the little column arrived at the
camp, and halted before the Commander-in-Chief's tent.

A guard of soldiers with bayonets bore back the curious
crowds, to allow space for the caravan within the circle. As
they came in there was a faint attempt at cheering heard, but
it died away in an indistinct murmur, as soon as he that
started it found no one to second him.

There were women and children amongst them, and when

the thin treble voice of a child was first heard, instead of cheering enthusiastically the great burly soldiers laughed.   Yes, actually laughed!

Assuredly there must have been something wrong.   This could not be the reception the captives anticipated before being released; this did not appear as the Salem of their hopes!

Neither was it, I confess, what I had limned. I had imagined that the whole army as one man would have lifted up their voices at the sight in loud and glorious huzzahs, amid loud-sounding brass and tinkling cymbals; that everybody would appear delirious with joy; that the captives would weep copious tears, and rush simultaneously towards Sir Robert's knees to hug them and kiss his hands in a perfect transport of thankfulness; and that Sir Robert, catching the spirit of the moment, would have delivered a lecture or a sermon upon a beneficent Providence; and afterwards with uplifted hands and upturned face would call down the blessings of God upon the liberated and the liberators.   English people have done such things; but it is passing strange they did not do it here!

The officers produced their eye-glasses, and one gazing upon the scene might have witnessed nine-and-ninety lustrous reflectors arranged upon nine-and-ninety eyes, belonging to nine-and-ninety different gentlemen, busily scanning the appearance of the captives, critically viewing their clothes, making dead sets at their buttons, coolly commenting upon the appearance of the women and children, and last of all, prosaically reckoning up how old the ' *ladies* ' were.

Generally the captives looked in much better condition than we had been led to expect.   Save three, all looked well; these were young Kerens, McElvey, an Irish boy servant of Cameron, and a middle-aged man named Schrieber, one of Theodore's engineers.   These latter looked very pale, and were clothed like Romans, in lengthy togas, and straw hats of their own manufacture à-la-Crusoe.   McElvey, the Irish boy, was exceedingly vivacious and merry.

In the hurry of departure Mrs. Flad had been left behind by Mr. Rassam, she being slightly ailing in health.

Mr. Flad had to go back to Magdala in the dead of night to get his wife and children, for he and P—— met the

captives coming down Fahla slope, as they were the second
time returning to Magdala.

These were the captives—sixty-one the total number of
them—for whom the Crusade had been undertaken ; these were
those whose graphic letters had drawn tears from Christendom.

Their accounts had led us all to believe they were about
to die; that they were flogged every day ; that they were for
ever in manacles ; naked ; a mass of sores and corruption ; and
yet here they were healthy, well-clad people, four of whom
sported those everlasting quizzing-glasses ! Extraordinary !

The second day dawned, after the pyrotechnical exercises of
Good Friday which proved fatal to so many of the Abyssinians.

The captives were all, with the exception of Mrs. Flad and
Bardel, who were ill, in our camp. Their black and white and
red tents outnumbered the army's. According to the list,
there should have been sixty-one persons ; there were forty-eight
tents !

With these captives there were 187 servants and 323
animals. Mrs. Flad arrived safely by 9 A.M. that morning.
The only one left in Magdala was A. Bardel, Artist and Professor, who was sick of a fever. It will be remembered that
this man was the one who was accused of translating to Theodore the letters and papers of the Missionaries, causing them
to be flogged for writing defamatory articles about the
Emperor. I have stated that there were sixty-one to be released, number sixty-two was Theodora Morritz, born at 10
o'clock the next morning after their arrival in the English
camp, and was named Theodore, after the Emperor.

The astonishment of Political Secretary Tweedy was
very great when he read the list of captives and followers, and
still greater when he found that about one-third of the
servants were Abyssinian women.

Such facts as these caused a revulsion of feeling amongst
all who heard of them. Many other interesting facts came to
light which proved that too much credence must not be placed
in those agonising letters written by some of the captives.
When they had beef, bread, honey, milk, houses, gardens, wine
&c., we wondered what more they wanted ?

Since we had taken position before Magdala, no water
could be found nearer than two miles ; but on the second day

that supply failed us, and the road thither was so beset by the thieving Gallas that mules were constantly being stolen from the muleteers by those unscrupulous natives. When it was perceived that the supply of water in the neighbourhood was totally inadequate for the large number of animals attached to the expedition, the whole 10,000 were taken to the Bechilo River, about seven miles off. 500 bheesties (bullocks for carrying water) were constantly employed in carrying water from the river for the use of the army.

Of such luxuries as grog, cigars, tobacco, sugar, tea, coffee, rice, &c., we had none, as all had been left at our camp near Lat.

It was mid-day when 1,000 beeves and 500 sheep were sent by Theodore to Sir Robert Napier, with the hope that, as this day was Easter Sunday, the British soldiers should eat their fill, for were we not all Christians?

As soon as Sir Robert heard this message he sent an officer up to Magdala, to say that he could not think of accepting anything from His Majesty, until himself, his family, and the fortresses, were surrendered to the Queen of England.

In the afternoon more escalading ladders were made out of the dhoolie poles, the rungs being pickaxe handles. Powder charges, hand grenades, &c., were also prepared ready for use.

The elephants, forty-two in number, having brought up the Armstrong battery to the camp, conveyed the guns in the evening to the ridge which divided our camp from the plain below Fahla. The mules were brought up from the Bechilo with two days' rations of grain, so that all would be in readiness for the assault upon Magdala, which was expected to take place at noon of the next day.

Before going to sleep, Sir Robert Napier, unwilling to be harsh when there was the least chance for an amicable settlement, despatched a native courier with the final summons to surrender. He explained in his letter how useless and vain was resistance, and promised Theodore that he would be well treated if he delivered himself up, winding up with his declaration that if an affirmative answer was not received by him by 9 o'clock next morning, he would move forward to the attack.

CHAPTER X.

RESPLENDENTLY glowed the proud summits of Fahla and Selasse
in the sunshine on Easter Monday morning, April 13, 1868.

Directly the *réveille* was sounded, with bugle blare and
drum beat, teeming life sprang forth. Soldiers shook themselves outside their tents, with many a sleepy yawn and growl,
mingled with jest and laugh. The respective companies
mustered themselves before their quarters ; the roll was called,
names were answered to ; ranks were then broken, and the
soldiers retired to breakfast.

Down the slope of Fahla, about the hour of eight in the
morning, came eight Abyssinians apparelled in all the frippery
of chiefs of high rank—silken boddices and brocade gowns,
with hair variously gotten up, according to taste, in plaits,
curls, or frizzles, necks heavy with pendent ornaments, breasts
glittering with silver decorations. They advanced before Sir
Robert's tent, and were immediately surrounded by a ring of
soldiers, who were summoned to preserve order and keep an
open space clear for the council.

On the English commander presenting himself before them
they salaamed lowly before him, and then announced to him,
through their spokesman Aytoo Samuel (the favourite of the
Emperor), that they came down to deliver Fahla and Selasse
into his hands upon the simple condition that they be allowed
to depart unmolested. Their conditions met with the approval
of Sir Robert, and he accordingly gave solemn promise that he
would protect them until they crossed the Bechilo on their
way home.

On being questioned as to the whereabouts of His Majesty, the chiefs made known the fact that Theodore had departed soon after the receipt of the last summons from Sir Robert Napier, which must have been about midnight. They said that they expected that he had either gone to Gojam or to the camp of the Galla queen, Mastevat, in which latter case, though the Mahomedan queen was his deadly enemy, still, according to Mahomedan custom, his life was perfectly safe while in her camp.

This was a thing that General Napier had never dreamt of; at least that he would be able to get away clear of the cordon of cavalry pickets that he had stationed rearward of Magdala, in the valley of the Melkaschillo, another mountain stream running parallel with the Bechilo. He immediately issued a proclamation that he would pay the sum of 50,000 dollars for Theodore, dead or alive; and messengers were at once despatched with it to the camps of Wagshum Gobozye Mastevat, Mahomedan queen of the Wollo Gallas, and Walkeit, Christian queen of the Asubo Gallas.

Sir Robert next determined upon occupying Fahla and Selasse, and moving upon Magdala, after strengthening himself in the former fortresses, and Sir Charles Staveley was commanded to muster his division on the plain below Fahla.

There was created a buzz of excitement among the soldiers; much talk among the officers; servants and sowars flitted about with astonished faces; specials, electrified at the news of the prospective battle, ran hither and thither rapturous.

The great event upon which the campaign hinged was about to transpire; that for which we had all toiled drew near its accomplishment.

Then there was heard a concatenation of bugle sounds answering to each other as the different regimental bugles repeated the commands, and the grey hills, majestic and vast, flashed them abroad and across to the far away ranges; and then there was long rolling drum-beating, angry trumpet-blaring, and stern, brusque, military orders; and lastly, regiments forming columns of fours debouched past their respective camps towards Fahla plain, preceded by a small army of musicians.

The 33rd Irish Regiment, 'Duke of Wellington's Own,'

now restored to favour by their subsequent good conduct and prompt obedience since leaving Dildee, of awkward memory, marched to the exhilaratory quick step sounds of 'Yankee Doodle.' This regiment was destined to play an important part in the last scene of the trinoctial act of the drama. It was to lead the assaulting column.

This day the regiment knew their conduct was to be watched closely by the Chief, and their walk and manner seemed to say, 'Never mind, we'll show you that fighting and marching are two different things, and that we aspire to the reputation of being the best fighting regiment in the world !'

Following the 33rd Foot went the 4th King's Own, the band playing 'Garry Owen.' Then went the 45th Foot (European), the band playing 'Cheer Boys, Cheer.' After these splendid well-equipped regiments, numbering in the aggregate 1,600 Europeans, went the Sepoy regiments, the Belooches, 800 strong, the Punjaubees, 800 strong, two companies 10th Native Indian Infantry, the Sappers and Miners, followed by Penn's Mountain Train Battery, with two mortars, and the Naval Brigade in charge of the Rocket Battery. The cavalry numbered two companies 3rd Light Native Cavalry. The 3rd Dragoon Guards were sent rearward of Magdala as pickets to prevent the retreat of Theodore.

The road up to the low gorge or connecting ridge between Fahla and Selasse was very steep; probably it inclined one foot in three, but, like all roads that the Emperor had made, it was too steep a gradient for an easy ascent. Presently Sir Robert Napier and staff made their appearance, and I made bold to ascend with them. From a thousand points on the heights the road up Fahla slope was commanded, and all the officers admitted that a few Europeans stationed behind the projecting masses of rock to contest the approach, would have materially changed the aspect of affairs.

So, we defiled by fours past enormous boulders, as we went toiling up; dipped into paths between echoing rock, and emerged sharply in full view of a sloping gradine bossy with toppling stones; round a projecting butment of a hill we wound upward, and then at steady pace tramped along a bit of level road, to fetch up at the base of another steep acclivity; and then up, up, and still up, until we found ourselves full under the noonday sun on the crest of the gorge between Fahla and its sister-fortress of Selasse.

The 3rd Light Cavalry, under Colonel Loch, had been despatched, soon after the chiefs came down to surrender the two fortresses, to take possession of them, and drive the natives out.

When the 33rd Foot arrived upon the heights, six companies were detailed to aid the cavalry. They were marched to the extreme eastern end of Selasse, where they faced right about, and deploying across the height, advanced westward, steadily, with fixed bayonets, driving the natives before them out of the fortress as they came towards the road, on the top of which the English army halted. Captain Speedy was with Colonel Loch riding hither and thither, assuring the chiefs and principal men that they would not be harmed, but that they must leave at once for the Fahla plain. Near the mouth of a pass, through which the natives must travel, a company of soldiers was seen drawn up on each flank of it, for the purpose of disarming them as they emerged out of it.[1] After descending through this pass, the natives appeared upon the terraced ridge where the army was being halted.

At sight of the long lines of soldiers coming up, the natives recoiled in dismay, but reassured, they again advanced and halted on the terrace, until the road should be cleared for them to descend to the plain. Upon being satisfied by their chiefs that the Feringhees were their best friends, they came on the terrace between the army and the scarped sides of Selasse, and raised the welcome Li-li-li-li, which proved immediately evocative of other welcome chants—until nearly thirty thousand people caught the joyous diapason, and wafted it skyward in a grand and exciting chorus.

Men, women, and children, eager to greet the conquerors, rushed downward from Selasse, adding force and power to the mass until it covered the whole area between the twin fortresses !

It was noonday, and the hottest sun that we had experienced rested directly centrally in the azure concave.

So soon as the elephants loaded with the Armstrong battery had surmounted the ridge, at the word of command from a mounted officer, the soldiers urged the mob onward with their

---

[1] It will be seen that Napier's conduct at Magdala was different from that which Sir Garnet Wolseley followed at Coomassie six years later.

bayonets. The road became filled with the hurrying fugitives, then the slopes began to teem with them. In the place of the welcome Li-li-li, rose high and shrill a sorrowful threnody, which thrilled even our souls as we gazed on the departing multitudes.

While these scenes were passing in view of the ridge whereon we halted, Colonel Loch and Captain Speedy were manœuvring and bye-playing at the extremity of Selasse, on the road which encircled that fortress, and hence led to Magdala.

A number of men were seen careering about on the plateau of Islamgee, which connected Selasse with Magdala. They were Abyssinians, mounted on ponies, dressed like chiefs; one was specially noticeable from the gorgeousness of his robes and the majesty of his person. This one rode a white horse very gaily caparisoned. When they beheld the cavalry rounding the angle of Selasse they retired slowly towards Magdala, firing as they went.

Once fairly upon the isthmian plateau of Islamgee, lynx-eyed Speedy discovered that the object of that detachment of Abyssinian cavaliers was to secure a number of cannon and mortars lying at the Selasse end of the isthmus a few score of yards to his right. Coming towards them from the direction of Fahla, were some artillery officers and two special correspondents of the London Press. Colonel Loch's cavalry and this other detachment of officers and specials met each other at the artillery park, or arsenal of the Emperor. The cannon were seized at once, and declared to be the property of Her Majesty Queen Victoria. Most of the guns were of French and English manufacture; others there were evidently made by the Russian *employés* of Theodore.

After retiring to the base of Magdala, a few of the Abyssinian cavaliers made demonstrations as if they were anxious to charge, but were dubious of the result, and contented themselves with galloping about at the Magdala end. Standing by, apparently irresolute, were a hundred native foot-soldiers; and a line of dark heads gloomed darkly along the brow of the famous fortress, and now and then musket shots woke loud echoes.

Suddenly the horsemen made a dash towards Speedy and his companions in charge of the artillery park, and among them figured conspicuously the princely rider of the white horse. After coming within three hundred yards of them, the natives halted; and then Speedy discovered that the gorgeous rider of the white horse was no less a personage than THEODORE, EMPEROR OF ABYSSINIA!

Captain Speedy affirmed that the savage potentate shouted out a challenge: ' Come on; are ye women, that ye hesitate to attack a few warriors?'

This, of course, made all minds easy at once concerning his whereabouts, which was very satisfactory to the army. A messenger was despatched to Sir Robert Napier to apprise him of the fact. No attention was paid to the challenge until he manifested a disposition to make an advance upon them, together with the foot-soldiers and horsemen that were on the plateau. Then some soldiers of the 33rd were signalled to approach and take position, so as to command all paths leading to the valleys on either side of the plateau, and a company of the 33rd, who had ascended Selasse to plant the colours thereon, were invited by signals to come and aid them in the defence of the captured artillery. Six of the guns were also loaded, and a few balls were sent hurtling amongst the natives, who had commenced a desultory firing. This had the desired effect; for they immediately retired behind some low booths, temporarily erected for the accommodation of his people when the Emperor first returned from Debra Tabor. A few more cannon-balls ousted them from their hiding-places there also; and the little force of soldiers and Press gentlemen had the satisfaction of seeing Theodore and his men make a precipitate retreat up to Magdala Amba.

But from his rocky eyrie Theodore kept sputtering his rage with sharp musket-balls at all that came within range of his mountain citadel, and the detachment of the 33rd Infantry were finally teased into murderous activity by the sullen shots from Magdala. An aide-de-camp, however, shortly galloped up with express orders from Sir Charles Staveley to cease firing.

It was the noise of this firing that induced Sir Robert Napier to ride forward to reconnoitre. The 33rd Infantry were ordered to hoist the colours above Selasse, and the 45th to do the same above Fahla. The British flag was hoisted in both citadels at the same time, and though it was a bloodless affair, still the event being honourable, naturally elicited admiring cheers, and so was this sight also evocative of much enthusiasm. Two of Theodore's strongholds had thus fallen.

Crossing over to the southern side of the intermediate ridge, the real magnitude and strength of Magdala became instantly apparent.

Suppose a platform of rock, oval in shape, a mile and a half in length, and from a half to three-quarters of a mile in width, rising 500 feet perpendicularly above a narrow plateau which connected its northern end with Selasse. The rock was Magdala—the plateau was Islamgee. On the western and southern sides Magdala towered above the valley of the Melkaschillo 2,000 feet ; the eastern side rose in three terraces of about 600 feet in height one above another. Its whole summit was covered with houses straw-thatched of a conical shape. The extreme brow of the fortress was defended by a stone wall, on the top of which a hurdle revetment was planted ; but the side fronting Islamgee was defended by a lower wall and revetment, constructed nearly half way up the slope. In the centre of the revetment was a barbican, up to which led the only available road to the fortress.

Leaving only a sufficient garrison to guard Fahla and Selasse, all the other troops were withdrawn to Islamgee, where they were massed behind the captured artillery.

Sir Robert Napier had taken much pains to ascertain the exact strength of the enemy and its defences. From one point only was Magdala attackable, and that was the side which fronted us, as we stood upon Islamgee. Longer and stronger escalading ladders than the army had made would have been requisite. But it was believed that though no attack was meditated upon any other point than the northern front, that all avenues of egress were blocked ; the investment being made by regular European cavalry, the 3rd Dragoon Guards.

Penn's Battery of Mountain Train Artillery was ordered to take position to the right of the captured pieces at a distance of 800 yards from Magdala. Twiss' Battery of Mountain Train Artillery was posted on Selasse a full thousand yards from the object of attack. The Armstrong Battery was posted on the intermediate ridge between Fahla and Selasse at a distance of

2,500 yards from Magdala ; thus the Armstrong battery and Penn's were posted each on the horn of a crescent curving towards Magdala, which would give a convergent fire. The eight inch mortars had also been posted to the left of the Armstrong Battery.

At 2 P.M. Penn's battery piped for battle, and its notes rang out clear and loud. In the meantime Sir Robert Napier was on Selasse height to observe what means were taken by the besieged for defence, as the firing went on.

Following the howitzer shells flew a tentative Armstrong conical shell, singing a tune an octave higher with a slight quaver in its tone as it went twirling and bowling through the air, bursting directly over Magdala. Then followed the mortar's shell, wailing mournfully as if it detested its errand, but yet was bound to perform it by inexorable fate.

The Artillery then warmed into action ; and soon twenty guns of all calibres were thundering with might and main at the gates of Magdala, while a chorus of fierce hisses burst from the Rocket Battery.

During the energetic bombardment one of the correspondents asked me if I had seen Theodore's last handiwork. I replied in the negative. He then said that he could show it to me ; leading me, at the same time, to the edge of the precipice, when he pointed downward, and, looking in that direction, I saw a sight which for ever beggars description.

Of a verity, reader, I am no lover of the horrible or the disgusting. But if you can conceive 308 dead people, piled one upon another, stripped naked, in a state of corruption, with gyves and fetters round their limbs, you will save me the unpleasant task of describing the scene !

Is it possible to award fitting punishment to such a wholesale murderer ? A man who takes another man's life is sacrificed at once by the avenging laws of civilisation. But what can be done to him who murders over 300 people ?

The late captives said that they were present during the execution ; that the prisoners were manacled hand and foot, and that they were sabred and shot by Theodore and his men as they lay helpless on the ground. Many of them, it is said, struggled to their knees, and entreated in the wild ' fervour and passion of prayer ' for mercy, but it was of no avail. They were butchered to the last soul. This took place the day we arrived upon the Dahonte Dalanta plateau, April 9.

Sir Robert Napier perceived during the bombardment, which lasted two hours, that the defenders were weak, and that his troops would suffer no great loss in the assault. He therefore ordered the Royal Engineers, the 33rd, the 45th, and the ' King's Own,' to prepare for the final work.

The musketry from the fortress had ceased soon after the British artillery had opened fire. The troops destined for the storming were drawn out in battalions at quarter distance, across Islamgee, facing Magdala. The Engineers, under the command of Major Gordon Pritchard, were in the front of the battalions deployed as skirmishers. Soon signals for rapid firing were made to the artillery, and under the furious cannonade which now commenced, the troops began their march along the plateau with trailed arms. Upon arriving within fifty yards of the base of Magdala the artillery ceased firing, and the Royal Engineers at once opened fire with their Sniders, which was taken up by the 33rd and 45th, who plied their breech-loading rifles with admirable rapidity, raining a storm of leaden pellets, for ten minutes. The volleys of musketry were as continuous and rolling as ever an army of 20,000 men, armed with the ancient muzzle-loaders, produced.

Theodore and his faithful adherents had lain concealed during the artillery fire ; but so soon as it ceased, up he sprang, sounding his war cry, and with his followers he hurried to the gates and defences, determined to give the advancing columns the benefit of a reception worthy of an Emperor who was about to conquer or die.

So long as Theodore was wedded to the idea that there was no forgiveness for him, so long would he have manifested the same implacable rancorous spirit towards the English had the siege lasted as many months as it did hours. Now, seeing no avenue open for him to escape, with his faithful men willing to shed their blood for him, nerved with the deepest courage born out of his extreme despair, he arrived at the barbacan and lowest revetment, posted his men at the loopholes and along the wall topped with the wattled hurdles. As he saw the soldiers still firing while they faced right, and wheeled into columns of fours for the purpose of ascending the path that led up to the barbacan, his signal was given, and their presence was known to the English by sharp shot falling amongst them, wounding several.

Instantly the British fire was concentrated on the barbacan and the suspicious revetment, through the interstices of which wreaths of pearly smoke issued, indicating the presence of the dogged riflemen. At this time, as on Good Friday, lightning and thunder played and rattled, as fit concomitants of a battle, and the rain fell in large warm drops, splashing heavily on our heads. But through the pelting shower the soldiers, invincible seemingly, undreading certainly, advanced, making their Sniders keep fatal music to their steps, scrutinising suspiciously granite embrasures and natural basalt battlements that shot up here and there on the brow of the cliff of Magdala.

Still preceded by the Engineers, and still sowing the deadly missiles over every inch of the superjacent slope, they arrived at, and halted near the barbacan.

For a minute there was a pause, and again a dozen shots hurtled amongst the more advanced of the Engineers, wounding Major Pritchard and three or four of the Engineers; but they were immediately replied to by a thousand directed at each spot from whence the reports came, and Major Pritchard and Lieutenant Morgan—the latter a most enthusiastic officer—made a dash upon the barbacan to effect an entrance. They found the gate closed, and the inside of the square tower completely blocked up with stones to the depth of ten feet, so that a passage through was not readily available.

'Hasten up with the powder!' shouted the Major.

'Hasten up with the powder! hasten up with the powder!' was passed along the contiguous columns.

But who was to bring up the powder? Why were those canvas bags with fuses attached made? Were they not for the purpose of blowing up this place? Where were the heavy hammers and iron wedges for breaking open the barrier and knocking off the bolts or drawing the hinges? Where the handsaws and the axes to cut the hurdle revetment down? Where the pickaxes to demolish or sap the wall? Where the escalading ladders, constructed for the purpose of scaling the walls and stockades or any raised defences there might be?

Echo answered, 'Where? oh where?' No mortal voice answered the eager question.

Private M'Guire of the 33rd thought he would climb up the cliff wall. Surmounting a ledge, without pause he ascended another in the same way, and then turned round and shouted aloud that he had found an opening. With a fierce cheer which was faintly heard on Islamgee, he faced the wall on the height, hotly seconded by Private Bergin of the same regiment.

The whole regiment, now urged by the bold example of their daring comrades, scrambled up the almost perpendicular slope, and after a few minutes of breathless work they surmounted the ledges, and seeing men suspiciously moving about on the summit of Magdala they opened fire, at once sweeping them away as with a breath.

Intruding their rifles into the interstices of the hurdle fence which topped the wall, they lifted it up, and in a second had passed over the lower defences. Scattering themselves over the ground, they made simultaneously for the other defence, which was seventy-five feet above them, passing over several ghastly relics of the battle.

Shortly we heard them firing quickly, eagerly, as they discovered their enemies moving about. Then were heard their clubbed muskets beating a fierce 'rat-tat-tat' upon the gates. Big rocks were thrown by them with a crashing force against the gates, and now and again they pertinaciously fired through every crevice and loophole. Neither gate, nor fence, nor stone wall, nor brush heap, nor even sheer rocks and strong barriers could stop the excited Irishmen; and no sooner had their bayonets gleamed through the fence than it was laid prostrate, and a wild 'huzza!' 'hurrup!' and 'hoorah!' was shouted out as they leaped over. Forward before them they flung their bristling rifles, and fired volleys into the very faces of the Abyssinians!

But we must not forget the Icarian charge of Drummer M'Guire and Private Bergin upon Magdala. The two men were advancing onward, a few paces from each other, to the upper revetment, when they saw about a dozen fellows aiming at them. They instantly opened fire, and so quick and so well delivered was it that but few of their assailants escaped. Seeing a host of redcoats advancing upward, the others retreated precipitately. Over the upper revetment both men made their way, and at the same time they observed a man standing near a haystack with a revolver in his hand. When he saw them prepare to fire he ran behind a haystack, and both men heard plainly a shot fired. Marching on with their Sniders on the present, they came to the haystack, and saw the man who had run behind lying prostrate on the ground, dying, with the revolver still convulsively clutched in his right hand. To their minds the revolver was but their proper loot, and without any ceremony they took up what they considered their own; but on a silver plate on the stock, during an examination of it, they perceived an inscription which read thus :—

PRESENTED

BY

VICTORIA

QUEEN OF GREAT BRITAIN AND IRELAND

TO

THEODORUS

EMPEROR OF ABYSSINIA

AS A SLIGHT TOKEN OF HER GRATITUDE

FOR HIS KINDNESS TO HER SERVANT PLOWDEN

1854

' F''what d'ye think, Pat; can this be that unblessed deevil of a Theodorus, the No-goose as they call him?' asked M'Guire.

' Meb-be, Mac; can't say; but we had better shtop near him till the Sergeant shows his phiz inside Mag-dá-la. Och, here he comes!'

They saw a swaying line of Irish soldiers advancing, and at once these two heroes raised their sun helmets, and, swinging them round their heads, they shouted the warlike cry of the Anglo-Saxons, ' Hurrah!' with the strength of unusual lungs, to greet their comrades.

With heads bent low, like charging bison, the ' Duke's Own ' came surging up almost intact; the colour-bearer in the centre; officers cool and martial like to the rear of their companies, all striding audaciously forward, alert, keen-eyed, and prompt as tinder, to burst into a white-heat blaze upon the slightest provocation.   Near the spot where the dying man, who had been drawn out to the open, lay, the centre of the regiment halted.

At this moment the rain ceased, and the sun shone forth into the full power of his departing splendour.

Eagerly stepped out the standard-bearer at the word of command, and high and triumphant, in all its silken bravery, streamed the ' Wavy Cross ' emblem of Britannia's majesty and power, above the surrounding world of mountains—an omen to all beholders that the tyrant Emperor had been humbled, and that his proudest stronghold, MAGDALA, had passed into the strangers' hands.   As it fluttered and rippled in mid air, the ' Duke's Own ' doffed helmets, and simultaneously, in the acme of enthusiasm, they raised their voices in cheers, which sounded to those on Islamgee, 500 feet below, like the deep roar of an ocean's tide.   The cheers were recognised, caught up, and flung from Magdala to Selasse, thence to Fahla, and that grey crag sent it quivering far below ; finally the British camp nearly two miles off caught the sounds, and strengthened the universal " Hurrah " by their own exuberant voices.   Strains of music burst from the martial bands.   The National Anthem of England, ' God Save the Queen,' was never played or sung with greater effect or vigour than when the hoary crags of Magdala responded to its notes in an overwhelming chorus of echoes !

A few unarmed Abyssinians, attracted by the clamour of music and shouting, mustered courage enough to approach the standard, which waved so gaily in the mountain gale ; and, on beholding one of their countrymen on the ground, they bent over the body, but quickly recoiled with fearful dismay on their faces, exclaiming ' Todros ! Todros !'

The words attracted the attention of every one, and together they strode towards the body, jostling each other eagerly in the endeavour to obtain a glimpse of him the natives styled ' Todros, Negus, Negashi of Itiopia !'

And what did they see ?   The body of a native seemingly half famished ; clad in coarse upper garments, dingy with wear, and ragged with tear, covering under garments of clean linen !

The face of deep brown was the most remarkable one in Abyssinia ; it bore the appearance of one who had passed

GOING UP TO ATTACK MAGDALA.

through many anxious hours. His eyes, now overspread with a deathly film, gave evidence yet of the piercing power for which they were celebrated. The mouth was well defined and thin-lipped. The lower lip seemed well adapted to express scorn, and a trace of it was still visible. As he gasped his last, two rows of whitest teeth were disclosed. Over his mouth two strong lines arched to a high aquiline nose. The nostrils expanded widely as he struggled to retain the breath which was rapidly leaving him. The face was broad, high-cheek-boned, with a high, prominent forehead, and overhanging eyebrows. The hair was divided into three large plaits extending from the forehead to the back of the neck, which latter appeared to be a very tower of strength. The body measured five feet and eight inches, and was very muscular and broad-chested. There was a character about the features denoting great firmness or obstinacy mingled with ferocity; but perhaps the latter idea was suggested upon remembering the many cruelties ascribed to him. And thus was it, that we saw the remains of him whom men called THEODORUS, EMPEROR OF ABYSSINIA, THE DESCENDANT OF MENI-LEK; SON OF SOLOMON, KING OF KINGS, LORD OF EARTH, CON-QUEROR OF ETHIOPIA, REGENERATOR OF AFRICA, AND SAVIOUR OF JERUSALEM, now dying—dead, by his own hand![1]

Fitting punishment was it that the red right hand, which had bereft so many hapless ones of their lives, should have de-prived that of its outlawed owner! Fitting was it also that the banner of St. George should first shadow his body, as it first proclaimed his downfall!

The Irish soldiers took hold of his legs, and roughly dragged him to a hammock, where, after two or three gasps, he breathed his last.

Curious remarks were passed upon the body by the dense groups which surrounded it. One man, with a spice of Latin in him, uttered sententiously, 'Sic semper tyrannis,' to which many a one responded heartily 'Amen and amen!' Another Celtic

[1] At the post-mortem examination the surgeons found that, excepting a slight flesh wound in the right leg, he was uninjured by foreign missiles. The palate was destroyed and the roof of the mouth scorched, and a hole was found through the back of the head. The united opinion of the medical authorities was that a pistol fired in the mouth had caused the death.

warrior hoped the scoundrel would trouble ' nobody no more;' and another with some regard for decency covered up the bared abdomen, evened the nether limbs, and folded the arms upon the breast.

Larger grew the crowds around the body. Officers and privates as they came up hastened to get a glimpse of it. The released captives hurried to obtain a farewell glance at their dead captor, and when they recognised him all doubts as to his identification were at an end. Theodore had been fighting in disguise, knowing that bright colours attracted England's marksmen. The Commander-in-Chief with his staff rode up to view the corpse, but not one kind word of sympathy for the dead Emperor's fate was uttered. He who had been merciless to others was not deserving of sympathy.

Not until the last moment, when on the threshold of certain defeat, did he surrender his life. Seeing speedy death in the levelled muskets of the advancing soldiers, he quickly retired behind the haystack, and with the revolver—the Queen's gift— into his mouth the Imperial Suicide had fired, and died.

The advance was sounded, and the regiments with arms aslope filed off in columns through the narrow streets of Magdala, the Commander-in-Chief and staff following. Passing through a long lane flanked on each side by store- houses with conical thatched roofs, they came to a large open space on the southern side, which was probably used by the Emperor as a parade or drill ground.

At the upper extremity of this open area, near a group of black tents, were seen several groups of armed natives, who fired two or three shots as they perceived the soldiers. The 23rd formed line, deploying across its whole width, and opened fire upon them, steadily advancing the while. Before the determined approach of the soldiers and their withering fire, the last remnants of Theodore's army incontinently fled down the cliffs, led by his illegitimate son Dajatchmatch Masheshai.

When the Irish regiment arrived at the eastern extremity, having traversed its whole length, another flag was raised to announce the complete capture of the fortress; and the Prince with his flying warriors must have heard the glorious cheers that greeted the flag as it was waved, as well as the martial strains of 'Rule Britannia;' the Gallas on the alert on the opposite mountains of the Walla Galla country must have heard it also.

When Sir Robert Napier appeared, the music was changed to 'See the Conquering Hero Comes ;' and other rousing cheers and anthem chants, amid mutual congratulations, closed the STORMING of MAGDALA.

CHAPTER XII.

THE EMPEROR'S HOUSEHOLD—LOOT—THE GENTLEMEN IN BLACK—THE
DEAD TYRANT.

AFTER the important incidents recorded in the last chapter
the soldiers strayed away from their respective regiments by
twos and threes, and commenced an assiduous search for loot.
They moved off slyly in squads or singly, peering into every
house, overturning household furniture, and recklessly destroying the Lares and Penates of the departed garrison.

In their wanderings the soldiers came to the Imperial residence—a barn-like dwelling, two storied—very strong, and
thatched with straw. Close by were the Imperial kitchens,
the harems, the stables, the store-houses, and the great prison,
which enclosed a quadrangular area.

In the open space a multitude of females were discovered
gathered together. These females composed the women of the
Emperor's household. Surrounded by the ladies of her court,
apparelled and non-apparelled, was found the Empress Etegie
Torenachie—daughter of Prince Ubie of Samen and Tigre—
a nut-brown lady of six-and-twenty, or thereabouts, the loveliest
of all, but pale and consumptive-looking. Her character,
bruited abroad already by the captives, caused the soldiers when
they perceived her to pay her every respect. For a long time—
at least so long as Theodore was sober—the Empress was as a
guardian angel to him ; advising him like the ' Good Book,' as
a soldier pertinently expressed it ; watchful and assiduous in
his interest ; brave and courageous as her husband, when her
duty called her attributes forth. While Theodore was at Debra
Tabor, preparing for his expedition to Magdala, she resided
at the latter fortress. Twice during the time it was besieged

by that mythical being, King Menilek of Shoa, with an army of 40,000 (perhaps an extravagant number), the chiefs and elders wished to surrender the place; but the Empress was ubiquitous, and untiring in her devoirs to her Imperial husband; and Menilek's demonstrations were set at naught, and evil counsels were confuted.

Playing about her maidens, was the heir-apparent to the throne of Ethiopia—Prince Ala Mayu—a sturdy, intelligent little boy of eight years old. So soon as Sir Robert Napier was made aware of their presence in the fortress, Captain Speedy received orders from him to take them in charge and pay them every attention.

As the soldiers perambulated about, a party of them stumbled upon the store-house, where scores of twenty-gallon jars were found brimful of tej of exquisite flavour; and the joyful cries which they uttered brought down fifties of others to join in the ' liquoring up.'

Over all Magdala were scattered different kinds of military mobs : mob appropriative, mob reckless, mob bibulous, mob demoralised, mob humane, mob decently respectable, menadic mob, and prisoner mob. To describe all these mobs would, of course, be tedious, so I shall only touch upon those most interesting.

The mob bibulous was gathered in the Aulic quarters, prepared to quaff the grateful beverage bountifully provided by his Majesty for his own and harem's necessities. Several jars of very fine arrachi, as strong as brandy, were found, and of course eagerly seized and tasted by men who were connoisseurs in the business of tasting liquor. They tasted and tasted again and again ; and, as might be expected, the effects of the copious drinks taken by the men became distinctly visible in a very short time ; but not to any extravagant extent, as military law is rather strict in the English army, and severe penalties are always awarded to drunkenness.

Another party of soldiers, diving into the barn-like dwelling yclept Imperial Palace, perceived innumerable articles of furniture strewn about. One thing upon another the men picked up ; then, examining the article, pocketed it or threw it down : to be picked up, examined, and pocketed or thrown away by others coming after them. Little of the whole was worth

COOMASSIE AND MAGDALA.

carrying away by soldiers who would be obliged to convey luggage on their backs hundreds of miles.

Opposite the Palace was the Penitentiary of Magdala—a strong double-storied dwelling—from whence cries were heard. With lengthy leaps—quickened, no doubt, by a drop or two or three of arrachi—a score of soldiers sprang at the doors, and in a twinkling or so battered them down. As the rays of the declining sun fell slantingly inwards, a chamber of horrors was revealed!

The long disused room was crowded to overflowing with native prisoners. Some of them were of high rank in the different provinces. Princes, Generals, peasants stood in expectant and surprised groups, heavily manacled and fettered. Appalled for a moment, the soldiers, no less bewildered than the prisoners, gazed at the scene. They were recalled to their duty by a mute appeal to their sympathies—an exhibition of manacled limbs. With desperate haste the soldiers tore their fetters and manacles off, and with hearts full of gratitude the people stepped out into the glad sunlight, once more free!

Many of them were so weak they could scarcely stand. For several days they had cowered in horrible suspense in the loathsome dungeon, hoping against hope, longing passionately for the end, while expectant that the great day of deliverance was at hand. It was so suggested to them through hearing bye conversations between jailors, who, ignorant of the strength of the British army, dubious of the result, yet gossiped about it as foolish gossips generally do. Prisoners of all classes approached and kissed the hands of the soldiers.

Could gratitude be better expressed? Could any words equal this silent yet touching conduct? The glad mass wended their separate ways as best they could, gathering strength as they inhaled the fresh breezes. According to their usual custom upon a joyful event, they clapped their hands, quaveringly chanted a welcome anthem, and lifted on high their voices in the joyous 'Li-li-li,' in honour of the Feringhee Liberators.

Some provisions found in the store-rooms were distributed amongst the people, and they were then ushered out of the gates.

These scenes are really what passed around the Aulic portions of the fortress. But the tabular surface of this mountain plateau contained other sights, though not as virtuous, yet as interesting. There were an endless variety of dwellings— silken and canvas tents, Koord-like domiciles, cotes, and such like; and each of these had around it a knot of men commenting, gossiping, pocketing, analysing, breaking into pieces, or tearing into shreds, whatever thing their vision or fancy lit upon.

Upon approaching nearer I saw, not a Bedlam broken loose, but an extempore pandemonium enacted round the Koord tents, which I learnt, to my astonishment, were the treasure-tents. The ground was strewn with an abundance of seemingly costly things. The glitter certainly could not be surpassed, were all the treasures of Windsor Castle and the Tuileries exposed on their velvet lawns, side by side, by what were here. If all that glittered among these tons of treasure were gold, the English Government need not have entertained much concern about the cost of the expedition. But taking up a shining candlestick—massive and curiously gotten up—I perceived the worthlessness of the material; and though there were a very considerable number of wares manufactured out of the precious metal, the major part of them were gilt.

Over these trivialities the people sprawled and cuffed; and prominent amongst them, with their voices raised to an unpleasantly high pitch, and their coat-tails fluttering in the breeze, were the gentlemen in black coats, endeavouring to pack off the silken marquees. They would for ever have remained unmentioned by me, had they not sought unenviable notoriety by their unseemly conduct and screeching voices of aroused avarice. These gentlemen in black, with fluttering coat-tails—hear it, oh, ye philanthropists !—were the captives! Not all of them, but several of them. Some of the British subjects; three missionaries; the Prussian, German, and Russian mechanics.

Several of these gentlemen had secured this loot long before the soldiers had arrived. It lay in piles at their feet, or had been already spirited away under the envious coat-tails. It was reported that one of these wretches, who had neither the fear of God nor of the devil before his unholy eyes, had the audacity to break open the coffin of the most holy Abuna, and snatch from the neck of the putrid corpse a diamond cross worth many thousand dollars.

The fame of the treasures brought new comers by the hundred, and General Sir Robert Napier, Generals Staveley, Wilby, Snyder, and their respective staffs, came to visit them; and these gentlemen, together with the members of the press —to the honour of the Fourth Estate be it said—were the only ones exempt from the picking and pocketing mania which had taken possession of all heroes.

To enumerate even the one-tenth of the articles scattered about would be a task as tiresome as it would be fruitless. In one of the tents was found the Imperial standard of Ethiopia— a lion rampant, of the tribe of Judah, worked in variegated colours. In another was found the Imperial seal, with the same distinctive figure of a lion engraved on it. A chalice, of pure gold, was secured by Mr. Holmes, on which was engraved in ancient Ethiopic :—

THE CHALICE OF
KING ADAM LEGUD CALLED GAZOO
THE SON OF
QUEEN BEHAN MOGUSSA
PRESENTED TO KOSKWAN SANCTUARY GONDAR
MAY MY BODY AND SOUL BE PURIFIED

15th Century.

The Abuna's mitre, 300 years old, of pure gold, probably weighing six or seven pounds troy weight; four royal crowns, two of which were very fine specimens of workmanship, and worth a round sum of money; were worthy things to be placed in a niche of the British Museum. A small escritoire, richly ornamented with mother of pearl, was found also, full of complimentary letters from European sovereigns, and state papers; besides various shields of exquisite beauty. There were also an infinite variety of gold, and silver, and brass crosses, and censers, some of extremely elegant design; golden and silver pots, kettles, dishes, pans; cups of miscellaneous descriptions; richly chased goblets, of the precious metal; Bohemian glasses, Sèvres china, and Staffordshire pottery; wine of Champagne, Burgundy, Greece, Spain, and Jerusalem; bottles of Jordan water; jars of arrachi and tej; chests full of ornamental frippery; tents of rose, purple, lilac, and white silk; carpets of Persia, of Uschak, Broussa, Kidderminster, and Lyons; robes of fur; war capes of lion, leopard, and wolf skins;

VIEW NEAR THE KING'S HOUSE, MAGDALA.

saddles, magnificently decorated with filigree gold and silver;
numerous shields covered with silver plates; state umbrellas
of gorgeous hues, adorned with all the barbaric magnificence
that the genius of Bejemder and Gondar could fashion; swords
and claymores; rapiers, scimetars, yataghans, tulwars, and
bilboes; daggers of Persia, of Damascus, and of Ind, in scabbards of crimson morocco and purple velvet, studded with
golden buttons; heaps of parchment royally illuminated;
stacks of Amharic Bibles; missals, and numberless albums;
ambrotypes and photographs of English, American, French,
and Italian scenery; bureaus, and desks of cunning make.
Over a space growing more and more extended, the thousand
articles were scattered in infinite bewilderment and confusion
until they dotted the whole surface of the rocky citadel,
the slopes of the hill, and the entire road to camp two miles
off!

From this scene I strolled away to the northern gate, to
where the dead body of the late Master of Magdala lay, on his
canvas stretcher. I found another mob—mob indiscriminate
—of officers and men, rudely jostling each other in the endeavour to get possession of a small piece of Theodore's blood-
stained shirt.

No guard was placed over the body until it was naked, nor was
the slightest respect shown it. Extended on its hammock, it
lay subjected to the taunts and the jests of the brutal-minded.
An officer, seeing it in this condition, informed Sir Robert
Napier of the fact, who at once gave orders that it should be
dressed and prepared for interment on the morrow.

From this place I went to examine the dead bodies of the
faithful followers who had died at their posts, defending to the
last their strange master, who had elements in his composition
that, surrounded by happier circumstances, might have made
him a loved, revered, and renowned monarch. The dead chiefs
were doubled in all postures, but with singularly serene countenances. Their wounds were many and ghastly. The head
chiefs were seven in number, and they lay close to the barbacan
which they had so determinedly contested. Their garments of
silk and brocade were steeped in their life-blood, and the rocks
around them were marked with the same evidences of the
struggle.

Along the lower revetment which flanked in parabolic curves the barbacan, were nearly a dozen warriors, some with brains unroofed and nether limbs sheared, and others bearing frightful wounds. Near the upper barrier were thirteen dead natives, and dotting the interior of Magdala were perhaps a dozen more.

The barrier that defended the ingress to the fortress, though rudely constructed, was as well designed as the best military engineer of Europe or America could have done it; that is, of course, taking into consideration that it must be built with the tools and materials which Theodore used. The passage to it ran below two walls of sheer basalt rock, rising from 15 feet at the embouchure to 30 feet over the barrier. The gate was 6 feet high, constructed of solid wood, composed of two upright stiles and two horizontal rails, strengthened by diagonal braces on each side of the stiles. It opened internally, and was fastened by a bar of wood running from post to post and from wall to wall. It was 8 feet wide at the clear.

In the interior of the passage were built two or three short stone traverses, and along the summit of each basalt wall was erected an abattis.

Taken altogether, it was a very defensible place, and one 12-pounder would have rendered it impassable; but Theodore had planted his cannon upon Fahla and Selasse, and, defeated at Aroje, he had delayed their removal to Magdala because for two days he was not without hope that peace was being negotiated by his dear Armenian friend Rassam. When finally he came to the conclusion that peace was impossible, he had commenced their removal on the third morning; but, as I have described already, he was surprised in the act by Speedy and Loch in command of the 3rd Cavalry.

The number of dead, as described variously, amounted to about sixty natives, and double that number wounded. The British loss was seventeen wounded, none mortally. The bombardment of the fortress continued for over two hours, and it is wonderful that, out of the 4,000 people which inhabited it, innocent lives were not sacrificed. But out of the wounded of Aroje battle there were hundreds who had taken shelter in the straw cotes on Islamgee. Every alternate house was a hospital, housing one or more meagre wretches.

The rockets had wrought utter desolation among some of the villages below Islamgee, eastward.   One rocket in particular diverged widely from its intended mark, and darted through one of these combustible houses, setting fire to it instantly, and the wind fanned it until it reduced the whole group of villages in a short time to ashes.

Bardel, the French professor, the only European who had not arrived at camp with the others, because he was sick with fever as I have already mentioned, was found in high delirium, in a little booth at the Selasse end of Islamgee, surrounded by offal and filth, and covered with verminous rags.   He was taken so soon as found to the military hospital, where he received every attention, and finally, after a week's severe illness, he recovered his health.

All that I have related took but three hours.   It was optional with civilians either to remain upon Magdala and rough the night out in some hut, or proceed to the camp below Fahla, two miles off; but, after an examination of several huts, we decided to proceed to camp.

On Islamgee we met the camp followers and the equipage of glorious war rolling slowly upward to the soldiers encamped on the summit of Magdala—mules and horses, dhoolies and stretchers, in never-ending continuity.   Below Islamgee we met the elephants with their enormous packs on their backs, looking like phantoms in the dark ; and in the friendly shadows of the night the flitting forms of the Magdala fugitives pressing forward to the valley of Aroje, without regard to the militant current of England and India.

Once past Fahla, we stumbled down the slope with reckless force, until we came in full view of the camp of the Abyssinians —who left Fahla and Selasse early in the morning—all aglow with fires.   Far away in the eastern horizon a long belt of flame was visible—a warning beacon to native stragglers; for there the versute and ferocious Gallas looked wide-eyed and wondering upon the sea of fire about our camp.

We passed right through the Abyssinian camp, which was pitched on the plain, where the battle of Aroje first began— around the graves of hundreds of their people—and half a mile beyond was our own.   On approaching it we heard bacchanalian songs, and shouts for lost friends, and a combination of novel sounds. While travelling through, at the imminent risk of crushing some one to death under our horses' feet, we came suddenly upon the white snowy tents, rearing themselves in ghostly silence from the ground, around which we saw whole multitudes, who, adopting the bivouac, had stretched themselves on the earth and were already deep in slumber, dreaming, no doubt, of acknowledgments by grateful England of their conduct at the fall of Magdala !

CHAPTER XIII.

INTERMENT OF THEODORE—THE RIVAL QUEENS—FAREWELL TO MAGDALA—
GENERAL NAPIER'S ORDER—THE AUCTION OF LOOT—OFFICIAL REPORTS.

BEFORE quitting the neighbourhood of Magdala there were
various things to be performed : Theodore's body had to be
interred ; his Empress was to be consoled ; the young Prince's
fate had to be decided ; the hungry hordes of Gallas, called into
unpleasant activity, and now loud-mouthed for prey, had to
receive their quietus; the Abyssinian fugitives had to be
escorted across the Bechilo ; the revetments, barbacans, and
watch-towers, palace, prison, and peasant houses, had to be
demolished ; and, lastly, the triple mountain citadels had to be
delivered over to some responsible authority.

The first day was occupied in marching the victorious troops
back to camp, with the exception of a regiment which was left
as garrison ; in burying the dead chiefs who fell within Magdala; in distributing provisions to the homeless Abyssinians—
numbering over 30,000—who were camped a bullet's flight
from us.

The Imperial relict's wishes concerning the interment of
her husband were ascertained before Sir Robert Napier decided
upon Theodore's funeral. She replied that she was willing
that the Commander-in-Chief might do as he felt disposed,
but hoped the prayers of his own church might be read over
him by his favourite chaplain. To this the General made no
objection, and accordingly directed a party to decently swathe
the body in silken cerements preparatory to burial. That
task done, the body was extended on a rough bier, and conveyed
to the church wherein Sir Robert Napier had directed he should
be buried. A shallow grave had been dug in the centre of the

building. The chaplain—an old decrepit monk—was called
in and requested to read prayers over the body. Apparently
very distasteful was the task, for he used exceeding haste in its
performance. The body was then lowered by two stout soldiers
into the grave; the monk took and scattered a handful of soil
over it; shovels deftly handled rattled the clods over the coffin
until it was hidden well out of sight.

A messenger was despatched that day to Wagshum Gobazye,
proffering him Magdala. The General received an answer on
the third morning from that wily prince that he did not want
it; that he had been smart enough to keep out of Magdala
while Theodore was alive, and now that he was dead he did not
care to make himself a voluntary prisoner in it, as he would be,
he thought, surrounded as the Bechilo valley was by the
Galla forces. He added, moreover, that if Sir Robert would be
kind enough to present him with Theodore's cannon, as well
as the fortress, he might think better of the gift, as he would
then be furnished with the means for its defence.

So reasoned the Wagshum in his epistolary correspondence
with the British General. Gobazye was distinguished for nothing
else than masterly retreats, or acting the part of a detective from
a distance, always from safe positions. During two months he
kept himself engaged in viewing Theodore's camp from some
inaccessible peak or another, and the last movement on the part
of the Emperor was sufficient to send the bombastic Wagshum
winging a precipitate flight to furthest regions. Napier could
not see the benefit accruing to him or to the country by a
change of masters. Wagshum Gobazye was more unprincipled
than Theodore, and but a trifle less cruel, so neither Magdala
nor cannon was given to him.

On the second day the two rival queens, Walkeit and
Masteevat, came forward as candidates for Magdala, under the
ciceroneship of Meer Akbar Ali, who had been despatched to
evoke their aid in hemming Theodore in. Masteevat, the
younger of the two—fat, fair, and forty—was preferred by Sir
Robert. During her stay in the British camp she was received
with all due honours, and apportioned a silken tent as a residence. She and her little son, a boy of six, very handsome, of
the colour of a Spaniard, were arrayed in the most gorgeous
finery that native genius could manufacture.

The General's cooks had their culinary knowledge taxed to
the utmost to provide dainties for the Royal family; and the
Queen, nothing loth to show her appreciation of such talent
and skill in the *cuisine*, ate like a *gourmande* of each and every
article, never deigning to examine the materials, but disposing of
what came before her, without regard to the horrified looks of
the Political Secretary—pudding before beef, blancmange with
potatoes, drinking coffee before finishing her fricandeau,
emitting labial smacks like pistol cracks when a more savoury
dish than common pleased her palate.

Her Majesty often vented her satisfaction by hearty, boisterous guffaws, and proceeded deliberately to wipe her unctuous
mouth with the tail end of her turban, for which she specially
untied it; while a tawny-faced Hebe, with the bust and form
of a Venus, brought her a horn of tej strongly mixed with arrachi
(her usual quantum being a quart), which she quaffed with
infinite gusto, looking more imposing after it than before.

Some Iago of an officer, possessed of a tongue that would
charm the d——l, managed to persuade the Queen to sit for her
portrait; and after three or four trials, during which she looked
awry and unnatural, a very good one was taken, and copies of
the same have been sent to all the Governments of Europe, as
well as to the United States.

The third day was employed in destroying the cannon of
Theodore, thirty-three in number.

All this time we were suffering severely from want of water
and scarcity of provisions. However, the hearts of all were
gladdened on the fourth morning at the gladsome tidings that
the army would leave the Bechilo valley for Dalanta plateau on
the next day.

On the fourth morning after the fall of Magdala, the
Abyssinians, to the number of 30,000, commenced their march
for Dalanta. The Gallas, it seems, were aware of the fact; for
like vultures, keen-eyed, and eager for the prey, they hovered
on their flanks and rear, sometimes dashing boldly into the
main body, and capturing many prisoners.

At 3 p.m. I rode up along with several others to see the last
of the fortress. Two hours afterwards, every living soul being
driven out by the Royal Engineers, those deputed for the task
of destruction began their work. The Kafurbar, or southern

gate, was blown up first ; then the eastern towers and magazines ; and, as the wind was from the eastward, fire was set to the houses at the extreme eastern end of the fortress, and, retrograding quickly, the nimble Engineers touched each house with their fiery wands. The spectators made haste to secure good positions, from whence the mighty conflagration, soon destined to appear, could be seen to advantage ; and the best place they could select was the southern end of Selasse, which looked into Magdala from a distance of 1,000 yards.

The easterly wind gradually grew stronger, fanning the incipient tongues of flame visible on the roofs of houses until they grew larger under the skilful nursing, and finally sprang aloft in crimson jets, darting upward and then circling round on their own centres as the breeze played with them. A steady puff of wind levelled the flaming tongues in a red wave, and the jets became united into an igneous lake !

The heat became more and more intense ; loaded pistols and guns, and shells thrown in by the British batteries, but which had not been discharged, exploded with deafening reports, and projectiles whistled ominously near us. Three thousand houses, and a million of combustible things, were burning. Not one house could have escaped destruction in the mighty ebb and flow of that deluge of fire.

Cheerily sounded the bugle, merrily rattled the drum, loudly piped the fife, gaily sang the birds, brightly dawned the day on the welcome morning of April 18, 1868.

We had all seen the drama of the Abyssinian Campaign enacted. On this day we turned our faces northward for the homeward march.

At 8 A.M. the vanguard defiled down the slope of the plain of Fahla towards the Aroje ravine and the Bechilo. The baggage trains were then started, and by 10 A.M. the road was clear for the army.

The regiments were drawn up as if on parade—Infantry, Cavalry, Artillery, Engineers, Sappers and Miners, in order. As the chief mounted his horse the order ' March ! ' was heard, and preceded by its band the 45th Regiment, with sloped muskets, trod proudly down the slope to martial tunes ; then the Native Indian regiments, horse and foot ; then the Artillery and

THE END OF KING THEODORE.

Naval Brigade ; and lastly went the 'Duke of Wellington's Own.' As the most rearward of the regiment took up the step and got firmly under way, cheer after cheer burst forth from over six thousand voices, which went reverberating among the deserted and desolate heights, until the Gallas, on the *qui vive* ten miles off, were made aware by these sounds that the soldiers had said their ' Farewell ! '

Magnificently stern and sombre the triple citadels raised their lofty heads far above the mountains forming the southern horizon, which, mingling with the morning sky, glowed in radiant colours, and would have served for a rare background for any landscape. Magdala, the far-famed one, was covered with a haze, the smoke of the still smouldering embers rising upward like that of an incense offered to appease Divine wrath !

It was determined to allow the animals of the transport trains to recruit their strength with the aid of the plenitude of provender and other supplies gathered on the plateau during our absence before Magdala. Accordingly, a gentle depression in the centre of the plateau was fixed upon as the camp site.

An important matter remained to be attended to. The mass of trophies collected at Magdala were to be sold to those gentlemen who had either a superfluity of cash, or who leaned to antiquarian tastes, for the benefit of the soldiers engaged at the front. Commanding officers were requested to select what things they thought most appropriate as mementoes for their regiments or troops. Fifteen elephants and nearly 200 mules were loaded with these miscellaneous articles.

At 9.30 A.M., to the solemn beat of drum, the different regiments marched out of their respective camps to a level eminence 200 yards off, where they formed one large hollow square, with the captives, officers, and generals in the centre.

There was hardly a dry eye in the vast assembly when the clergyman finished a discourse which referred feelingly to the position of the captives and the triumph of our cause. It made a deep impression upon all who heard it ; and the tones of the voice, the glad sunlight, the wall of mountains in the far distance like a dark shadow, the warblings of the birds above and around, and the drifting clouds, will be remembered by those present as a Sunday unequalled in the calendar !

On the next day there was a brilliant and impressive general review of the army of Abyssinia.

The division lastly formed a hollow square six deep; the captives, foreign officers and civilians, were invited within. The Adjutant-General, Colonel Fred. Thesiger, then read aloud the General Order, reviewing in glowing language the objects and the incidents of the campaign:—

'SOLDIERS OF THE ARMY OF ABYSSINIA,—The Queen and the people of England entrusted to you a very arduous and difficult expedition—to release our countrymen from a long and painful captivity, and to vindicate the honour of our country, which had been outraged by Theodorus, King of Abyssinia.

' I congratulate you with all my heart on the noble way in which you have fulfilled the commands of our Sovereign.

' You have traversed, often under a tropical sun, or amid storms of rain and sleet, 400 miles of mountainous and difficult country.

' You have crossed many steep and precipitous ranges of mountains, more than 10,000 feet in altitude, where your supplies could not keep pace with you.

' When you arrived within reach of your enemy, though with scanty food, and some of you for many hours without either food or water, in four days you passed the formidable chasm of the Bechilo, and defeated the army of Theodorus, which poured down upon you from their lofty fortress in full confidence of victory.

' A host of many thousands have laid down their arms at your feet.

' You have captured and destroyed upwards of thirty pieces of artillery, many of great weight and efficiency, with ample stores of ammunition.

' You have stormed the almost inaccessible fortress of Magdala, defended by Theodorus with the desperate remnant of his chiefs and followers.

' After you forced the entrance, Theodorus, who never showed mercy, distrusted the offer of mercy held out to him, and died by his own hand.

' You have released not only the British captives, but those of other friendly nations. You have unloosed the chains of more than ninety of the principal chiefs of Abyssinia.

' Magdala, on which so many victims have been slaughtered, has been committed to the flames, and remains only a scorched rock.

' Our complete and rapid success is due, first, to the mercy of God, whose hand I feel assured has been over us in a just cause. Secondly, to the high spirit with which you have been inspired.

' Indian soldiers have forgotten the prejudices of race and creed, to keep pace with their European comrades.

' Never has an army entered on a war with more honourable feelings than yours : this has carried you through many fatigues and difficulties; you have been only eager for the moment when you could close with your enemy.

' The remembrance of your privations will pass away quickly, but your gallant exploit will live in history.

' The Queen and the people of England will appreciate your services.

' On my part, as your commander, I thank you for your devotion to your duty, and the good discipline you have maintained. Not a single complaint has been made against a soldier, of fields injured, or villagers wilfully molested in person or property. We must not forget what is due to our comrades who have been labouring for us in the sultry climate of Zoulla and the pass of Komaylee, or in the monotony of the posts which have maintained our communications. Each and all would have given all they possessed to be with us. But they deserve our gratitude.

' I shall watch over your safety to the moment of your re-embarkation, and to the end of my life remember with pride that I have commanded you.

' R. NAPIER, Lieutenant-General,

' Commander-in-Chief.

' Camp, Dalanta, April 20, 1868.'

(The above Order is true in each and every particular. The General has not overstrained himself whatever in its composition. It is as simple and as modest as himself.)

On the third day of our stay in our cantonment upon Dahonte Dalanta plateau, the articles for sale were ready for the auctioneer.

The perambulatory roll of the drum, which in all well-governed and systematical military encampments announces a new move or new event, assembled all the officers and crowds of on-lookers around the piled trophies of Magdala, which covered half an acre of ground.

The auctioneer selected was Lieutenant S., an indefatigable officer who had been gifted by nature with the physiognomy characteristic of the first Napoleon. Fathoms of finest carpets of all countries were spread about, and all the paraphernalia of a thousand churches glittered in the morning sunlight; and jostling each other in the characteristic confusion of mobs (and the most belligerent mob in the world is an English one), were the gentlemen buyers, who for the nonce were connoisseurs in antiquities, and displayed remarkable acumen in discerning tinsel from real, glitter from gold. The many things scattered round about, evidences of Theodore's pampered selfishness or harem folly, evoked queer remarks from the vulgar-minded and the rude.

Bidders were not scarce. Every officer and civilian desired some *souvenir* of Magdala. One bought a cross of silver or brass, another a censer, another chose a sword. Goblets and cups, pixes and chalices of silver, there were in plenty; silks, umbrellas, saddles resplendent with golden filigree, and starred with rare-coloured stones; tents, carpets, richly illuminated Bibles and manuscripts, trinkets, and jewellery, found ready purchasers.

Mr. Holmes, as the worthy representative of the British Museum, was in his full glory. Armed with ample funds, he outdid all in most things; but Colonel Frazer ran him hard because he was buying for a wealthy regimental mess—11th Hussars—and when anything belonging personally to Theodore was offered for sale, there were private gentlemen who outbid both.

When Theodore's shield, used by him in his younger days, was offered for sale, though only garnished by a few silver plates, the bidding became very energetic, and from ten dollars it speedily went up to 200 dollars, for which sum it was purchased by Colonel Frazer.

BURNING OF MAGDALA.

Mr. Holmes secured many interesting articles.

The auction lasted two days. The total receipts amounted to 5,000*l.* This sum was divided among the non-commissioned officers and men who were southward of the Bechilo, which gave each man a trifle over four dollars.

On the fourth morning after our arrival upon Dalanta the late captives started on their route home, under the charge of 'that best gentleman in the world,' Captain S., commissariat officer.

From Dalanta camp Lieutenant-General Sir Robert Napier dated his luminous despatches, which made known to the world the rapid and conclusive events that had transpired before Magdala.

Sir Robert Napier's Military Despatch.

Commander-in-Chief's Offices, Head-quarters, Camp, Antalo :
May 12, 1868.

RIGHT HON. SIR,—On April 3, when encamped on the Wadela plateau, I received intimation from the chiefs of Dalanta that Theodore, having moved from Magdala, and encamped on the plain of Aroje, was preparing for an expedition. Letters from the captives also warned me to be on my guard. Between the British force and the plain of Dalanta lay the Jedda ravine, 3,400 feet deep. As the passage of this formidable obstacle, so easily defensible, could not have been effected in the face of an enemy without serious loss, I made a forced march of eighteen miles, crossed the Jedda, and established myself on the plain of Dalanta. The mere distance, in miles, gives little idea of the labour and fatigue of the march; the excessively steep descent and ascent, and the great heat, were very distressing for troops heavily weighted. Theodore, however, did not cross the Bechilo, but plundered and burnt the villages between that river and Magdala, which had always been faithful to and trusted him. From the edge of the Dalanta plain, I obtained a distant but clear view of the position of Magdala and its approaches. I was able, with a good telescope, to appreciate the formidable character of the whole position, and became aware that I should require all the infantry that I could possibly collect to make the attack effective, and that every cavalry soldier that I could bring forward would be necessary for the investment. Even with all the force that I could hope to gather up, I felt that I could not complete the investment by sending a column to close the Kaffurbar or southern gate of Magdala, but I deputed an officer of the Intelligence Department, Meer Akbar Ali, to Masteevat, the queen of the

Wollo Gallas, to engage her to bring every man she could muster to close all escape on that side. Meer Akbar Ali's report will be forwarded, and will show how effectually he accomplished his mission. Reluctant as I was to incur any delay so near to Magdala, these considerations of the necessity of having supplies sufficient to carry me through the operations against that fortress obliged me to defer crossing the Bechilo for several days. I had not overlooked the probability of the unstable Abyssinian people despising the small postal detachments and the pacific demeanour of our troops; but the various difficulties of our transport, and scarcity of our supplies, kept me without sufficient troops to make posts of communication as strong as the circumstances required. Each day, however, was bringing forward some accession of strength, and in the meantime I had endeavoured by liberality and every means of conciliation to engage the petty chiefs between Antalo and the Takasse River to maintain their friendly assistance in forwarding native convoys for supplies. It was the only course that gave chance of success; unfortunately it succeeded but partially. Relieved from the pressure of our main force, the chiefs commenced to interfere with the Abyssinian carriers of our supplies, and to make attacks on our posts and convoys, so that the local carriage, which had enabled me to advance from Antalo, was suspended just at the time when its maintenance was most important. Thus it happened that on April 4 I had only a few days' supplies to depend upon. This force had left all its baggage at Lat, 100 miles in the rear, taking on merely the clothes in which they marched, and carrying great coats, blankets, and water-proof sheets; they had no other incumbrance than a bell-tent for twelve officers or twenty soldiers; the daily storms which we experienced rendered this shelter indispensable; the carriage so released was sent back for provisions to the points where native transport was doubtful, and gave me some hope of ultimate relief from my commissariat difficulties, but the immediate urgency was pressing. On the 2nd instant I deputed Brigadier-General Merewether to the Takasse, to arrange with the chiefs there to bring in supplies of flour. Major Grant was directed to return to Lat, and Captain Moore to Lake Ashangi, to remove obstructions which had arisen at these places. Captain Speedy and Mr. Munzinger proceeded, the former to Dahonte, the latter to the borders of Dalanta, that had been ravaged by Theodorus; and through the exertions of these officers I was enabled to feed my cattle and to obtain flour enough for eleven days' supply at 8 oz. for each soldier. The native followers received wheat in the grain instead of flour.

Relying on the admirable spirit of my force, I was prepared to commit myself against Magdala with these means. Besides the view which I had obtained of Magdala and its approaches, I received most valuable information from a chief named Beitwuddun Hailo, who had recently deserted from Magdala. Having engaged in some intrigues with Menilek, King of Shoa, he knew well that his lot would be instant death on his master's arrival. It is difficult to give, by description alone, a sufficient idea of the formidable position which we were about to assail.

The fortress of Magdala is about twelve miles from the right bank of the Bechilo, but the great altitude and the purity of the atmosphere exhibited the whole outline distinctly. The centre of the position is the rock of Selasse, elevated more than 9,000 feet above the sea, and standing on a plateau called Islamgee, which is divided into several extensive terraces, with perpendicular scarps of basalt ; a saddle connects these terraces with the hill called Fahla. Fahla is a gigantic natural bastion, level on the top, entirely open, and commanded by Islamgee. It domineers completely, at an elevation of 1,200 feet, over all approaches to Islamgee ; the sides appeared precipitous, and the summit, surrounded by a natural scarp of rock, accessible only in a few places, and from eighteen to twenty feet in height. Nearly concealed from view by Selasse and Fahla, the top of Magdala was partially visible. The road to Magdala winds up the steep sides of Fahla, subject to its fire, and to the descent of rocks and stones. One part of the road is so steep that few horses, except those bred in the country, could carry their riders up or down it. The whole road is flanked by the end of Selasse and the broadside scarp of Islamgee. Altogether, without taking into account Magdala itself, the formidable character of its outworks exceeded anything which we could possibly have anticipated from the faint description of the position which had reached us. The refugee chief, Beitwuddun Hailo, was very anxious that I should try the south side, at the Kaffurbar (gate), from the opposite range called Lanta, saying, ' If you want to take Selasse go from hence; but if you want Magdala you must go from Lanta.' This, however, would have been impossible. I had not force enough to divide, and I could not place this vast combination of natural fortresses between me and my direct line of communication. I also perceived that the real point to be taken was not Magdala, but Islamgee, where Theodore had taken post with all his guns, and that Fahla was the key to the whole.

On the 7th I descended to the bed of the Bechilo, and reconnoitred the crossing. The ordinary approach to Magdala is by the Aroje ravine, which commences under Islamgee, and is bounded on its right by a spur which extends from Islamgee, in a serrated ridge, to the Bechilo. A similar spur from Fahla stretches to the water of the Bechilo, and bounds the ravine on its left. The highest point of this ridge is about 2,000 feet above the bed of the Aroje ravine. The grand features of the ground rendered it impossible for me, with my small force of infantry, to hold both sides of the ravine. I considered Fahla the key of the position, and determined to occupy the ridge bearing in different parts the names of Gimborgi and Affijo, which leads to that imposing outwork. Established on this ridge, I could operate on either side of Fahla, as might seem expedient on closer examination.

Between the 4th and 9th inst. my force was increased by six companies of the 45th Regiment, under Lieut.-Colonel Parish, which, though long delayed for want of carriage, had marched from Zoulla in twenty-five days. A wing of the 3rd Bombay, No. 1, under Lieut.-Colonel Campbell, was detained seven marches in rear, owing to the urgent want of carriage; and I thus lost the services of an able officer and an excellent body of soldiers, on whom I had calculated for the attack. On the 9th the whole force concentrated on the edge of the plateau overlooking the Bechilo, which flows 3,900 feet below it. Major Chamberlain, with the 23rd Punjaub Pioneers, supported by a wing of the Beloochees, occupied the bed of the Bechilo, and repaired Theodore's road. The signallers of the 10th company Royal Engineers maintained communications. The Royal Engineers and the Madras and Bombay Sappers, under Captain Goodfellow, made up the necessary provision of sand bags, scaling ladders, and bags filled with powder for the demolition of gates, stockades, &c. As the only supply of water between the Bechilo and Magdala was under the enemy's fire, all the water-carriers of the force were organised under command of Captain Bainbridge, Transport Corps, with two subalterns, and Lieutenant Ramsbottom, Transport Corps, for the purpose of carrying forward regular supplies from the Bechilo. The bandsmen and a party of Punjaub muleteers were also organised under command of Captain Griffith, aided by Lieutenant Gaselee, Transport Corps, and furnished with stretchers for the removal of wounded men from the field. All preparations having been completed, I placed the cavalry under Colonel Graves to hold the Bechilo, but ready to advance, and moved the remainder of the force across the river under the immediate command of Sir C. Staveley; the 2nd Brigade, under Brigadier-General Wilby, to remain in the bed of the Bechilo in support; the 1st Brigade, under General Snyder, to occupy the Gimborgi spur, and advance to a suitable place for encampment, and also to cover a reconnaissance by the Deputy-Quartermaster-General of the enemy's position. The Deputy-Quartermaster-General reported that the ascent to Gimborgi was extremely steep and difficult, and that the King's road up the Aroje ravine was easy and secure for the mountain guns and baggage. They were therefore ordered to take that route.

When the leading part of the column had reached Affijo, I arrived at the front. The King's road emerges from the Aroje pass at a distance of 1,200 yards from Affijo, and 700 feet below it. I ordered Major Chamberlain's Punjaub Pioneers to be sent immediately to cover the head of the pass, and the remainder of the brigade to be closed up as soon as possible. The men were greatly distressed by the heat, the severe ascent, and want of water. Shortly after Major Chamberlain had taken up his position, the Naval Rocket Brigade, under Captain Fellowes, appeared rising from the pass, followed by Lieutenant-Colonel Penn's Steel Battery, escorted by detachments of infantry. At this time the enemy opened his guns from Fahla and Islamgee, making good practice at the Punjaubees and at the position of Affijo. Notwithstanding the distance, which was more than 3,000 yards, the enemy's shot ranged well into the positions, owing to the great command, and probably to excessive charges of powder ; but, the fire being a plunging one, no casualties ensued. Almost simultaneously with the opening of the enemy's artillery, a large force was seen pouring down from Islamgee and the side of Fahla, descending at speed the steep road and the faces of the mountains, until they filled the whole plain of Aroje. Many of the enemy were dressed in red, and almost bore the appearance of our own troops in the distance. About 500, principally chiefs, were mounted. The Naval Brigade hastened up to Affijo, and as each rocket-tube came into position it opened on the advancing masses of the enemy, who were startled, checked, and driven back at some points, but only to press forward at others. I directed Sir Charles Staveley to bring forward the remaining infantry, which by this time had closed up, to repel the attack. The 4th (King's Own) regiment, under Colonel Cameron, closely followed by Beville's Beloochees and the Royal Engineers commanded by Major Pritchard, and the Bombay Sappers under Captain M'Donnell, R.E., descended rapidly the steep path leading down to the Aroje plain, with unrestrained expressions of delight in having at last their enemy before them. Opening into skirmishing order, they ascended a small slope which separated them from the plain of Aroje, and immediately came in contact with the enemy, drove them back in spite of the efforts of their leaders, in masses on which the fire of the Snider told with terrible effect. Several gallant attempts were made by the Abyssinians to rally, but many of their chiefs fell, and they were driven down the slopes of Aroje towards the ravines on our left front. A portion of them withdrew up the sides of Fahla, and, taking cover in a thicket of cactus trees, opened a teasing fire on Staveley's right, causing some casualties. Captain Fellowes, having maintained the fire of his rockets until masked by the advance of the infantry, had been sent to support Sir Charles Staveley. The fire of the rockets, together with some volleys from Beville's Beloochees and the Royal Engineers, supported by two of Penn's guns under Lieutenant Taylor, cleared Staveley's flank from further annoyance. The rockets were then turned on the summit of Fahla. They were well directed, and, as I subsequently learned, produced a very great effect. A party of the enemy attempted to pass round the sides of Affijo to turn our right, but were checked by a few rockets, and dispersed by the K company, Madras Sappers, under Major Prendergast, V.C., Lieutenant-Colonel Loch, with a detachment of the 3rd Bombay Cavalry, accompanied by the infantry in support. Towards the left Colonel Milward ascended from the Aroje pass with Penn's Battery, escorted by detachments of the 4th (K.O.) Regiment, under Captain Kittoe, and the 23rd Pioneers under Captain Paterson, at the time when the guns opened from Fahla and Islamgee.

On perceiving the troops of Theodore descending from Islamgee, Colonel Milward took up a strong position and opened fire from Penn's Battery. Major Chamberlain, who was holding the Pioneers in hand to cover the head of the pass, moved to his left and joined Colonel Milward. A considerable body of Abyssinians bore down upon Milward's position. Notwithstanding the evident effect of Penn's guns, they continued to advance with much determination and order. Chamberlain with his Pioneers met this attack in a most prompt and spirited manner, driving them with great slaughter into the ravines to his left front, not, however, without gallant resistance on the part of the Abyssinians, who closed fearlessly with the Punjaubees, the spear-wounds received bearing witness to the closeness of the conflict. On the extreme left the enemy pressed in large numbers towards the head of the Aroje ravine, where the baggage had arrived. The baggage-master, Lieutenant Sweeny, 4th (K.O.) Regiment, with great readiness massed the baggage in a safe position; and the baggage-guards, consisting of two companies of the 4th (K.O.) Regiment and one of the 10th (N.I.), under Captain Roberts, of the former corps, until disabled, and subsequently under Lieutenants Abadie (11th Hussars), and Sweeny, were brought forward and most effectually checked the attempt of the enemy to penetrate into the Aroje ravine. Arrested at the head of the ravine, and driven back by the baggage-guards, closed in upon by Chamberlain's Pioneers and two companies of the 4th (K.O.) Regiment, whom Sir C. Staveley had wheeled on to their flank, the enemy suffered most severely. Large numbers were seen to fall from the admirably directed fire of the mountain guns. Theodore's troops had advanced with the full confidence of men accustomed to victory; they had cast themselves off from their vantage-ground, to which there was no return. They had been promised by Theodore that they should be

enriched by the spoils of the English ; and it was not without a stout
resistance that they were finally driven off the field. A heavy rain
continued through the greater part of the action. The troops, thoroughly
wet and tired, but highly elated with their victory, bivouacked for the
night, covering the road to Aroje pass, and before daylight had re-
occupied their commanding position on Affijo, from which they had
descended to meet the enemy. The wounded were promptly attended
to, under the direction of Dr. Currie, C.B., Inspector-General of Hospitals. Many wounded Abyssinians were also carried off the field by
our troops, and were carefully attended to in our hospital. The 2nd
Brigade, which came up in the night, occupied the ground which
had been held after the action by the 1st Brigade. According to the
best information the probable number of the enemy was not less than
5,000, of whom at least 3,000 were the regular musketeers, and the
remainder less efficiently armed. Theodore distributed new arms to
the troops on the day preceding the battle. The loss of the enemy
cannot be correctly estimated ; 349 dead were buried in front of
the left of our position alone, and exclusive of those who fell in
Staveley's first attack. Thirty very badly wounded Abyssinians were
carried to our hospital. Theodore's lieutenant, Fetaree Gobazye,[1] and
many chiefs of note, were amongst the slain. Nearly all night the calls
of the Abyssinians to their wounded friends were heard, and the greater
number of the latter were carried from the field. We saw a large
number of wounded when Theodore's army surrendered. The British
loss was only twenty wounded, two mortally. This disparity of loss
resulted from the determined and persistent attack of the Abyssinians
against a better-disciplined and better-armed force ; not better armed,
however, as regarded the 23rd Pioneers, whose smooth-bore is hardly
equal to the double-barrelled percussion gun of the Abyssinians.
There was no hasty flight ; the enemy returned again to the attack
wherever the ground favoured them. I issued orders to provide
against the pursuit being carried too far up the hill, which could
only have ended by our retiring, and giving renewed confidence to
the enemy.

On the morning of the 11th Lieutenant Prideaux and Mr. Flad
arrived in my camp, accompanied by Dajatch Alema, a son-in-law
and confidential chief of Theodore, with a request for peace. I replied
that if Theodore would bring all the European captives to my camp,
and submit to the Queen of England, I would promise honourable
treatment for himself and family. Lieutenant Prideaux returned to
Magdala with the letter containing these terms. In the course of the

[1] Feet Ourary Guvrie.

forenoon he returned again to the British camp with Mr. Flad, but without Dajatch Alema. He brought a letter without seal or signature from Theodore, refusing my terms. My letter was returned. I sent back Lieutenant Prideaux and Mr. Flad to intimate that no other terms would be granted. I considered that a fuller atonement than the surrender of the captives, when they could be retained no longer, was absolutely required, and must be exacted ; and, painful as was the thought of the possible consequences to the captives if Theodore's rage should become excited, I relied for their safety on the apprehension of a renewal of the conflict, which demoralised Theodore's troops, and from which Theodore himself was not free, as he was involuntarily betrayed by Dajatch Alema. I relied also on my threat, which I impressed on Dajatch Alema, of unrelenting pursuit and punishment of all who might in any way be concerned in the ill-treatment of the European captives. I pointed out how the power of Great Britain had already reached Magdala, and that no corner of Abyssinia, however remote, could screen any one whom we wished to punish. Lieutenant Prideaux was met on his return to Magdala by Mr. Rassam and the remainder of the British prisoners, and several of those of other nations, all of whom arrived in my camp before evening. At the request of Dajatch Alema, I had promised to abstain from hostilities for twenty-four hours. After the lapse of forty-eight hours Theodore had not surrendered himself. Reliable information reached me that his army was recovering from their defeat, that many soldiers who had been unable to return to Magdala on the night of the 10th had since rejoined their ranks, that fresh defensive arrangements were being made, and that Theodore and his chiefs even contemplated a night attack on the 2nd Brigade, encamped on the lower ground. I therefore prepared to attack the enemy's position. I had originally intended first to assault Fahla from the side which fronted our camp, and was screened from the fire of Islamgee and Selasse. But under the altered condition of the enemy, Theodore having by death, wounds, and desertion lost half of his army and his bravest chiefs, I determined to attack Islamgee by the King's road. All arrangements for this had been considered, and the positions for the artillery reconnoitred and fixed upon, when information was brought to me that Theodore had left Magdala, and that many of the chiefs, with their followers, wished to surrender. I agreed to accept their submission, and ordered Sir Charles Staveley to advance on Islamgee, relaxing no precautions that I had considered necessary for the attack.

The scarcity of water rendered it impossible to retain any considerable body of cavalry before Magdala. My personal escort, under a native officer, only remained, and, with a few details of other corps, was sent under command of Lieutenant Scott, A.D.C., to watch the west side of Magdala, where they took up a good position, until the arrival of the cavalry, under Colonel Graves, who completed the invest-ment up to the Kafurbar gate, which was watched by the Gallas. The Bechilo was held by the head-quarter detachment of the Scinde Horse, under Major Briggs, and detachments of the 3rd Dragoon Guards, 3rd and 12th Cavalry, under Major Miller, to secure that point and provide against the escape of the enemy in that direction by the Minjerra ravine. A detachment of the Beloochees, under Lieutenant Beville, ascended by the spurs of Fahla, and occupied that im-portant position, where they were reinforced from the 2nd Brigade by the head-quarters wing of the 10th N. I. under Colonel Field. The artillery was placed in position, and the troops advanced, preceded by Captain Speedy, of the Intelligence Department, with a small escort of the 3rd Light Cavalry, under Lieutenant-Colonel Loch, to communicate with the chiefs who wished to surrender, and to prevent any misunderstanding. No resistance was offered. Sir Charles Staveley effected an entrance to Islamgee and Selasse through a difficult crevice in the rocky scarps. It would be impossible to arrive at any correct estimate either of the number of armed men who laid down their weapons, or of the masses of people—men, women, and children—whom we found on Islamgee. It was necessary to collect and guard the arms that were surrendered. It was also necessary to send down all the disarmed soldiers, and the miscellaneous multitude that followed them, to the plain below, before I could proceed actively against Magdala. Theodore himself, having abandoned his attempt to escape, was making preparations for defence, and offering us defiance in front of Magdala. By three o'clock, the Abyssinians having nearly all cleared away from Islamgee, I ordered the attack upon Magdala to be at once carried out. The entrance of Magdala is 300 feet above the terre-plain of Islamgee, and the ascent is by an extremely steep and rugged path. Viewing the very difficult nature of the approach I made the attack as strong as possible, and massed the whole of my artillery fire to cover it, in order to overpower the enemy's resistance, and prevent the heavy casualties which I should otherwise have incurred. The assaulting force consisted of the 2nd Brigade, led by the 33rd (Duke of Wellington's) Regiment, accompanied by detachments of the Royal Engineers, and Madras and Bombay Sappers and Miners, to clear away obstacles, the 1st Brigade to be a close support. I concentrated the fire of the artillery on the gateway and the north end of the fort, which were crowded with the houses of the soldiers, avoiding, as much as possible, the higher part of the interior, occupied by the Abyssinian prisoners and non-combatants. The enemy carefully concealed them-selves from view, so that the place seemed almost deserted, though, when entered by our troops, it was found to be thronged with soldiers who had thrown away their arms, released prisoners, and the numerous voluntary and involuntary followers of Theodore's fortunes. The artificial defences consisted of stone walls, loopholed and surmounted by strong and thick barricades of thorny stakes, with narrow stone gateways; the lower one built upon the interior, the higher one being seventy feet above the lower, and approached by a very steep, narrow path winding amongst the soldiers' huts. The attack was ably conducted by Sir Charles Staveley, whose report is annexed, and gallantly carried by the troops. Fortunately the defences were very unscientifically constructed ; and though the attack was met by a stout fire from the enemy, yet they could not direct it on the head of the storming party without exposing themselves to the rapid and fatal fire of the Snider rifles, and our loss was, in consequence, very small. The Royal Engineers and Sappers, and leading sections of the 33rd Regiment, were long before they could force an entrance, and during this time nine officers and men received wounds and contusions.

At length an entrance was forced by means of the ladders near the gate, and by the leading men of the 33rd, who scaled a rock and turned the defences of the gateway. The enemy were driven to the second barricade, and when that was carried all resistance ceased. Amongst the dead near the outer gateway were found a number of Theodore's most devoted chiefs ; one of them, Dajatch Enjeda, had urged Theodore to murder all the captives, a course from which he was dissuaded by others. Close to the second gateway lay the body of Theodore. At the moment when the barricade was forced by the 33rd, Theodore fell, as I have since learned, by his own hands; the troops immediately fled, some by the Kafurbar gate, which we found choked with arms that had been cast away in their flight. Of these fugitives, the greater part fell into the hands of the Gallas, and the remainder seeing the fate of their comrades, and hearing the vaunting invitations of the Gallas, returned to Magdala and surrendered.

The command of Magdala was entrusted to Brigadier-General Wilby, who held it with the 33rd and wing of the 45th Regiments. So thickly was the fortress inhabited, and so great was the crowd of people, that it was no easy matter to establish order. Guards were placed at the gates and at such places as required protection.[1] The family of Theodore were committed to the care of Mr. Rassam, who was requested to do all that was in his power for their comfort and protection. The Abyssinian prisoners were released from their chains, and the numerous body of Abyssinians, whose histories and condition

---

[1] Napier's conduct at Magdala is in broad contrast to that of Wolseley's at Coomassie.

it was impossible at the time to investigate, were collected in an open space in the centre of the fortress, where they could be protected, and where they quickly threw up huts for themselves, and remained until their final departure. On the 15th the 4th (K.O.) Regiment relieved the 33rd in Magdala, and the 45th were moved to Islamgee to reinforce the detachment of the 10th N.I., under Colonel Field, for the protection of the captured arms and ordnance, and to furnish working parties for their destruction. The inhabitants of Magdala were collected at Aroje, where great vigilance was necessary to protect them from the Gallas, who were lying in wait both day and night for opportunities of plundering and destroying them.

Notwithstanding the friendly relations with the queens of the Gallas, their people were so little under restraint that it was frequently necessary to fire upon them to drive them from molesting our water-parties and carrying off our mules. A party of them in search of plunder even dared to make their way into Magdala, where they were captured by the guard of the 33rd Regiment. On the 15th and 16th the disarmed soldiers and people of Magdala made their exodus from Aroje. Every consideration was shown them, and they were allowed to take all their property. The Aroje defile was guarded by infantry, and their procession after crossing the Bechilo was guarded by cavalry patrols until they reached Wadela. No doubt many of these people deserved little mercy at the hands of the peasants of Dalanta, who had suffered so much misery from Theodore's troops; but, having surrendered to the British forces, it was incumbent on us to protect them until they reached a point of safety from whence they could go to their native districts. On the morning of the 17th orders were issued to clear every one out of Magdala by 4 P.M. At that hour, the whole of the captured ordnance having been destroyed, the gates of Magdala were blown up, and the whole of the buildings were committed to the flames. The wounded Abyssinians, who had no friends to take charge of them, were conveyed to our hospital. The elephants and heavier ordnance having been sent in advance on the 15th, on the 18th of April the force recrossed the Bechilo on its return to the coast. I have the honour to enclose plans and photographs to illustrate the course of the operations. I regret that I have not been able to complete my despatch earlier; but owing to the marching in a difficult country, which has occupied the greater part of the days, and the many demands upon my time, it has been quite out of my power. By the next mail I hope to submit the conclusion of my despatch, which will include the re-embarkation of the greater part of the forces, and I shall then have the satisfaction to represent for the favourable consideration of Her Majesty's Government the services of the officers and men of the force which I have had the honour to command, of which it is impossible for me to write in too high terms, and to which I could not now do adequate justice.

I have the honour to be, Sir, your most obedient servant,

R. NAPIER, Lieutenant-General,
Commander-in-Chief.

Return of Ordnance captured in the Magdala Fortress on April 13, 1868.

Three brass 56-pounders 5 in. smooth bore, one brass 18-pounder ditto, four brass 6-pounders ditto (Turkish), two brass 6-pounders ditto (English) cast at Cassipore, one brass 6-pounder ditto (French), five brass 24-pounder howitzers (native and French), three brass 12-pounders ditto, one brass 3-pounder, four iron 1-pounder howitzers.

Appended to the despatch of Sir Robert Napier is Sir Charles Staveley's report to the Commander-in-Chief, narrating the capture of Magdala. It goes over the incidents already recorded by Sir Robert, and ends with a list of names of officers brought under the notice of His Excellency.

I wish to bring the names of the following officers to the notice of His Excellency :—Brigadier-General Wilby, who commanded the 2nd Brigade and led the assault. He mentions the very effective service rendered him by Captain Hicks, his Brigade-Major ; Captain James, Deputy-Assistant Major-General ; and Captain Capel, 12th Bengal Cavalry, A.D.C. ; Major Pritchard, R.E. (who was wounded), the senior officer with the Royal Engineers and Sappers, and who mentions the able assistance he received from Captain Elliot, commanding K company, Madras Sappers. Captain Foord, Lieutenant Bird, Cornet Dalrymple, and Surgeon Pearl of the Madras Sappers ; Sergeants Harrold and Dean, R.E., Corporal M'Donagh, and Sapper Bailey ; Major Cooper, commanding 33rd Regiment, who led this regiment to and over the gate. And I beg here to bring especially to notice No. 3,691, Drummer Michael M'Guire, and No. 949, Private James Bergin, 33rd Regiment, the two men who first forced an entrance to the extreme and turned the gate ; Captain Elliot, who commanded the Madras Sappers ; Captain M'Donnell, commanding Bombay Sappers ; Lieutenant-Colonel Parish, commanding 45th Regiment ; Colonel Field, commanding 10th Regiment Native Infantry ; Brigadier-General Snyder, commanding 1st Brigade ; Lieutenant-Colonel Cameron, commanding 4th (King's Own) Royal Regiment ; Major Beville, commanding wing Belooch Regiment ; Major Chamberlain, commanding 23rd Punjaub Pioneers ; Brigadier-General Petrie, commanding Royal Artillery ; Lieutenant-Colonel Wallace, commanding division Royal Artillery ; Lieutenant-Colonel

Milward, commanding division Royal Artillery; Lieutenant-Colonel
Penn, commanding A-21, Royal Artillery; Commander Fellowes, com-
manding Naval Rocket Batteries; Captain Twiss, commanding B-21,
Royal Artillery; Colonel Graves, 3rd Brigade Light Cavalry, command-
ing Cavalry Brigade; Lieutenant-Colonel Tower, commanding 3rd
Dragoon Guards; Major Gough, V.C., commanding 12th Bengal Cavalry;
Major Briggs, commanding Scinde Horse; Lieutenant-Colonel Loch, commanding details of Cavalry; Captain Moore, commanding 3rd Brigade
Light Cavalry; Lieutenant W. Scott, A.D.C., commanding the Commander-in-Chief's Escort; Lieutenant A. Le Messurier, Royal Engineers; Captain Leslie and Lieutenant Leacock, Bombay Sappers;
Captain Bainbridge, commanding Land Transport Corps, who had the
very difficult duty of providing water for the troops throughout the
operations since leaving the Bechilo river; Captains Ross, Griffiths,
and Twentyman, 20th Hussars, commanding divisions Land Transport
corps; Major Mignon, Assistant-Commissary-General, 1st Brigade, and
Major Goldsworthy, Brigade-Major of Cavalry, acting as Assistant-
Commissary-General 2nd Brigade; Dr. Mahaffy, principal medical
officer 1st Division, whose arrangements for the field hospitals were all
that could be desired; Surgeon-Major Wyllie, in charge of the field
hospitals, 2nd Brigade.

I desire also to bring to notice the officers of my staff: Lieutenant-
Colonel Wood, Assistant Adjutant; Major Baigrie, Assistant-Quarter-
master-General; Lieutenant Saunders, 20th Hussars, Aide-de-Camp;
and Captain George Arbuthnot, extra Aide-de-Camp.

These officers rendered me great assistance in carrying out the
orders of His Excellency the Commander-in-Chief during the day, and
have served with me since my arrival at Zoulla with the first arrived
brigade.

*Lieutenant-General Sir R. Napier, G. C. B., G. C. S. I., to
the Secretary of State for India.*

Commander-in-Chief's Office, Camp, Koomaylee:
June 1, 1868.

RIGHT HON. SIR, — In continuation of my despatch, No. 40,
dated May 12, I have the honour to report that I this day passed the
Sooroo defile with the last column of the Abyssinian Expeditionary
Force, the 25th Bombay Native Light Infantry, and the 27th Belooch
battalion.

The march from Dalanta to Antalo was trying, from the frequent
severe storms of rain which appeared to accompany us, and from which
our troops, in some degree, and more especially the followers and transport animals, could not fail to suffer.

CHAPTER XIV.

HOMEWARD—MARCH TO TAKAZZE—DEATH AND BURIAL OF LIEUTENANT
MORGAN—THE ARK—THE PRIESTS' SONG OF THANKSGIVING—THE HEAD-
WATERS OF THE GREAT NILE—PARTING WITH THE FUGITIVES FROM
MAGDALA—STILL HOMEWARD REJOICING—QUEEN VICTORIA CONGRATULATES
THE CRUSADERS—DEATH AND BURIAL OF THE EMPRESS OF ABYSSINIA—
PRINCE ALA MAYU—KUSSAI'S GOOD FAITH AND ITS REWARD—HIS PART-
ING WITH NAPIER—THE GREAT FLOOD IN THE SOOROO PASS AND A
MIRACULOUS DELIVERANCE — MY ARAB HORSE—HOSPITABLE FRIEND—
MORNING—THE TENDER FAREWELLS—ON THE SEA—HOMEWARD BOUND—
THE LAST OF THE ABYSSINIAN EXPEDITION.

On April 24 the army broke up its encampment on the plateau
and took up the line of march for home.

The British army, after the plenty that reigned in the camp
the last four days, with the good rest, so necessary, that they had
enjoyed, felt as blithe and merry as could be desired for a
rapid march to the sea-coast.

And with all dangers, all anxieties removed, with home and
its delightful associations rising before us, so

> ' No foe, no dangerous pass we heed,
> Brook no delay,—but onward speed
> With loosened rein ; '

and with the lively tune of ' Garry Owen ' ringing in their ears,
what could the gallant Irishmen of the ' Duke of Wellington's
Own ' do but spur

> ' O'er the hills like hayroes fighting
> And tayring all before them,'

followed by their war-comrades the ' King's Own,' the 45th, the
3rd Dragoon Guards, the thousands of native Indian soldiers,
and the thousands of camp followers.

The army was in excellent health when it started from Dalanta. There were but eighty in the dhoolies, inclusive of those wounded at Magdala and Aroje.

But the marching *en masse* by brigades soon exhibited the weakness of the animals, superinduced by the sufferings they had undergone in the Bechilo valley. We passed the Jeddah ravine, and terrible were the effects the next day. Officers and men began to suffer from a relaxation of their whole bodily system, occasioned by the removal of the objects for which the campaign was undertaken.

From Sindee we passed on unhalting to Abdecoom; thence to Gazoo and Santarai; thence downward to the Takazze, on the banks of which we encamped on the 28th April.

At this place the British army suffered a loss in the death of one of its most promising officers—Lieutenant Morgan, of the 10th Company of the Royal Engineers. He was a young, able, courageous, and energetic officer at all times, besides being a true Christian gentleman. He was buried on the banks of the Takazze river, under the shade of pleasant, umbrageous trees which surrounded the church of St. Gabriel. All the officers and civilians attached to the 1st Brigade followed the body to the grave, to pay the last sad respects to one who was entitled to the goodwill of all who knew him. The services read over him were very affecting, and those who stood by during the interment felt that indeed a bright soldier had departed. Under the waving trees his grave was made, which in future years, when another generation shall have come and gone, will be an object of veneration to European travellers, as the resting-place of one of the famous Crusaders who marched to Magdala.

The Commander-in-Chief issued an order commendatory of his high excellence, his unflagging zeal and energy, and expressive of his regret at the loss the British army sustained in his death. Lieutenant Morgan was interred with all military honours.

In the evening we had signal evidence of the feelings with which the natives regarded the conquerors.

While we were marching southward they held their fingers to their ears and their hands before their faces whenever the soldiers spoke to them of what they intended to do; but now

that they saw that the soldiers had reduced the fortress to a
scorched rock, and dispersed Theodore's grand army of In-
vincibles to the four winds, they came out to tender ovations.

Their pride and arrogance gave way before the stupendous
fact that they beheld the conquerors of ' Theodorus.' They
approached in masses to sing our praises, taking for the subject
of their psalmody David's thanksgiving psalm for ' powerful
deliverance and manifold blessings,' which commences with
' The Lord is my rock, my fortress, and my deliverer; the God
of my rock; in Him will I trust; He is my shield and the horn
of my salvation, my high tower and my refuge, my Saviour;
Thou savest me from violence ' (2 Samuel, xxii).

It was chanted with due Davidic fervour.   ' David played
before the Lord on all manner of instruments made of fir wood,
even on harps, and on psalteries, and on timbrels, and on
cornets, and on cymbals,' while ' leaping and dancing,' as
Michal, daughter of Saul, saw him, girded with the linen ephod
—with the people shouting with all their might and with the
' sound of trumpets '—so the Abyssinian priests, deacons, elders,
and children of the church, brought forth an imitation of the
Ark.

Five yards was its length, a yard its breadth, and a foot was
the height of this Ethiopic image of the Ark.   It was covered
with a scarlet cloth embroidered with gold, and on the top of
it there was a semblance of the Mercy Seat—not of pure gold,
but of wood covered with crimson silk.   Above it there was a
canopy of this silk, like unto a crown with rich embroidery
round about it, with a border of two hands' breadth pendant
from its sides.

On the ark there were neither cherubim nor the sheckina
visible.   In these things the ark was deficient, and came not up
to the original.   High above the heads of the psalmodising mass
were held the garments and the vessels, the candlesticks, the
laver, priests' robes, hyssops and communion cups, pixes and
chalices, crosses of brass, and of silver, and of gold, together
with other officiating instruments denominated ' holy things.'

The head priest wore his ephod—a short vest without
sleeves—over his stole, with the semblance of a mitre enfolded
by an enormous turban round his head.   The breast-plate of

this sub-hierarch was a richly decorated square piece of tanned ox hide.

The neophytes, clothed with their cotton togas, held the belts and kept up with him—to the intense admiration of the boys—a deafening jingling clamour; their instruments of juniper wood and brass, one-stringed banjoes, clanking brass cymbals (stuck on the players' numerous joints), keeping symphonious time, while high rose the epopeæ as they advanced towards head-quarters.

Amongst them, chanting with vigour, were several arrant rogues, who, with unparelleled mendacity indexed on their grotesque features, beseeched the mighty Feringhee Sooltans to give them backsheesh.

In the mazy crowd of dusky men, women, and children there was a remarkable paucity of honest faces. I fear I should be flattering them if I admitted that there was even one who possessed a soul which could have resisted the temptation of picking and stealing if opportunities had been forthcoming.

The inherent human failing to acquire things by superior cunning was stamped by nature on each tawny physiognomy. After all the romance of the land had become stale with familiarity; no one in the army, I believe, could ' abide ' to see one Abyssinian near his tent.

However, as they came to sing a lay in the Feringhees' honour, no one had the heart to object, provided they kept at a respectable distance from the tent doors. As Sir Robert Napier in his triumph over them and in his large-heartedness believed them to have been faithful auxiliaries in the matter of furnishing the various camps with provisions, the soldiers and officers put as good faces on the matter as they could, and found humour enough left in their bosoms to enjoy the singing, and applaud the Terpsichorean performances.

As the priests command by far the greatest respect among the natives, they of course took the front of the assembly. When they were all grouped in their respective positions—fifty paces in front of head-quarters—a strong-lunged priest, evidently the choragus of the native choir, struck up, with a serio-ludicro air, the first note of the impromptu stanzas about to be sung in our honour, descriptive of the Feringhee conquest over the Negus.

As he warmed in his theme, and his voice rose to enthusiasm, the motley assembly, at the waving of a crucifix, chimed in with chorus, which, sung with stentorian lungs, had a tremendous effect.

After the chorus, six priests clad in cotton stoles, headed by the sub-hierarch, took the eulogy up at a very low key, which soon, however, rose so high and shrill in a protracted continuity of sound, that one momentarily expected to hear their lungs crack, ending with a stormy chorus as before.

Then, forming themselves into a circle, a hundred of them commenced a dignified sailing round their neighbours to the right and left, their togas getting inflated with the movement, weaving each into another, until it might have been imagined they had manufactured some complicated knot on the Gordian principle ; but, soon taking the reverse method, they reached their former positions in time. The singing went on louder and louder, and, the choragus giving the warning clap, the whole multitude clapped their hands ; the women and children struck up the silver-toned ' Li-li-li,' performing a dance similar to the Chinese hop and skip.

They then formed triplets, and massed themselves together, when a shrill note from the boys sent them all into a confused whirl, round and round, the sub-hierarch and his six assistants going faster and faster, as they acquired momentum, clapping their hands, singing louder than ever, the head priest ducking his body lower and lower, and more energetic, until the dance, in the excitement which they all laboured under, assumed the appearance of a Jubilee medley, composed of waltzes, Dervish dances, sarabands, fandangoes, pirouettes, and chassés—the three latter performed by the most youthful of the assembly.

It must not be forgotten that all this time the Ark and Mercy Seat—minus the cherubim—stood on the ground near the priests, while a choice number of infantine neophytes manfully rung the merriest chimes, and the instruments of juniper wood, the one-stringed banjoes, and cymbals, made as much discordant music as was possible under the circumstances. The Ethiops, before concluding the entertainment, raised once again the *Canto Trionfale !*

Situated as our camp was, near the fountain-head of the

mysterious Nile, on the banks of the Takazze, a dreaming stream, flashing and murmuring at our feet in the bright evening—the men's voices resonant among the sinuosities of the heights, the children and women's voices toned in harmony with the deep bass, like the piping of a living gale through a ship's shrouds, to the deep roar of the storm—one may well imagine that there was a charm in the scene, that it was another event to be remembered in the drama.

Perhaps the thought of heroic Morgan's grave, the magic tints of the sunset, or the pleasurable emotions which the weird music aroused, or all combined, tended to heighten the enjoyment.

One hundred and twenty-three mules, which were crippled and galled till past service, were presented to these people for their favours, besides an abundance of corn and grass which had been collected during our absence, and which the army transport train could by no manner of means take with them.

At this camp we parted from the last of the fugitives from Magdala—they continuing their march, Lake Tzana-ward, *viâ* Lalibala, while we jogged on northwards. While in company with the army they were fed out of the abundance gathered at each station.

There was no scarcity of food on the return. The garrisons, during the absence of the army, had collected stacks and heaps of provisions of all sorts; for when once the silver dollar had touched the palms of the natives, whatever they possessed became immediately vendible. It was really marvellous how the supplies rolled in, considering the insufficient tillage of the ground, their natural timidity and distrust of strangers, as well as the narrow band of country from which the provisions were raised.

Forward the army went rejoicing by the way from the Takazze to Muja—Muja to Wandach, over cloud-fenced Mezgar Amba, bidding adieu to the Takazze valley, and downward to Dildee, where we recovered our clothes and effects, and once more revelled in Bohea, whose essence and aroma were as water to parched lips ! From Dildee to Marowar was a sore march ; the mules died by hundreds. From Marowar to Lat, where the 1st Division had stopped for action on its way to Magdala— thence to Muzzageeta—and next to beautiful Lake Ashangi, where the wide Syrtus on its further side was smiling with every green thing !

At Ashangi we received news that the floods in the Sorroo Pass and Senafe Ghaut had begun; that the military road through the pass had been destroyed ; and that there was every appearance that, now the rains had commenced in earnest, the road could not be rendered available for the passage of troops to the sea-coast.

Miles of telegraph wire had been swept away, and much hay had been destroyed. Various suggestions were of course started for the amelioration of the evil.

It was proposed to build huts on Senafe, wherein the troops could stay during the rainy season ; and again it was proposed that the camps should remain as they were in the ghaut, and that each station should cook provisions for the soldiers by the time they should arrive—the telegraph to warn the respective stations of the hour the regiments might be expected. But these suggestions fell to the ground, on account of the difficulty of decision until Sir Robert was thoroughly posted as to the extent of the damage done by the floods.

That night at Ashangi we experienced a heavy shower which made the tents very wet and cumbersome ; but by 10 o'clock next morning they were perfectly dry, and the army continued its march. There was a large amount of forage burnt at the lake, the General preferring to burn it rather than distribute it to the wretches who had repaid his kindness with so much ingratitude. They had taken every opportunity to retard the conveyance of supplies to the posts beyond while Colonel Milward was practising his battery, the day before, upon points near the lake and on its waters.

The groves of Haya, on the southern slope of Mount Masobo, were reached three or four hours after we left Ashangi.

Here we met, on our return, the officers of the Trigonometrical Survey, Lieutenants Carter, Dummler, and Holdich, Royal Engineers, who, by the most strenuous exertions and at the cost of great fatigue and privations, had succeeded in surveying nearly 6,000 square miles, carrying their work from the sea-coast to the interior, as far as Dalanta.

At Haya, though surrounded by wet dingles and mist and

fog, our stomachs were gratified with port wine, brandy, and the pure Jamaica—the results of a necessary distribution of the hospital stores, as their carriage demanded more animals than any benefits that could accrue from their preservation were worth.

From Haya to Pilago was a short march, owing to the rains and mire. Five of the transport elephants were shot here—a necessity that could not be obviated because of their extreme weakness.

The next day the army left Pilago for Mukhan, thence to Atzala. It was thought that Sir Robert Napier, upon arriving at the latter place, would have thrown a few shells at the amba of Aleggie out of compliment to its owner, who had been assiduous in his efforts to retard the native carriage of supplies; but Welda Yassous had fortunately taken wings and deported himself away, in which case any demonstration made against the amba would have been deprived of its effects.

From Atzala the prudent Napier marched his heterogeneous army still northward, unhalting, homeward. The pleasant glades and bubbling stream near Mesheck were reached and passed ; the wild and weird amphitheatre before Musgee was seen as soon as we passed the portals of Aleggie ravine; and the next day, to our great relief, the fortified camp and plain of Antalo dawned upon the view. We were half way to the sea-coast. The rearmost brigade arrived at Antalo at noon, May 12, 1868.

The march from Magdala to Antalo proved very fatal to the animals of the transport train. The road was literally strewn with them. In a proportionate sense the retreat from Moscow could not have proved more adverse to animal life than this march. Hundreds of mules died daily. Camels succumbed the most readily, and even the elephants suffered. The cavalry horses died by dozens.

The crossing of the Jeddah ravine and the Dildee mountains cost the lives of 700 animals. A single glance at a route chart will at once explain how trying so many consecutive ranges of mountains would naturally be to loaded animals.

From Antalo to the sea-coast, instead of the bungling disorderly mass that the army presented on the previous marches, it was to be divided, each regiment of foot and horse having its appointed days of marching.

Upon the arrival of the army at Antalo the following General Orders and telegrams, received from the Queen of England and the Royal Duke of Cambridge, were read to the regiments by their commanding officers :—

'Camp Antalo, May 12, 1868.

'The Commander-in-Chief has the highest satisfaction in conveying to the army of Abyssinia the following message received this day by telegraph.

'Sir Robert Napier heartily congratulates all under his command on the flattering recognition of their services by Her Most Gracious Majesty Queen Victoria :—

'"The Queen sends hearty congratulations and thanks to Sir Robert Napier and his gallant force on their brilliant success."'

'Camp Antalo, May 12, 1868.

'The Commander-in-Chief has much satisfaction in publishing to the troops under his command the following messages received by telegraph from His Royal Highness the Field Marshal Commanding-in-Chief, and from the Right Honourable the Secretary of State for India, respectively :—

'"We all rejoice in your great success, and in that of your gallant and enduring army."

'"I congratulate your Excellency with all my heart. You have taught once more what is meant by *an army that can go anywhere and do anything. From first to last all has been done well.*"'

Thus were the first blasts of the universal Jubilee that convulsed all England wafted to the interior of Abyssinia. Well might Her Gracious Majesty Queen Victoria be pleased to congratulate Sir Robert and his Crusaders; well might His Royal Highness the Duke of Cambridge transmit greetings by cable to meet the army of Abyssinia. Though a little war, it was a great campaign; and not in Abyssinia—mountain-locked though it were—was there a revulsion of feeling experienced; the fame of it sounded with loud reverberations over wide Asia. Princes and potentates, scattered far apace, heard the noise of it and trembled; a peaceful epoch, they saw, had not vitiated England's strength; and *this last, this best, this greatest of victories,* seemingly made so little of, established her prestige on a firmer basis than ever.

From Antalo to Senafe the army continued its way unfalteringly and without intermission. The only event on the route was the death of Empress Egenie Torenachie, relict of the Emperor, from consumption. Very genuine grief was exhibited by her dependants at her funeral. She died on May 15, at Haikhullut, where she was buried. The military band of the 'King's Own' played the 'Dead March in Saul,' preceding the coffin to the grave. Near the coffin were her maids of honour, weeping violently at what they considered an irreparable loss. Behind came the officiating priests of her church; the young Prince, accompanied by Mr. Rassam, besides one of the Commander-in-Chief's aides-de-camp, and several of his staff, followed by a great number of Abyssinians.

She had purposed to go to Bombay with her little son, believing that course the wisest, since such was the advice of Sir Robert Napier.

The young Prince Ala-Mayu, which name means 'I am seeing the world,' was now left an orphan; but the Commander-in-Chief's intentions concerning him were in no wise changed— rather strengthened than otherwise.

The last of the army of Abyssinia arrived at Senafe on May 24. In spite of the monotonous journey from Antalo to this first highland post, it every day grew more and more evident that the campaign was drawing to its close; that the sea was nearing us, or that we were nearing it; and that once on the sea, Home, Sweet Home, lay not far distant.

Six months before the army had left Senafe with bosoms steeled for their duty, and nerves braced for what was to be done. Two harvests had been reaped since we had left this initial post, and it was the vernal period with every blade of grass and stalk of corn for the third time.

Prince Kussai of Tigre, whom I have unintentionally left in silence for so long, deserves one more notice from me. Loyal ally of a loyal man he was, and had remained so since first he plighted his faith by the side of that stream near Haussein. Whatever prince could have done was done by him. Constantly assiduous in furthering the British General's intentions, deftly energetic in raking supplies of forage and provender for both man and beast, not a particle of dereliction of duty could be charged to him; but what thanks could be clothed in words were certainly due to him. Wagshum Gobazye had been as full of promises as the Prince; but his promises were, as a rule, but fruitless words for the entertainment of the British General. While in his country of Lasta and Samen, Sir Robert Napier could not say that he knew thoroughly the insincerity of Gobazye's pseudo-friendships; but when the open, manly, sincere face of the Tigrean Prince and the recollection of his courtesy flashed across his mind, it is of course unnecessary to say that Kussai could claim the preference in his favour.

The second day after the British army arrived at Senafe, Prince Kussai, with a chosen body of 300 cavalry, came to bid adieu to the British army. He was rapturous at the great success of Napier, and complimented him with all the courteous phrases that the Abyssinian vocabulary admits of. He said he wished Napier ' all the happiness that Paradise can bestow, the applause of his countrymen, all and highest honours, plenty of wives and children, years of life, and a comfortable seat at the right hand of the patriarchs—Abraham, Isaac, and Jacob.' Amen! says the author.

It is an indubitable fact that Sir Robert Napier, though a British General, discarded all frigid conventionalities, reciprocated the glowing compliments, and lovingly embraced him for his fealty.

There was another review held at Senafe in honour of the Queen's birthday, and the admirable and intelligent manœuvring excited the Prince's warmest admiration. He, also, desirous of appearing in the best light before Sir Robert Napier, the gentlemen of his staff, and the groups of foreign officers, drew his men up in two lines and went through the evolutions of a sham *melée*.

The Prince performed his part very well, his horsemanship being by far the best. Once the signal was given, the intended sham battle became, as I term it, a sham *melée*, where each man dashed against his opponent without regard to the order of the formation, fiercely assailing with the blunt end of his spear whoever came across his path. The retreating party, which soon

made itself recognisable, were almost as dangerous in their retreat as in their onset, for, Parthian-like, they shot imaginary
bullets while flying from the field.

For his good faith Sir Robert Napier was pleased to bestow
a valuable guerdon on Kussai—especially valuable for one in
his straits—namely, 1,500 muzzle-loading muskets, 'Brown
Besses,' 400,000 rounds of cartridges, and six pieces of artillery
with shot and ammunition to match. With this magnificent
present there was no doubt that he would be able to make his
own terms with the depredating Egyptians who had too long
been allowed to perform nameless crimes upon Abyssinians
unchecked. Kussai, with the aid of several Europeans who
wanted to go back to him, might now dictate terms to both
Wagshum Gobazye and the Egyptians.

I started down the Senafe Ghaut with five animals loaded
with baggage and trophies, attended by as many servants, three
or four days prior to the advent of the army within its confines
on the return. The pass appeared as when I first started on
the campaign ; but I thought the birds sang blither notes of
welcome, and the steep slopes, garnished with numberless shrubs,
looked greener and more joyous in the light. Rayra Guddy
was reached by noon, and after an hour's rest we continued our
way downward, threading patiently the numerous angles and
labyrinth of curves, until we arrived at Undel Wells.

The next day, after despatching a breakfast cooked long before dawn, we continued our journey, arriving at Upper Sooroo
station at noon. Here we rested a couple of hours while the
animals were feeding and our lunch was being cooked.

About 2 P.M. my little kafilah was on the road again, followed in the rear by two or three officers and their baggage
animals, who were as eager to reach Zoulla as I was.

The heat at this time was intense, increased as it was by
the incandescence reflected by the white walls of quartz,
granite, marble, and limestone, between which the pass ran its
serpentine course. Two miles from Sooroo the extreme width
of the pass is about 30 feet, and for five miles afterwards it
contracts itself at intervals to within 20 feet, the walls of
sheer rock rising to a dead height of 800 feet.

As we advanced, the heat became more intense, and I
noticed that the sun had become obscured by drifting murky

masses of clouds. Whole gusts of torrid wind occasionally swept towards us.

When about three miles from Sooroo, these premonitory symptoms of a storm had been noticed with much anxiety by the party in our rear. As we rounded an abrupt curve of the gorge we met a multitude of officers and men, numbering somewhere about a thousand, advancing to meet us on the run, as for dear life.

As we neared each other the officers shouted out for us to return to Sooroo, saying that it was impossible to proceed further because of the deep pits the cataracts had formed in the road, which had been left unfilled.

They said they had been busy with the working parties to try and make the road passable by the time the army should arrive from Senafe, but were obliged to retreat up again to the station; for during the rains nothing could live in the pass, as it filled up in three hours to the depth of ten feet, sweeping on its impetuous course ' boulders weighing hundreds of pounds.'

The lurid clouds, the close atmosphere, and general gloom that prevailed, predicted beyond a doubt the approach of the diurnal rain-storms; and, said they, ' It is impossible even now to reach Sooroo before the water will be two feet deep in the pass.'

Their earnest expostulations and the evident danger to be apprehended had the desired effect upon the officers who were in my company, and they retreated back to the depôt; but I was determined to proceed, and my only reason for doing so was the fact that the mail steamer departed from Zoulla for Suez on the next evening, and a delay of half a day would prevent me from reaching Zoulla in time for making my preparations for departure, the distance being nearly thirty miles of roughest road. The officers hurried on their way, satisfied that I was a doomed man.

The Irish soldiers, at all times ready with their jibes, saluted me with numberless witticisms, to all of which I replied good-humouredly. They made up their minds that in about ten hours I would find myself in ' Davy Jones's locker.' Some of them seriously remonstrated with me; but, I suppose, remembering the adage that ' a wilful man must go his own way,' they left me. The Coolies sung out in chorus, ' Rah Naheen

Sahib, Sahib—Rah Naheen (there is no road, sir); but when they saw that I was deaf to their cries, they also felt assured that they saw the last of the Yankee Sahib, and went on their way saying, ' Khooda hafiz Sahib '—' Sir, may God be your protector!' I thanked them for their prayers, but nevertheless whipped my animals to a quicker pace.

The riband of sky, seen betwixt the two walls of the pass, assumed the blackness of a velvet pall, and, before the last of the working party had barely passed beyond view, the rain began to fall.

Down came the drops slowly, of almost the size of a dollar as they fell on the stones and pebbles; then thicker, larger, faster. The confined thunder, low, muttering at first, like the rumble of artillery over a wooden bridge, suddenly broke forth in loud peals. The forked lightning became vivid and fierce. The clouds unfolded themselves; the ' windows of heaven ' were opened; and the rain poured in an unbroken flood.

Forward for our lives we travelled, almost on the gallop, plunging into already formed puddles, leaping into pits still left unfilled, over round boulders, and smooth, slippery rocks, looking keenly around for some sheltering rocky ledge convenient for halting during the storm.

We were drenched to the skin in an instant, and the water dripped from us as if we were mermen. The luggage was also rendered so wet that it soon became evident that the animals could not carry their loads. The mule which carried the tent was already staggering under the weight, and the one which bore Theodore's carpet looked, as it were, suddenly broken-backed. The road became choked with streams. A current was created which swelled and raved and washed deep ruts in it. As yet there was not a sign of any place that could afford shelter to a mouse—the walls on either side of us were as upright as if constructed by line and plummet.

From each summit of these walls a few tiny slender strings of water poured downward of the colour of molten glass; but before a second thought could have been taken, the tiny waterfalls had become cataracts which fell from either height along the whole extent of the walls, deafening to the ears by the terrific roar which filled all space with its thunder. This new danger made itself apparent directly; Zahoorka, one of my

Hindostanee servants, carelessly went too near the base of one of the cliff walls, and was caught by one of the cataracts, which instantly bore him to the earth, and the current, now ominously strong, swept him with the speed of an arrow over a lot of great rocks into a pool of water a few paces ahead of us.

By timely aid he was saved, but not before he had been severely wounded on the head, and had received serious bruises on the body. The pool of water halted my party ; we could go no further. Upon examination it was found to be 8 feet deep and 50 feet long, extending across the pass. A rocky platform, whose side and surface were worn smooth as a flag- . stone by the action of many a century of torrents, jutted about ten feet into it, leaving a space of five feet between the sides of the projection and the pass walls. The ground at its upper end was composed of gravelly soil, the loose friable nature of which offered no impediment to the powerful torrent that now flowed towards the sea irresistibly.

On this rock I massed my party. It was to us a very rock of refuge, unless the debacle, as the British officers said, should fill the whole pass to the depth of ten feet, in which case there would be nothing for us to do but to swim our horses adown it at the rate of twenty knots an hour, or—drown !

While our minds were excited by the dangers which beset us, their nature and magnitude were not so apparent as now that we had found a halting-place, and were condemned of a necessity to inactivity upon our friendly eyot. Our knowledge of them hitherto partook of vague formless presentiments, which animal instinct served to convey to minds already all absorbed in the problem of how to evade the effects of the flood. But now that we were providentially rescued from instant death, my mind was soon engaged in the analysis of the causes of this diluvial phenomenon. Here is not the place to enter into a detailed description of them ; let it suffice for the general reader that this sudden flood so powerfully disruptive—as we may be assured of that when we know that it had formed a chasm of 800 feet in depth—is the collected rain from an area of a hundred miles in extent, and the unaccountable cloughs and ravines that disembogued their contents into this great pass which is the main duct for the passage of the water to the sea.

Naturally, man cannot long philosophise when surrounded

by an element risen to the destructive pitch, and mind and
body suffer from its effects. We were led to note the action of
the current fast rising into a deluge ; and the more we gazed
upon the scene the more certain was our fate defined should
the rain long continue. How bitterly I repented not having
conformed to the prudent admonitions of the working parties,
will be known only to myself. The time and place were very
adverse to vain regrets. What remained for us to do, was to
be as patient as possible ; to watch with aching eyes the progress of the flood, listen with awe to its roar, and mutely admire the sublime grandeur of the scene.

Downward poured the rain unceasingly, and the thunder
roared above, the echoes of which deafened our ears, multiplied
ten thousandfold as they were flung from wall against wall,
while the lightning illumined with its ghastly green flames the
rifted edges and pinnacles of each colossal height.

Still downward plunged the watery sheets, now silvered with
foam. Onward adown the pass swirled the turbulent current,
growing higher, ever higher, bearing everything before it but
our own firm granite eyot, tearing the friable gravelly soil away
by tons, rioting round abrupt curves and angles in raging
confluxes, foaming over obstructions, forming powerful rapids,
whitened cascades, and cataracts, which roared like ten thousand
bass drums beating the battle-call !

These infinite sounds and multiplex echoes were enough
in all conscience to appal the stoutest man ; and my Arab and
Hindoo servants were far from laying claim to such distinction.
They expressed their fears and the wild alarm which had taken
possession of them in prolonged shrieks.

Still the rain fell unceasingly, plashing tumultuously in our
faces, blinding us with the spray. The current rose higher,
ever higher, every instant waxing mightier, every drop adding
more force to it, and increasing its violence. There were three
feet of water in the pass.

The current became a river—a mighty one— all powerful—
sailing towards us in an irresistible flow, its placidity only ruffled
when it came surging against our rocky islet ; for it had been
made one long before, the bank of gravel by which we came
having been eroded and torn away so soon as the current had
gathered any strength. Well for us it was, and yet we could

not by any means be deemed safe. The water was fast rising; it was even then within eight inches of the surface, and half-an-hour's more rain would cover the rock three inches, according to computation made.

Engirt as we were by those looming heights, shedding powerful cataracts which resounded like the advent of doom, and shrouded as we were in the deep gloom, made darker by the spray and vaporous masses which sailed majestically upward, our condition was truly pitiable.

To add to the horrors of the scene, corpses of men rolled by with outstretched arms appearing above the turbid waves, and carcases of oxen, horses, and mules tumbled headlong, colliding heavily with rocky obstructions; telegraph poles shot past with lightning speed, dragging long lines of wire after them; carts and wheels crushed against our granite islet; bales of hay and yellow straw, clumps of bushes, and again corpses of men, all commingling in the turgid stream.

These ghastly evidences of the ruin and death created by the wild element increased the terrors of my servants, and their shrieking prayers to 'Allah' were heard far above the tumult of either tempest or flood.

After raining steadily for an hour and three-quarters, there was an instant's pause, when it recommenced more furious than before. The water became level with the top of the rock, and was still rising. We looked upward; the mist was gradually clearing. Then a few sun-blinks stole through the dark vapour tinging the rain-drops with a mysterious shimmer, transforming them as it were to showers of brilliants, and the dripping projections glistened with a breathing lustre; and there was revealed also to us a few scant bushes, which dotted the Alpine summit, quivering and flapping under the battery of the storm, like forlorn souls tossing on the verge of eternal death. There was something inexpressibly grand in this sudden lighting of the gloom.

The water touched our feet and washed our baggage resting on the rock. The animals became restive and whinnied piteously, the muleteers prayed with unusual volubility, and I gave up hope. I felt certain that in a few seconds at most we should be battling for life in the waters, which were still rising higher and higher. But suddenly the rain ceased. The mist dissolved before the splendour of the returning sun, and a segment of the Bow of Promise was seen spanning the zenith! We were saved!

Within five minutes the cataracts had diminished into a few weakly jets, which soon ceased altogether, and within ten minutes the water had receded two inches.

At five o'clock there were but tiny rivulets where but an hour previous a river had flowed.

But how were we to descend from our rocky pedestal, the surface of which was four feet above the water? I had five baggage animals and my riding horse, with all the luggage, on an island in the centre of a pool of water 8 feet deep. Most uncomfortable position to remain all night there, shivering in wet clothes; and, unless matters were hurried, it would be dark before we could get to the Lower Sooroo station, three miles below. As it was evident I should have to sacrifice the greater portion of the baggage, I at once set about selecting such things that I could best dispense with. So Theodore's carpet, which I had bought at the auction for fifty rupees, now of a weight of nearly 800 pounds, was tumbled over into the pond, though with many regrets at the loss of my money, the dissolving of my romance, and the futility of the trouble I had taken with it over a journey of 350 miles; then the servants' tent and half their bedclothes were discarded; a trunk full of clothes, curiosities, a lion skin or two; all superfluous camp-kettles. Two days' rations, and two other trunks, with their contents, I divided in light loads among the animals. My tent was next cut in two, and the poles thrown away, rendering the carriage of the canvas easy enough.

It now remained for me to get the animals away from the rock. They could not go of themselves, because the sides of the rock were perpendicular and as smooth as glass, and coaxing them into it proved futile. The plan had been tried and it would not answer.

I was therefore obliged to have them pushed in so that they could swim for the road, which was but 15 yards below us. The first one upon which we tried the experiment was an Abyssinian mare. She fell in with a heavy splurge, but soon came up swimming bravely for the shore below, which she reached in good order. The next one was a mule, a thick-skulled animal, and decidedly muleish in every way. It required the united efforts of all my servants to push it from the rock. In falling it turned a somersault, fell with a heavy sound, and never came up again—so much for incorrigible stupidity. The other three animals reached shore in fine style. Then tying a hitch rope to the baggage, it was all thrown into the water, and dragged through safe to land.

My indispensable adjunct Ali of happy memory, my Arab charger Sayed, and myself, were all that now remained on the rock. It therefore devolved on Ali and myself to remove Sayed from his rocky pedestal whereon he stood. Unsaddling him, and fastening the bridle well around his neck, we led him close to the edge of the rock, and tried to push him over ; but this was an ignominious style of treatment which Sayed's noble spirit could ill brook, and with his heels and teeth he at once set about to inform us that he could not and would not put up with it.

After debating that question with Ali—who was a wiseacre concerning horses, especially Arabian ones—it was decided that the 'Yankee Sahib' should ride him in. Now, I am a capital rider on a smooth, level prairie, or even on an Abyssinian road ; but it is quite a different thing to leap a horse into a deep pool of water from a height, and it was with considerable misgiving that I mounted him—after, however, first seeing well to saddle gear—while Ali, in pure Arabic, had bestowed any number of endearments upon him.

Cautiously taking up the bridle reins and requesting Ali to jump in and swim away, I urged Sayed gently to the edge of the rock, and permitted him to smell and snuff and snort as much as he pleased for three or four minutes. Then backing him two paces and setting myself well in the saddle, I dug my spurs simultaneously into each flank, lifting him up at the same time with the bridle. The effect of the spurs on the blooded horse was instantaneous ; he couched himself for a spring for one-eighth of a second, and like a panther he bounded off the rock clean into the air and into the turgid water. Sayed, after flapping his ears, headed for the shore, half swimming.

When he landed the lively Ali warmly hugged Sayed, and in honour of the triumph which he felt the noble Arab horse had won he extemporised an eulogistic lay of his high courage then and there.

We had soon packed up again, and resumed our journey, the pass being now almost dry. After struggling over two miles of boulders, and pits, and flat rocks, the chasm widened into a ravine, flanked by sloping ridges, wet crags, and dripping dingles stretching away in vague and mystic perspective.

We arrived at the Lower Sooroo station about 7 o'clock in the evening, just as twilight became night. An officer who was in command of the post came out very much astonished to find that we had survived the dangers of the pass. The bodies of five men had been found drifting when the flood was at its maddest, from which the garrison had presaged greater calamities. After I had concluded my tale, the officer said he was 'distwessed beyond meashaw,' but would I 'take a peg—something warm, you know; or a glawss of bwandy and sodawh; do you good, you know—'pon word, vewwy sowwy, vewwy sowwy.'

The hospitable officer's invitation to 'take a peg,' need I say, was accepted with many thanks; but it was not until we had pitched upon our camp and built a roaring fire, and had gorged ourselves to repletion with a bountiful supper, drunk quantities of hot coffee, and had lit our pipes and bubble-bubbles, that we could talk and laugh at our ease at the walled Pass, the mighty deluge, and the terrible dangers of the Great Sooroo.

Warmed by the glow of the cheerful fire, and fortified against rheumatic fevers by the hot liquids we had taken, we sat and talked together in broken English—for I always make friends of my servants when they are inclined to be good. For hours we chatted in the lurid shadows of the shelving rocks, until, unconsciously, we were wafted between the portals of the ivory gate into the phantom realm, 'the wide pathless desert of dim sleep.'

At 4 A.M. the next morning we were on the road for Koomaylee, and we reached that camp at the embouchure of the Sooroo Pass, about an hour later.

Breakfast despatched, we defiled past Koomaylee, and emerged into the fulness of the day with the fervid plain of Zoulla expanding to the right and left, and for twelve miles in our front; and beyond the snowy cantonment gleamed the azure waters of the Red Sea, with the leviathans of the deep riding on the surface in magnificent array !

About noon we all arrived at Zoulla, glad at heart at the prospective rest before us.

Zoulla appeared very different from the complicated turmoil and chaotic disorder that reigned there when I first landed. Immense warehouses had been erected, which contained tons upon tons of stores and every known requisite of a well-ordered campaign. Bazaars were numerous ; and a half restaurant and half grog shop, euphoniously named 'Annesley Hall,' had been built to accommodate the officers, suffering from the atmospheric fervour which averaged 120° Fahrenheit, with delectable ' pegs,' 'bwandys and sodawhs,' ' smiles,' and such like refreshers. All the hospitals had been removed to Koomaylee, which was undoubtedly as good a location for a sanatorium as could be procured between the highlands of Senafe and the sea.

By 6 o'clock in the evening I had sold my baggage animals for a mere cypher, discharged my servants, and seen Ali and my Arab horse Sayed[1] safe on board the transport steamer ' Indore.' She was to leave early next morning for Suez, with Colonel Milward of the Royal Artillery, who had been chosen by Sir Robert Napier to convey his long delayed despatches and the four crowns of Theodore to the Queen of England.

Our ' mutual friends' of the Commissariat, and Captain Dawes, Bunder-master, did the honours of a *recherché* table to the returned Crusader. Before I embarked there were uttered some solemn promises of friendship between us, and we each covenanted over a last brimming ' peg' to correspond with one another while life lasted ; and then there were affecting leave-takings ; and, would you believe it, before I had gone ten yards from the door I had to return to take one more ' peg' for a *bon voyage*?—and so was it done, after which we all breathed our final ' Farewells.'

Next morning at daybreak the steamer ' Indore' got under weigh. The sun rose above the eastern horizon to accompany us on our way, tinging the clouds above and around with an intense carmine tint. The low sandy shores of Annesley Bay,

_____
[1] The Arab horse Sayed was shipped to New York, where he still lives in luxury after his Abyssinian toils, a honoured inhabitant of a luxurious stable.

the vast camp of Zoulla, and the proud fleet of England's marine receded in the distance, and finally vanished ; but it was three hours afterwards before the gigantic heights of the Hǝddas and Hazorta ranges, through which the Senafe and Sooroo Ghauts run, were lost to the view as they blended indistinctly with the vast mountain ranges of Nubia.

The army arrived safely at Koomaylee, and came by the railway to Zoulla on June 1, having lost only ten men in the floods altogether. Eight days afterwards the different regiments embarked for Suez and Bombay, according to their respective destinations.

A month later the great camps had disappeared from Abyssinia ; the transport fleet had spread its wings, departing for ever from Abyssinian waters ; and thus the Modern Crusade became numbered with past events, to be remembered of all men, in all lands, among the most wonderfully successful campaigns ever conducted in history. *Finis coronat opus.*

THE END.

APPENDIX.

—••◦••—.

STATISTICS RELATING TO THE TRANSPORT SERVICE.

The following statistics in connection with the Transport Service were courteously supplied by Captain Tryon, R.N., the able Director of Transports :—

Troop Ships.

Number of arrivals in Annesley Bay	669
„ „ „ at Aden of ships that did not proceed to Annesley Bay	19
Ships chartered	688
Number of vessels arrived at Aden that did proceed to Annesley Bay	465
Number of departures from Annesley Bay . . .	669
Total tonnage taken up at Bombay, Kurrachee, Aden, &c.	219,299
Total tonnage taken up at Calcutta	45,427
„ „ „ „ in England	30,225
„ „ „ „ at Suez	6,972
	301,923
Tonnage of Bombay and Bengal Marine Ships, including 'Semiramis,' 'Hyderabad,' tugs, and steam tank vessels, but not lighters	9,605
Total tonnage	311,528
Cost per month of transports, supposing that all the transports that were engaged were under employ at one time	£433,028 17 6
Amount paid to freight, excepting 'Vine' and 'Columbine,' chartered in England, as far as Annesley Bay, the particulars of which are not known	£10,710 12 0
Amount paid for line of bugalows at Bombay, and Bushire, to convey animals and fodder . .	£55,585 16 0

Coal.

		Tons
Coal supplied at Aden		62,647
„ „ in Annesley Bay		19,651
Total . . .		82,291

Coal expended in condensing in Annesley Bay on board
transports 'Semiramis' and 'Hyderabad' .   .   . 8,020½
When the expedition was at its height, coal was discharged from colliers at the rate of, per day,   .   . 191

Water.

Gallons

Water supplied at Aden by P. and O. Company, from a
return furnished by Mr. Davis, their agent .   .   . 854,796
Water condensed at Annesley Bay by Transports specially detained for the purpose, and by 'Semiramis' and
'Hyderabad'   7,286,926

Total   .   .   . 8,141,722

Water supplied in Annesley Bay, condensed by transports and condensing ships   28,918
Water supplied by men-of-war, taken out of sailing
transports, sent from Suez and Bombay, about .   . 11,000

Total   .   .   . 39,918

Cost of water condensed by transports in Annesley Bay,
supplied by condensing steamers, including the cost
of the four steamers always detained there for the
purpose (per gallon) .   .   .   . 4·35d.

Actual number of troops, &c., that landed :—

Troops.

Officers	520
European troops	4,250
Native troops	9,447
Followers	26,214
Civilians	433
Women followers	140
Total	41,004

Animals.

Horses	2,538
Elephants	44
Mules	16,022
Ponies	1,651
Camels	4,735
Donkeys	1,759
Bullocks	7,071
Sheep	12,839
Total	46,659

Number of Troops, &c., that returned. The following is the number of soldiers, &c., conveyed to Bombay at the close of the expedition :—

Troops.

Officers	369
European troops	2,258
Native troops	9,606
Followers	21,896
Civilians	286
Total	34,415

Animals.

Horses	2,122
Mules and ponies	4,126
Bullocks	1,075
Elephants	39
Camels	40
Total	7,402

MEMORANDUM, showing the amount of Sickness and Mortality in the British Troops, Abyssinian Expeditionary Force, during the entire Campaign :—

Average strength	2,688·8
Average daily sick	156·62
Ratio per cent., sick to strength	5·8
,, deaths	1·3

Casualties.

	Officers	Men
Highlands	7	17
Lowlands	4	20
Total . .	11	37

Officers' Deaths caused as follows :—

Diseases	Number
Dysenteria Acuta	2
Insolatio	2
Vulnus Sclopetarium	2 [1]
Morb. Valv. Cordis	1
Apoplexia	1
Emersio	1 [2]
Febris Continua	1
Angina Pectoris	1
Total . . .	11

Men's Deaths.

Diseases	Number
Dysenteria Acuta	17
„ Chronic	1
Vulnus Sclopetarium	1
Insolatio	8
Febris remitt	1
„ intermitt	1
„ typhoid	1
„ Continua	1
Hepatatis Acuta	1
Aneurisma	1
Nephria	1
Nephritia	1
Verrenatio	1
Paralysis	1
Total . . .	37

[1] One accident, one suicide.   [2] Accidental drowning.